Switching to Renewable Power

Switching to Renewable Power

A Framework for the 21st Century

Edited by
Volkmar Lauber

London • Sterl

First published by Earthscan in the UK and USA in 2005
Reprinted 2006

ISBN-13: 978-1-84407-241-5
ISBN-10: 1-84407-241-X

Typesetting by FiSH Books, Enfield, Middlesex
Printed and bound in the UK by Bath Press, Bath
Cover design by Paul Cooper

For a full list of publications please contact:

Earthscan
8–12 Camden High Street
London, NW1 0JH, UK
Tel: +44 (0)20 7387 8558
Fax: +44 (0)20 7387 8998
Email: earthinfo@earthscan.co.uk
Web: **www.earthscan.co.uk**

22883 Quicksilver Drive, Sterling, VA 20166-2012, USA

Earthscan is an imprint of James and James (Science Publishers) Ltd and publishes in association with the International Institute for Environment and Development

A catalogue record for this book is available from the British Library

Library of Congress Cataloging-in-Publication Data

Switching to Renewable Power: a framework for the 21st century/edited by Volkmar Lauber.
p. cm.
Includes bibliographical references and index.
ISBN 1-902916-65-4
I. Renewable energy sources. I. Lauber, Volkmar. II Title.

TJ808.S89 2005
333.79'4–dc22

2005011832

Printed on elemental chlorine-free paper

Contents

List of Figures and Tables

Figures

Tables

List of Contributors

Peter Connor is lecturer in renewable energy policy at the University of Exeter in Cornwall, UK (BSc programme in Renewable Energy). Before January 2005, he spent two and a half years at the Centre for Management under Regulation at Warwick Business School. His research interests include policy on renewable sources of electricity and heat (UK, Europe) and renewable energy manufacturing industries.

David Elliott worked initially with the UK Atomic Energy Authority at Harwell and the Central Electricity Generating Board. He is now professor of technology policy at the Open University where he has developed courses on technological innovation, focusing in particular on renewable energy technology and related energy policy issues.

Frede Hvelplund is professor of energy planning at Aalborg University, Denmark. Trained in economics and social anthropology, he has written extensively about energy planning institutions and innovation. His main interest is in analysing how specific societal institutions determine the evolution of technological development, and in designing institutions which foster democratic and sustainable energy systems.

Staffan Jacobsson is professor in environmental systems analysis at the Department of Energy and Environment at Chalmers University of Technology in Gothenburg, Sweden. His current research interests lie in the transformation of the energy system to a more sustainable one, and in science and technology policy. He has been honoured with three teaching awards.

Volkmar Lauber is professor of political science at the University of Salzburg, Austria, after teaching for several years in the US and in Bologna, Italy. His main research interests concern comparative public policy in the fields of energy, technology and sustainability, with a focus in recent years on renewable energy policy in the European Union and several of its member states.

Ole Langniss is deputy head of the Systems Analysis Department of the Centre for Solar Energy and Hydrogen Research, Stuttgart, Germany. His work is focused on implementation strategies for renewable energies. Previously, he was guest researcher at Lawrence Berkeley National Laboratories (US), post-doc at Lund University in Sweden and senior scientist at the German Aerospace Center.

Kristian Hvidtfelt Nielsen is assistant professor in the History of Science

Department at the University of Aarhus, Denmark. He completed his PhD on the history and sociology of wind power technology in Denmark in 2002. He is currently working on the sociology and anthropology of markets for wind power and wind turbines.

Ian H. Rowlands is an associate professor in the Department of Environment and Resource Studies at the University of Waterloo, Canada. He is also director of the Environment and Business Program at the University of Waterloo. His research interests extend from energy management strategies to corporate environmentalism and international environmental relations.

Jörg Schindler is managing director of Ludwig-Bölkow-Systemtechnik GmbH (LBST) in Ottobrunn, Germany, a strategy and technology consultant for sustainable energy and transport concepts. There he participated in projects on market introduction of photovoltaics and other renewable energy sources, clean road transport systems, clean fuels (such as hydrogen) and the future availability of fossil fuels.

Giulio Volpi coordinates the WWF Climate Change Programme for Latin America and the Caribbean from Brasilia, aiming to integrate climate change into development plans with a view to developing clean and efficient energy services. Previously, he worked for the WWF in Brussels on European renewable energy policy, for the UK Foreign Office, for Ecotec and for the United Nations Environment Programme in Geneva.

Ryan Wiser is a scientist at Lawrence Berkeley National Laboratory, Berkeley, California. He leads research in the planning, design and evaluation of renewable energy policies and green power marketing opportunities and conducts economic analysis in the renewable energy field. Previously, he worked for Hansen, McOuat, and Hamrin, Inc, the Bechtel Corporation and the AES Corporation.

Werner Zittel is a scientist at Ludwig-Bölkow-Systemtechnik GmbH in Ottobrunn, Germany, a strategy and technology consultant for sustainable energy and transport concepts. He has been involved in projects on environmental aspects of energy supply, renewable energy technologies, energy efficiency, the future availability of fossil fuels (especially oil and gas) and hydrogen storage technologies.

Acknowledgements

This book was inspired by two Summer Schools on the Politics and Economics of Renewable Energy, organized at the University of Salzburg in July 2002 and 2003. The Summer Schools provided both a focus and an incentive to put our ideas into writing. They were made possible thanks to generous support from the European Union (HPCF-2001-00361), the Austrian Ministry of Science and the University of Salzburg.

I would like to thank the authors of the present volume for their care in preparing and repeatedly updating complex manuscripts, and all those who also contributed valuable teachings to the Summer Schools, even if their diverse contributions could not be accommodated in a single coherent volume. They are Valentina Dinica (The Netherlands), Irene Freudenschuss-Reichel (Austria), Antony Froggatt (UK), Harry Lehmann (Germany), Mark MacLeod (US), Niels Meyer (Denmark) and Rolf Wüstenhagen (Switzerland). Their very diversity helped to make the Summer School a livelier place.

My thanks also go to the students from all over Europe, East and West, whose interest in the subject matter played a decisive role in making the summer schools a stimulating experience. Finally, I want to express my thanks to Dieter Pesendorfer who excelled both in organizing the Summer Schools and in overcoming some of the technical problems associated with the preparation of the manuscript.

Volkmar Lauber
Salzburg, June 2005

List of Acronyms and Abbreviations

API	'grade API' density definition of the American Petroleum Institute, which characterizes different oil qualities (the higher the value, the better the oil quality)
AEUB	Alberta Energy and Utilities Board
Amoco	former American Oil company, merged with BP
Aramco	state oil company of Saudi Arabia
ARCO	Atlantic Richfield Corporation, former oil company, merged with BP
AWEA	American Wind Energy Association
AZ	Arizona
BBC	British Broadcasting Corporation
bbl	Barrel (1 barrel = 159 litres)
BDI	Bundesverband Deutscher Industrie (Association of German Industry)
BM	balancing market
BP	British Petroleum, second largest western private oil company
BSC	Balancing and Settlement Code
BVMW	Bundesverband mittelständische Wirtschaft (confederation of small and medium sized enterprises)
BWE	Bundesverband Windenergie
CA	California
CBI	Confederation of British Industry
CCGT	Combined Cycle Gas Turbine
CCL	climate change levy
CDM	clean development mechanism
CDU/CSU	Christdemokratische Union/Christlich-Soziale Union (christian democratic parties, with the CSU being the offshoot in Bavaria)
Ceq	carbon equivalent
CHP	combined heat and power
CNR Inc	Canadian Natural Resources Inc, Canadian oil company
CO	Colorado
CSD	Commission for Sustainable Development
CT	Connecticut
DEFRA	Department of Environment, Food and Rural Affairs
DETI	Department of Enterprise, Trade and Investment of Northern Ireland
DEWI	Deutsches Wind-Institut (German wind institute)
DG	distributed generation

DG	Direction Générale, Directorate General (administrative unit in EU Commission)
DG TREN	Directorate General for Transport and Energy
DM	Deutsche Mark
DNC	declared net capacity
DNO	Distribution Network Operator
DoE	Department of Energy
DTI	Department of Trade and Industry, UK
EC	European Community
ECA	Export Credit Agency
ECPR	European Consortium for Political Research
EEA	European Environment Agency (Copenhagen)
EEG	Erneuerbare-Energien-Gesetz (Act on Renewable Energy Sources – RESA)
EIA	Energy Information Agency
EOR	Enhanced Oil Recovery
EPIA	European Photovoltaics Industry Association
ERCOT	Electric Reliability Council of Texas
EREF	European Renewable Energies Federation
ESI	Electric Supply Industry
EU	European Union
EUR	Euro (currency)
EUR	Estimated Ultimate Recovery
EWEA	European Wind Energy Association
ExxonMobil	Largest western private oil company (after the merger between Exxon and Mobil)
FCCC	Framework Convention on Climate Change
FDP	Freie Demokratische Partei (Liberal Party)
FSU	Former Soviet Union
G8	Group of leading industrial countries; members are the UK, Canada, France, Germany, Italy, Japan, Russia and the US; the European Union is also represented
GB	Great Britain
Gb	gigabarrel (1 billion barrel)
GCN	Global Challenges Network
GDP	gross domestic product
GJ	gigajoule
GoM	Gulf of Mexico
GROWIAN	Grosswindanlage (large wind installation)
GW	gigawatt
HFC	hydrofluorocarbon
HI	Hawaii
HMSO	Her Majesty's Stationery Office
IA	Iowa
IAP	International Action Programme
ICS	International Council for Science
ICT	Information and communication technology

IEA	International Energy Agency
IG	Industriegewerkschaft (trade union for an industrial sector)
IGCC	Integrated gasification combined cycle
IHS-Energy	American consulting company which collects detailed oil and gas statistics worldwide
IMF	International Monetary Fund
IPCC	Intergovernmental Panel on Climate Change
ISET	Institut für Solare Energieversorgungstechnik (Institute for solar energy technology)
ITER	International Thermonuclear Experimental Reactor
JREC	Johannesburg Renewable Energy Coalition
kb	kilobarrel (1000 barrel)
kb/d	kilobarrels per day (flow rate of oil production)
kWh	kilowatt-hour
KW_P	kilowatt peak
LCIP	Low Carbon Innovation Programme
LEC	Levy exemption certificate (from climate change levy)
LFG	landfill gas
Lower 48	The US without Alaska and Hawaii
MA	Massachusetts
Mb	megabarrel (1 million barrels)
mcf	mille cubic feet (= 1000 cubic feet)
MD	Maryland
MDG	millennium development goal
ME	Maine
MEP	Member of the European Parliament
MIW	municipal and industrial waste
MMS	Minerals Management Service
MN	Minnesota
MP	member of parliament
Mt	megatonnes
MWh	megawatt-hour
MW_P	megawatt peak
NC-oil	non-conventional oil
NEB	National Energy Board of Alberta
NETA	New Electricity Trading Arrangements
NFFO	non-fossil fuel obligation
NFPA	Non-Fossil Purchasing Agency
NGL	natural gas liquids
NGO	non-governmental organization
NI-NFFO	Northern Ireland non-fossil fuel obligation
NIRO	Northern Ireland Renewables Obligation
NJ	New Jersey
NM	New Mexico
NMT	Nordic Mobile Telephony
NOF	New Opportunities Fund (National Lottery)

NPD	Norwegian Petroleum Directorate
NV	Nevada
NY	New York
ºC	Degrees Celsius
OCS	outer continental shelf
ODPM	office of the deputy prime minister
OECD	Organisation for Economic Co-operation and Development
OFGEM	Office of Gas and Electricity Markets
OPEC	Organization of Petroleum Exporting Countries
pa	per annum (= per year)
P/R	production to reserve ratio
P50	reserve estimate with equal probability of being too small or too large
P90	reserve estimate with 90 per cent probability of being too small
PA	Pennsylvania
PdV	Petroleos de Venezuela (state oil company of Venezuela)
PIU	Performance and Innovation Unit
PPA	power purchase agreement
ppm	parts per million
PTC	production tax credit
PURPA	public utilities regulatory policies act (of 1978)
PV	photovoltaics
PV-TRAC	Photovoltaic Technology Research Advisory Council
R&D	research and development
RAB	Regulatory Asset Base
RD&D	research, development and demonstration
RDD&I	research, development, demonstration and information
RE	renewable energy
REC	renewable energy certificate
REFIT	Renewable Energy Feed-in Tariff
Renewables 2004	International Renewable Energy Conference, Bonn 2004
RESA	Renewable Energy Sources Act, 2000 (in German: EEG)
RES-E	electricity from renewable energy sources
RET	renewable energy technology
RFP	request for proposals
RI	Rhode Island
RO	Renewables Obligation
ROC	Renewables Obligation Certificate
ROW	rest of world
RPA	Renewable Power Association
RPS	renewables portfolio standard
RRC	Texas Railroad Commission, state agency that collects oil and gas statistics
RWE	Rheinisch-Westfälische Elektrizitätswerke
SCO	Synthetic Crude Oil (upgraded bitumen which is purged,

	cracked and hydrogenated by a so-called upgrader)
SFV	Solarenergieförderverein, Förderverein Solarenergie (association for the promotion of solar energy)
Shell	British-Dutch oil company, third largest private western oil company
SPD	Sozialdemokratische Partei Deutschland (German social democrats)
SRO	Scottish Renewables Obligation
TAGS	Temperature Assisted Gravity Segration; EOR-method of hot steam injection to enhance the oil flow rate
TGC	tradeable green certificate
TNK	Tjumen Neftegaz Company, Russian Oil company in which BP acquired a 50 per cent share in 2003
TPES	total primary energy supply
TWh	terawatt-hour
TX	Texas
UAE	United Arab Emirates
UBA	Umweltbundesamt (Federal Environment Agency)
UK	United Kingdom
UN	United Nations
UNDP	United Nations Development Programme
UNEP	United Nations Environment Programme
US	United States of America
USD	United States dollars
USGS	United States Geological Survey (geological State Agency)
UVS	Unternehmensverband Solarwirtschaft (solar business association)
VDEW	Verband der deutschen Elektrizitätswirtschaft (association of German utilities)
VDMA	Verband der Investitionsgüterindustrie (Federation of the Engineering Industries), formerly Verband Deutscher Maschinen- und Anlagenbau
WALRUS	Wettability Alteration of Reservoirs Using Surfactant; EOR-method of chemical injection to reduce the viscosity of oil and to enhance its flow rate
WB	World Bank
WCD	World Commission on Dams
WHO	World Health Organization
WI	Wisconsin
WMO	World Meteorological Organization
WSSD	World Summit on Sustainable Development (Johannesburg 2002)
WWF	Worldwide Fund for Nature

1

Introduction: The Promise of Renewable Power

Volkmar Lauber

During the last quarter of the 20th century it became increasingly clear to all who cared to know that the prevailing energy system was unlikely to be sustainable, whether physically or economically. Instead, it began to look more and more as if it was heading down a dead end, plagued by such problems as nuclear risks, climate change and other high external costs, and the prospect of stagnating and then declining oil and gas extraction within a generation or two. The future was likely to be bleak, with costs and risks accumulating over time.

Within a relatively short time – mostly during the last years of the century – the outlines of an alternative energy system began to emerge progressively: one based on renewable energy sources. Few doubted that such a system would be more benign ecologically, but many questioned its economics. Despite those early doubts, it appeared increasingly likely that such a system could be as economically efficient as the previous one – and probably more so in the long run. As to the social dimension of sustainability, that is the social aspect of such an energy system, everything seemed to speak in its favour. Renewable energy sources are more evenly distributed around the globe than fossil or nuclear sources, their harnessing could bring economic vitality to languishing regions, and there was little question of accepting the new technologies, as one survey after another illustrates.

There remained one major question: how would the transition from the old energy system to the new one be organized? Such a transition was and is unlikely to come about by itself: it requires supportive politics and policies and these in turn require substantial vision and imagination, determination and skilful dealing with vested interests and the inertia those tend to produce. This book is about such politics and policies in a particular area – that of the switch to renewable power, with a focus on Europe. Electricity from renewable sources is indeed an area of the renewable energy system in which the new technologies are well on their way. This places particular urgency on the politics of transformation in this area.

Part 1 of this book deals with general aspects of the fossil and nuclear energy system: oil and gas depletion, the climate change issue and the rising demand from the developing world, which cannot possibly be accommodated by the current system, thus making transformation more pressing still. Part 2 deals with the main subject: frameworks for transition to renewable power as they were created in three European countries – Denmark, Germany and the United Kingdom (UK) – and also in Texas, the state that has one of the most successful regulations in the US in this respect, a regulation which in turn is also of interest to the European Union (EU). Finally, Part 2 also looks at the action of the EU in this area so far. Part 3 evaluates these and similar frameworks to assess how they deal with the challenge at hand – to promote the switch to renewable power with lasting success politically and economically.

The problems plaguing the current energy system

Humankind's use of non-renewable sources is historically relatively recent. It essentially started in the 17th century, when deforestation was spreading rapidly throughout Europe as a result of industrialization and urbanization. In response, energy use shifted from wood to coal, and later on to other fossil fuels. Oil became important in the 20th century, particularly in the second half of that century as far as Europe was concerned, and was soon paralleled by the rise of natural gas. Also at that time, nuclear power promised to supply a ready source of cheap electricity, so abundant that other technological options were often abandoned. The oil crises of the 1970s initiated a process of diversification, but mostly a reinforced emphasis on coal and nuclear power, both of which received massive support from politics. Natural gas was also favoured and its consumption expanded significantly. This response to the crises, particularly the plans to multiply nuclear power, drew strong resistance in many European countries. The resulting political pressure helped to strengthen support for programmes aimed at developing renewable energy sources. But financial support was relatively modest then and did not, for some time, produce results that were reflected in any significant market penetration. When oil prices came down again in the mid-1980s, renewable energy seemed to be set for decline again.

In the meantime, however, the intensified reliance on fossil fuels and nuclear energy had to deal with rising difficulties that affected its political base. These difficulties concerned import dependence and security of supply (for both fossil fuels and uranium), unsolved and perhaps unsolvable problems of nuclear power, climate change due mainly to fossil fuel use, accumulating evidence of other damage caused by fossil fuels in the form of external costs, and finally the prospect of stagnation and decline of oil and gas production within a few decades. Greater reliance on renewable energy began to supply a plausible answer to these problems, and not just at some distant point in the future when the technology would be ready, but starting there and then and to be phased in during the course of the 21st century, to allow for the inevitably slow pace that characterizes large infrastructure investments such as those of the power system.

Historically, dependence on energy imports was the first problem that the fossil energy system presented for Europe. This was already evident in both

World Wars, and emerged again with the Suez crisis in 1956. But vulnerability to oil imports in particular kept growing until the 1970s, partly as a result of cheap oil. The oil price increases of the 1970s led to greater efficiency, diversification and the activation of new discoveries in the North Sea. But efficiency gains began to stagnate when prices came down again in the 1980s, and production in the North Sea has already passed its peak. At the present time, the outlook for Europe is again sombre with regard to energy imports. From about 50 per cent in 2000, they are expected to increase to about 70 per cent by 2030 (European Commission, 2000). Also during this time period, the Organization of Petroleum Exporting Countries (OPEC) will probably achieve a much stronger position on the world oil market than currently since most other countries' reserves will have declined strongly by then; in many cases, former exporting countries will actually have become net importers (see Schindler and Zittel, in this volume). At the same time, demand for energy is likely to increase worldwide, especially in industrializing countries such as China and India where per capita consumption is still low but rising fast.

Table 1.1 *Primary energy sources for electricity generation and their ratios of import dependence: EU 1998*

	Total share (%)	Import dependence (%)
Solid fuel (mostly coal)	27	50+
Oil	8	76
Natural gas	16	40
Nuclear (uranium)	35	95
Renewable energy sources	15	near 0

Source: European Commission, 2000.

What can Europe do to reduce its energy dependence, particularly in the electric power sector that is at the core of this book? Nuclear power, coal and gas (oil plays only a limited role in Europe's power generation) all present considerable problems. The most unproblematic sources of power in the future are clearly efficiency increases and renewables. But let us look at the situation in detail.

Nuclear power has experienced a revival in the levels of political and media attention in several European countries. Some of the problems that have haunted it in the past have been neglected in this discussion. For example, the fact that a significant increase in nuclear power use would also increase dependence on imported uranium, and that known and conjectured reserves of uranium will only last for about 60 years by present counts (Traube, 2004). This can be changed by a second attempt to shift to plutonium on the basis of fast breeders and reprocessing plants. However, the fast breeders built in the 1970s and 1980s in the UK, France, Germany and Japan all turned out to be expensive failures; for this reason the technology, and the plutonium cycle for civilian use, were abandoned. Second, there is the problem of safety. Since the partial core meltdown at Three Mile Island in Harrisburg in 1979, and the actual meltdown in Chernobyl in 1986, nuclear power has a blemished record. Since then, prominent nuclear operators such as Sellafield or Tokyo Electric Power have been caught faking their safety records for a number of years, and the Davis-Besse incident in Ohio illustrated a similar mentality. Owing to these safety problems, the acceptance of nuclear fission at the political level is currently restricted to a small number of countries in the EU, several of them in Eastern Europe and relatively backwards with regard to their energy systems. Third, there is the unsolved problem of nuclear waste storage. And since 2001, a whole series of new questions have emerged for which there is no satisfactory answer: for example, the vulnerability of nuclear power plants to well-organized terrorists. In the past, security systems were notoriously easy to overcome by quite 'unprofessional' nuclear power opponents who dramatized their protest by taking along bazookas and other explosive weapons and remained undetected. Proliferation is another issue that probably cannot be solved in a satisfactory fashion if nuclear power is to be revitalized. And then there is the problem of the economics of nuclear power, particularly if external costs are taken into account, a point to be taken up below.

This leaves the hope of nuclear fusion – or does it? Work on fusion as an abundant energy source has been pursued for over 50 years. The next step currently being discussed is the plan for an International Thermonuclear Experimental Reactor (ITER) that will cost about US$5 billion to build and a similar amount to operate. This is only one of several steps with a similar price tag which by 2050 – so it is hoped by fusion's supporters – may result in a reactor capable of producing commercially competitive power. Whatever the merits of fusion (and the eventual outcome of current research is still open), it is evident from this schedule that it will not play much of a role – if any – in European power production before the last quarter of the 21st century.

Coal and gas are the main fossil fuels used in European power generation today. Their long term prospects have suffered a severe setback due to climate change. There is widespread scientific agreement on the basic facts. In pre-industrial times, the concentration of carbon dioxide (CO_2) in the atmosphere was around 280 parts per million (ppm). Currently it stands at 375 ppm, and will exceed the figure of about 500 ppm unless carbon emissions are stabilized at the present level of about seven gigatonnes per year for the next half century (Pacala and Socolow, 2004; see also Ian Rowlands, in this volume), with a decline thereafter. If the threshold of 500 ppm is exceeded, natural disasters on a major scale appear inevitable (in fact, the major insurance companies point to the fact that such disasters have already been on the rise during the last two decades, though not yet on the scale expected at higher emission levels). So the question is how to stabilize emissions at current levels.

The most plausible conclusion is to cut back fossil fuel use, or at least carbon emissions. The Kyoto protocol – entering into force in early 2005 – is a first step, requiring industrial countries that are signatories to the treaty to take the lead with these cutbacks, while 'developing' countries are expected to take on commitments at some later time. The EU, which for a long time took the lead in the Kyoto process, has taken the first steps by introducing an emissions trading system and requiring national allocation plans from all member states in 2004. At the expected price of 8 Euro per tonne of carbon,[1] once it is traded, this is expected to add about 8 Euro to the price of one megawatt-hour (MWh) of electricity from coal, and 3.2 Euro to the price of power from gas (Milborrow, 2005). This is not likely to pay for the full climate cost of those fuels – and the price is expected to rise later on – but it is a first step that will help to check further growth in the consumption of those fuels. At some point, however, a drastic reduction in their use will be necessary to meet global stabilization goals, considering that developing countries are allowed to increase their emissions.

Can fossil fuels gain a new lease of life thanks to technologies of carbon sequestration and storage? This avenue is already being pursued, but it is open to question on two counts. First, can storage be relied upon to last for many centuries and beyond, or is there a danger that it will leak into the atmosphere at some point? Second, how does the procedure make economic sense? While it is hard to predict specific values, the cost of one to two additional US cents per kWh of coal-generated electricity has been put forward by International Energy Agency (IEA) experts as a goal to be attained for capture. There would be additional costs for transport and storage in the range of US$5–10 per tonne of CO_2 (Gielen and Podkanski, 2004), roughly similar to the price a tonne of carbon is expected to be traded for; for this reason, only the cost of capture needs to be considered here as extra cost. Even at this rate, the process appears to be profoundly uneconomic in the long run, that is, with sufficient availability of renewable sources. The costs of capture will increase generation costs of coal-based electricity, from about 3.5–4 Euro cents, to about 5.25 Euro cents. If average external costs (e.g. of air pollution – see next paragraph) are added, coal's generation cost will go up by another four or five Euro cents on average and thus come close to 10 Eurocents. That is substantially more than wind power costs in Germany (even offshore) today, and corresponds approximately to the admittedly high rates currently paid under the British Renewables Obligation. Nonetheless, carbon capture may be interesting for a time of transition, until safer and more economic alternatives are available in sufficient amounts. It must be remembered again that large infrastructures (as in the power generation sector) only change at a slow rate, and that wind power and other renewable sources need many years – in fact decades – of high growth rates before they can replace coal and nuclear.

External costs of energy use are a relatively recent discovery, not in terms of economic theory but in terms of calculating specific monetary values. In his groundbreaking discussion of external costs in 1912, A. C. Pigou illustrated the concept by referring to the locomotives that frequently used to set fire to fields and forests, with the railroads making no efforts to avoid such damage as they were not held liable for it; this made these costs external to their accounting (Pigou, 1912). While many costs have been internalized since those days, in the energy sector external costs are still considerable. In the early 1990s the European Commission began to take them more seriously, fully expecting that this would lead to reorienting economic activities in more sustainable directions (European Commission, 1993). A large research project

(ExternE) was set up to explore this field. Originally it was conducted by both EU and US research teams; however, the US dropped out in the mid-1990s after data showed unexpectedly high costs (Krewitt, 2002). In 2003 the EU Commission published provisional 'consensus' results from this study. For electricity generation from coal, this publication (European Commission, 2004) shows figures ranging from a low of 2.4 Euro cents per kWh in Sweden to a high of 4–15 Euro cents in Belgium. For Germany, the figures were 3–6 Euro cents. The European Environmental Agency (EEA) calculated that for 2001, total external costs of electricity generation in the EU 15 amounted to 42–72 billion Euro. This is a low estimate due to the fact that the EEA (based on ExternE) practically excluded the cost of nuclear power: mortality and morbidity from exposure to high-level nuclear wastes as well as the contribution of civilian nuclear power to the risk of nuclear proliferation and terrorism were excluded as being too difficult to value, and the risk of nuclear accidents was not fully priced, in fact just about ignored (European Environmental Agency, 2004, p17, footnote 17). In addition to these external costs caused by conventional generation, there are substantial subsidies to electricity generation, again primarily for coal and nuclear for which no new technologies are produced by such expenditures. By comparison, renewable energy is subsidized on a modest scale; at the same time, the external costs of renewable powers are quite modest (ibid, 2004).

Now, if external costs were taken seriously, this would indeed make many renewable power technologies immediately competitive. As Jacobsson and Lauber show in this volume, including external costs in Germany means that wind power, which in 2004 came under heavy attack there for supposedly being uneconomical when compared with coal, is actually quite competitive with that fuel. Indeed, if subsidies for coal are factored in, it is already the cheaper option. The major difference between coal-generated electricity and wind power is not costs; it is in the allocation of those costs. In the case of wind power, they are borne by electricity consumers; in the case of coal, they are partially borne by the general public, regardless of causation; in other words, they are 'externalized'. Internalization of external costs – required in order to optimize welfare – will lead to a reorientation of the energy sector in a more sustainable direction. But serious internalization of external costs will also require a major change in politics.

The prospect of declining oil and possibly gas production during the coming years and decades is another powerful incentive to develop other sources of energy. Of course this is still a matter of controversy. But as Schindler and Zittel show in this volume, the estimates of oil deposits in the earth's surface have not changed much during the last several decades. The formation of oil requires very special conditions that were present in only a few places in the earth's crust; also, the extraction of each deposit – and of all deposits taken together – follows a certain curve, named the Hubbert curve after an American geologist who in the 1950s, on the basis of his theories, predicted that US oil production would reach its peak in 1970 and then decline. History proved him to be correct (Deffeyes, 2001). On the basis of the same theories, but much more sophisticated exploration methods and more recent data, many contemporary geologists have since concluded that world oil production is likely to reach its peak now or within a matter of one or two decades. This is supported by the fact that no really big oil discoveries have been made since the 1960s, and that exploration is becoming ever more difficult. At the same time, an increasing number of oil exporting countries are turning into net importers due to rising domestic demand, and energy demand in many developing world countries without notable resources of their own is rising at a rapid rate (see Volpi, in this volume). Such an interpretation is rejected however by a substantial number of economists who argue that if the prices are right, there will be no scarcity of oil as new reserves will be found at higher prices. They do not address the physical argument of the geologists however. Scarcity of oil is indeed likely to subside if market prices are high enough – if only due to a decline in demand.

Let us retrace for one moment the logic of the Hubbert curve and explore the likely characteristics of such a peak, and the consequences that such a decline of production would bring about, from prices to geopolitics. There is no reason to doubt that this peak will be reached at some point; the only question is 'when?'

It is certain that oil wells will not run dry all of a sudden, neither now nor in the future. At first, annual increases in production will become ever smaller. Later on, decline, modest at first, will set in and reach a slightly steeper downward slope after a decade or so (likely to be much steeper in the case of natural gas). On this slope it will continue, probably for another century. If demand is inflexible, this will mean a substantial price increase, until at least part of the demand can no longer compete and is eliminated. This will cause very severe effects in terms of economic disruption. In reaction (and perhaps even in anticipation), industrial and industrializing countries will scramble for oil and gas, and other sources of energy as well. But as the experience of the oil crises of the 1970s has taught us, it takes decades before new sources can deliver results – whether in terms of stepping up efficiency and conservation, shifting to other fuels (coal, tar sands, nuclear) or developing renewable energy sources. The time lags involved here are such that a deep crisis of adjustment will be unavoidable.

In any case, the matter is not likely to rest there. One of the foreseeable results of oil depletion is that the OPEC countries, and particularly those of the Middle East, will gain in importance as their reserves are the largest, and their demand comparatively modest. They will thus be in a better position to dominate the market than they have been in decades. The temptation to 'secure' continued access to – or control over – oil by military means will increase. In the early years of the 21st century, some US neoconservatives (who as a group dominated foreign policy-making at that time) argued well before the Iraq War of 2003 that the United States should secure a strong position in the Middle East so that it could control the distribution of the oil resource – not just to guarantee its own supplies, but because this would give it a handle over its 'friends' (and competitors – Europe and Japan) as well as its opponents (www.newamericancentury.org). Experience shows that intense conflict can hardly be averted in such a situation. The only alternative is to bring about a decisive change in energy policy, one that prepares the world for the decline of oil and gas. For the European power sector, natural gas is of decisive importance. Even though its regional distribution is not identical to that of oil, the structure of the conflict in case of gas depletion is likely to be similar. And those regions that are able to reduce their dependence on hydrocarbons are likely to be at an advantage.

Paradoxically, the need to increase the efficiency of energy use (required by the response to climate change as well as to an increased scarcity of hydrocarbons) is sometimes used as an argument against the development of renewable power. The argument is that the best way to reduce fossil fuel consumption, and in particular greenhouse gas emissions, is to use the cheapest options first; these are conservation and efficiency. At present, the cost of saving one ton of carbon by increasing efficiency is substantially lower than to avoid emissions by an equivalent amount of renewable power. While this is true, the reasoning is faulty as it ignores the time dimension. If renewable power is to play a significant role within a few decades, it is urgent to develop it now. Energy efficiency can play a significant role, but over time it will yield declining results (see Rowlands, in this volume). For this reason, it is imperative to develop alternative, low-carbon technologies right now.

The promise of renewable power

It is argued here that renewable sources are so abundant that they can easily supply European demand for electricity by the end of the 21st century and indeed earlier, provided that the transition is organized by appropriate policy and over an appropriate time period. This view is hardly dominant today, whether in business or government. (At the level of public opinion, renewable energy is viewed highly favourably at present in most western industrial countries). Sceptics – most often representatives of incumbent, conventional energy sources (particularly coal and nuclear), political institutions which co-developed with them and media allied to them – usually argue that renewable energy sources are not sufficiently abundant to satisfy energy demand, and that even where available, these sources are too expensive to be competitive. The fact that (with the exception of large hydropower) they have not been able to impose themselves on the market up to now is presented as proof that things will remain the same way in the future. In fact, things are changing already.

A different view is asserting itself with growing intensity. This view was long stifled at climate policy conferences, due to the insistence of the US, Australia and OPEC countries. In a key paper for the conference, Renewables 2004 in Bonn, this view was aptly summed up:

> Renewable energy flows are very large in comparison with humankind's use of energy. Therefore, in principle, all our energy needs, both now and into the future, can be met by energy from renewable sources (...) With decisive efforts to speed up (...) dissemination, all human energy needs could be met by rerouting

> a small fraction of naturally occurring renewable energy flows within a century (Johannsson et al, 2004).

Despite the dominance of fossil and sometimes nuclear power generation technologies in most European countries, the case in favour of a shift to renewables has been made several times in recent years. This will be illustrated by positions taken by the British and German governments and by the Danish energy sector, since only 2001.

In recent years the governments of both Britain and Germany – and undoubtedly other countries as well – have explored the potential of renewable energy sources' contribution to cover domestic needs for the next half-century. In the UK, Tony Blair in 2001 charged the Performance and Innovation Unit with a general review of British energy policy, with a particular focus on the role that renewables might play in this area. For the context, it should be remembered that (subsidized) British coal was largely discarded as a source of power generation during the Thatcher years, in the conflict that brought the government into confrontation with the National Miners Union (NMU). Coal was replaced by the 'dash for gas'. However, it has become evident that British North Sea gasfields are declining, so an alternative needs to be developed. At the same time, the current British government is among the most committed in Europe to taking climate change seriously. Under these circumstances, the review initiated by Prime Minister Blair in 2001 became 'the most wide-ranging energy review for 20 years' (Boyle, 2002) and was among other things entrusted with the task of formulating a vision for low carbon energy systems in 2050, and possible pathways towards such systems. One of its goals was also to look for ways to permit a 60 per cent carbon reduction as recommended by the Royal Commission for Environmental Pollution in 2000, a goal officially proclaimed by Blair in February 2003 (BBC, 2003). The review strongly emphasized efficiency and stressed renewables as the most important and most favourably priced low carbon technologies. It stated that Britain had the best wind and marine resources in Europe – onshore wind, offshore wind, wave and tidal stream, with energy crops and solar PV as further additions. Even though renewable power technologies were in most cases not yet fully competitive commercially, their costs had the potential to fall rapidly, and by 2020 should in most cases be able to outdo fossil fuels. Nuclear power was evaluated as being more expensive than renewables and as holding little promise of cost reductions in the future. For that reason the review recommended phasing out nuclear but keeping the nuclear option open in principle, in case renewables did not perform as expected. For renewable power, it recommended doubling the target of 10 per cent by 2010 to 20 per cent by 2020. A first step in this direction was taken in late 2003; the 20 per cent goal is still 'aspirational' at present (see Connor, in this volume).

Germany is also one of the countries most strongly committed to climate change policy, and is leading in Europe in terms of installed capacity of wind and solar power. Since 2002, the Environment Ministry has been in charge of renewable energy policy. In June 2001, the Environment Ministry commissioned three leading research institutes in the area of energy and the environment to conduct a study on the 'ecologically optimised increase in the use of renewable energy in Germany' (DLR et al, 2004), which it subsequently adopted as the basis of its policy. This study discusses the goal of reaching a 50 per cent share of renewable sources in total energy supply by 2050 by adding about 10 per cent each decade. For power, the goal for this date is 68 per cent from renewables, of which 54 per cent is expected to come from domestic generation (ibid, 29). This is only possible with a dramatic increase in the use of renewable sources. The study compares the cost of such a shift to renewables in the electricity sector with the cost of conventional generation factoring in foreseeable cost increases, including increases resulting from carbon sequestration and emissions trading (the fossil fuel price increases of 2004, however, are not included). On this basis it concludes that over the long term, that is by 2050, an electricity system relying on fossil fuels will be substantially more expensive than a system relying on efficiency and renewables. Early investments in renewable power will be paid back by then, and further advantages will accrue from the lower prices that cannot be matched by other forms of generation. If external costs and other problems of conventional sources are included, the advantages are even greater. Wind power (under the relatively unfavourable conditions prevailing in Germany) is expected to be commercially attractive – that is cheaper than fossil fuel technologies – by around 2020–2025. By 2040, renewable power is not expected to create additional costs; extra costs due to photovoltaics (PV) by then will be balanced by the competitive advantage of other technologies (ibid, 37). Total extra costs for renewable power (in the absence of a strong climate change policy, which would reduce such costs) are calculated at a total of Euro 55 billion for the period 2000–2020, the time of the highest expenditures for market creation and support. This figure is quite reasonable when compared to the much higher cost expended since the 1970s on coal subsidies, support for nuclear power, or to the external costs of coal-generated electricity in recent decades (see Jacobsson and Lauber, in this volume). The overall analysis outlined here is found in many similar studies and is very close to the views of the Environment Ministry and of the red–green government majority in parliament, although there is some opposition from coal interests among the social democrats.

In Denmark, a similar statement came from a source that will be less easily suspected of bias in favour of renewables – the industries of the entire energy sector. Ever since it turned down nuclear power in the 1970s and 1980s, Denmark has had a particularly innovative energy policy, at least until the current government, which came into power in 2001 (see Nielsen, in this volume). Such an energy policy had the broad support of Danish society – and it apparently still does. In November 2004, a report for the Danish Society of Engineers concluded that for the period ranging from 2004 to 2030, Denmark could reap substantial economic rewards if it converted its energy supply from oil and gas to renewable energy, even without taking into account the external costs of fossil fuels and using the low oil prices of 2002. Only a few days after publication of this report, a meeting of Danish energy executives published a unanimous declaration to the government, supported by the whole sector, including all electric utility companies and all forms of generation. This declaration proclaimed that government policy should aim for a share of renewables of 30 per cent of total energy consumption by 2025 (instead of 12 per cent today), and a share of 70 per cent in the area of electricity (Møller, 2004).

If we look at the growth potential of renewable sources for power generation, two sources currently stand out – wind and solar PV. To be sure, there are significant other sources. But hydro is unlikely to achieve high growth rates here, and biomass has many other uses outside of power generation, in particular in heating and transport. Wave energy is likely to become an important source but it is difficult to estimate its potential at present, given the early state of this technology. While Europe does not have ideal resources in the area of solar irradiation, a 2004 study by the European Photovoltaic Technology Research Advisory Council[2] stated that all of Europe's electricity needs could be covered by fitting the total surface of south-oriented roofs with PV equipment (PV TRAC, 2004, p4). Again, the time factor is important – the same study concludes that at least 20 per cent of Europe's electricity could be supplied by PV in 2050 (ibid, p24), given an appropriate policy. Deployment would be slow at first, but pick up speed once costs have become more competitive. As to wind, a study by the prestigious British wind power consultancy firm, Garrad Hassan, commissioned by Greenpeace and published in February 2004, puts the potential contribution of European offshore wind power to total electricity supply at about 30 per cent by 2020 (Garrad Hassan/Greenpeace, 2004). In both cases, a substantial political effort is seen as necessary, even though there would be very substantial side benefits, even in economic terms.

Needed: a regulatory framework for transition

Transition to renewable energy sources is unlikely to come about in time – in view of such problems as climate change or oil and gas depletion – unless it is supported by public policy in several countries. The task of this policy is to bridge the gap between the abundance of renewable energy sources and the scarcity of their use.[3] To refer again to wind and solar energy – in 2001, wind and solar contributed about 0.4 per cent to electricity generation in IEA countries (IEA, 2004). The figure is somewhat higher in Europe given Europe's current lead in wind power, but it will still be puny (European Environmental Agency, 2002). At the same time though, wind power in Europe more than tripled in the five years from the end of 1999 to the end of 2004 (from about 9300MW to about 34,000MW), while PV increased at a much more rapid rate still. On a worldwide basis, wind and solar together grew by annual average rates slightly above 23 per cent in both the 1980s and 1990s (Sawin, 2004; IEA, 2004). Growth for PV accelerated considerably beyond these values in the current century.

The tasks such a framework has to fulfil are several. For one, it has to remove the obstacles to renewable energy – in this case renewable power – resulting from the bias of the current regulatory framework in favour of incumbent energy sources. Second, such a framework needs to support rapid renewable power development and expansion at a reasonable cost to society. Third, a regulatory framework also needs to generate legitimacy and support.

Remove biases favourable to incumbent technologies, fuels and operators

A first set of biases under consideration is listed in the first section of this chapter. The power system resting largely on fossil and nuclear fuels has high social costs. The dependence on imports at a time of increasing conflicts over energy resources implies risks of very serious economic and social disruption. Not to prepare for oil and gas depletion is to risk economic collapse altogether – and also involvement in major military confrontations. The risks of nuclear power (from inadequate storage of wastes to accidents, from terrorist attacks to nuclear proliferation) are certainly real; insurance companies know what they are doing when they decline to cover them. Climate change is a survival issue with very high financial consequences, even though the finances are unlikely to be the most serious aspect. External costs are increasingly well estimated and in any case substantial. Current policy takes a view of all these issues best characterized as 'benign neglect', at least in terms of their financial consequences for the energy sector. In the name of international competitiveness, the costs caused by these aspects of the current energy system are rarely reflected in the bills of energy consumers. While it is often emphasized that nowadays such choices must be made by the consumer (an unconvincing argument in areas where collective action is indispensable), the consumer is not even confronted with full information or true prices. Disclosure on consumer bills regarding primary sources used in generation is a far cry from effectively pricing the risks resulting from them.

But there are other biases that operate in favour of incumbent technologies and fuels, more specific than the ones listed above. This concerns for example the regulatory preference given in many countries to large-scale thermal power plants, which though inefficient at using primary energy, arrive at lower electricity prices by neglecting combined heat and power (CHP); similarly for the organization of the grid in such a way as to serve large units, following a logic that puts small generators at a disadvantage. High and probably excessive charges for grid use are quite frequent in Europe; this too usually benefits conventional generators at the expense of renewable power sources. In many cases, large conventional generators own and operate the grid and use this position to inhibit access by small generators or traders, notwithstanding the unbundling rules of the directives on the internal electricity market of the EU. The penalties imposed on intermittent generation from renewable sources (as in the case of wind), and also the calculations of the cost of balancing energy, tend to reflect more the interests of large generators rather than the real costs these incur in the process of fitting together generation and demand; similarly for the cost of new transmission systems resulting from the use of renewables (Milborrow and Harrison, 2004; Knight, 2004a; Knight, 2004b). At the same time, there are few rewards, if any, for some of the special advantages offered by renewables – thus the advantages of embedded generation (supplying customers without need to use the high-voltage grid while at the same time increasing grid stability) or the fact that PV generation peaks at the highest point of the demand curve.[4] There is a multitude of other chicaneries by which transmission or distribution system operators have inhibited renewable technologies in a variety of European countries in the past (delays in linking up installations, delays in payment, excessive bureaucratic or technical requirements and so on). Although there are also positive examples, it is important to insist on fair competition, which – as experience shows – cannot be taken for granted.

Support rapid renewable power development at a reasonable cost to society

There are three aspects to this: technological development, market creation/ expansion and the search for maximum efficiency in such policies. Support for technological development involves public financial support to develop the knowledge base for new renewable energy sources that are not mature yet (and few of them are). This involves technologies in their early stage (currently for example wave power, or solar thermal power), but also at later stages when they confront new issues (offshore wind, PV). In this context it is important to remember that such support was given in the past to a series of new energy technologies in the power sector. It was comparatively modest in the area of renewable energy, as IEA statistics show. For comparison's sake, it will be useful to list estimates of development costs for several such technologies. The biggest chunk by far went to nuclear power in various forms (uranium-based reactors, plutonium-based reactors and related technology, fusion reactors). Notable amounts of research also went into fossil fuel technologies. None of these sources – with the possible exception of fusion – has a potential approaching that of wind or solar power.

Table 1.2 *Estimates of costs of various fuels/technologies (RD&D; other support, including cross-subsidies; external costs; accident costs)*

Fuel source/technology	Costs
Total support for nuclear power in OECD countries until 2001	About 150 billion Euro in RD&D[a] About 300 billion Euro in cross-subsidies[a]
Estimated cost of Chernobyl-style nuclear accident in western Europe	About 2,800 to 5,300 billion[a] Euro
Amount covered by insurance	Insurance covers 0.01% to about 1% (dependent on country)[a]
Annual external costs of fossil fuel power generation in EU-15	43–72 billion Euro (annually)[b]
Expected cumulative support for all renewable energy sources in Germany between 2000 and 2020 for scenario to reach over 50% of renewable energy generally, and 68% of renewable power by 2050	50 billion Euro in 20 years Lower if there is a strong climate change policy or if fossil fuel costs remain above 2001 levels[c]
Coal subsidies in Germany, 1975–2002	About 100 billion Euro in 28 years

Sources: a Rechsteiner 2003 (referring to PROGNOS and Swiss government figures); b Source: EEA 2004; c DLR et al, 2004; d Jacobsson and Lauber in this volume.

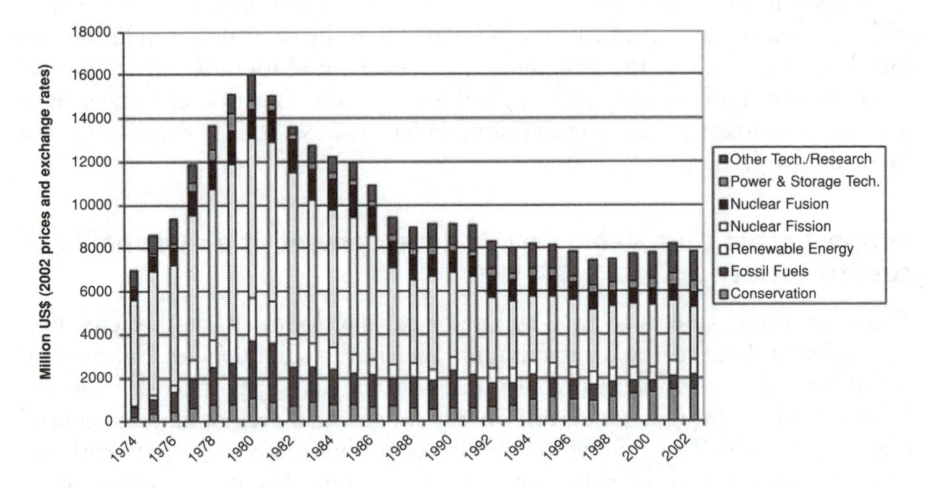

Source: IEA 2004

Figure 1.1 IEA government energy RD&D budgets

With regard to RD&D funding, the development of renewable energy technologies is characterized by a special situation when compared to large-scale technologies, which received massive public support. They cannot be developed in big steps, the way nuclear power is often presented as having been developed. There, the creation of a knowledge base was supposedly followed by the development of a few prototypes; after that, the technology was turned over to the market (in fact, this characterization does not even fit fission technology, as can be seen from Figure 1.1, above). In any case, unlike nuclear installations, most renewable power technologies are embodied in comparatively small devices whose efficiency can be improved at a very rapid rate with every new generation of products. This demands the creation of a special regime for sustained market creation of technologies whose maturation depends on a large scale of market introduction. At the same time, this market creation needs to introduce elements of competition in some form. This policy challenge is the main subject of Parts 2 and 3 of this publication.

Develop political legitimacy for renewable power

Public policy is not only about creating regulatory frameworks. Governments or other authorities also need to convey what is really at stake in the current energy transition (that is issues of basic security from disruption, war and disasters), develop visions of the opportunities offered by the transition to a truly sustainable energy system, put the cost of the transition into the context of catastrophes that may be avoided, and compare the cost of renewables support with the nature and volume of the advantages accruing to incumbent technologies, fuels and operators. Of course such action may require greater government independence from those operators than is often the case today. However, liberalization can be used to further reduce cliental relationships between governments and the power sector, and the growing importance of renewable power and the prospect of new opportunities may serve as useful counterweights.[5] To support this legitimacy, it is also important to establish an organization that promotes such a transition at the international level.[6]

Rationale of current renewable power policies

Clearly we live in a time of transition, possibly of a breakthrough. This is evident from the growth rates of new renewable energy sources, growth rates that would not have come about were it not for determined government policies. These policies are motivated by two sets of considerations, which may pull in the same direction but carry different implications.

The first set of motivations is to confront the major problems outlined in the first part of this chapter: climate change, pollution (external costs), import dependence and the risk of oil and gas depletion, in many cases the risks associated with nuclear energy – all this militates in favour of efforts to mobilize sustainable sources that promise to be sufficiently abundant to make a real difference. These efforts are checked however by several considerations. First, by the decline of public budgets at a time of austerity, a process that set in with the advance of neoliberal ideology; for the EU this is embodied in the Maastricht criteria. Second, by the fear that expenditures in this area are producing collective goods that benefit everyone – and that any one country that is investing disproportionately in such goods by moving ahead too far or too fast will do so at peril to its own competitive position. This is the traditional problem of collective action (Olson, 1965). The answer is seen in international cooperation to fairly distribute the costs of such action. From this perspective the current policy of the US appears as a major impediment to such collective action. Under President G. W. Bush – as previously under President Reagan – the federal government reduced its efforts in the area of renewable energy; it also took a 'benign neglect' approach to climate change, downplaying the very nature of the problem. Together with Australia and the OPEC countries, it also prevented renewable energy from becoming a major topic at climate conferences or at the Johannesburg summit in 2002.

There is however another set of motivations that inspire renewable energy policy in some of the pioneer countries. It is the prospect of becoming a leading exporter of technologies whose production is growing at dizzying rates already and which may become the object of immense investments in the future. For this strategy to succeed, it is enough if the market for such technologies keeps growing apace; there is no need for unanimous support by the international community, and if some countries choose to fall behind in this race this is actually a boon for the early movers. In some countries, the prospect of energy scarcity is perceived more intensely than in others; it is probably no coincidence that among the leaders in renewable energy we find

Germany (leading with wind installations) and Japan (leading in PV, with Germany catching up in new installations in 2004); both have practically no oil and not even a big national oil company. This reasoning may also have induced some of the newly industrialized countries in southeast Asia to plan enormous investments into PV programmes of their own. Thus South Korea announced a 100,000-roof programme in 2004 and appears to be aiming for a goal of 1300MWp of installed capacity by 2011 (Hirshman, 2004). Thailand was considering a similar programme in the fall of that year (Photon International, 2004).

Under the impact of neoliberalism, generous renewable energy programmes, supported by government or electricity consumers carrying the proclaimed goal of creating public goods will inevitably draw criticism from business, and perhaps from other groups as well. It is one of the ironies of the age that the same goal and high public purpose may be served as effectively by what is basically nationalist industrial policy or plain business strategy. But in the end the exact formulation of those rationales may be of secondary importance, as long as they contribute to solve one of the major problems mankind is facing today, that is the transition to renewable energy, in the present case to renewable power.

Notes

1 It was more than three times this price in July 2005.

2 PV TRAC was set up by EU Commissioners Loyola de Palacio and Philippe Busquin in 2003 and consists of the major European PV stakeholders, including the PV industry, the research community, the construction sector, utilities and government agencies.

3 In some areas of renewable energy, use of existing resources is not scarce in Europe, especially in the area of large hydro, and to some extent in biomass.

4 A 'thought experiment' operating with historical data conducted in 2004 showed that if 1000MW of PV had been installed in the New England states in 2002 (corresponding to 0.7 per cent of supply), average wholesale electricity prices would have been reduced by 2–5 per cent, due to the fact that the price spikes that occasionally emerge at peak time would have been substantially reduced. In Germany and Japan, such levels of PV penetration are expected by the end of the current decade (Martin, 2004).

5 In 2004, worldwide investments just in wind and solar power were in the range of Euro 17 billion, more than half of which was in Europe. This amount can be expected to multiply in the near future.

6 This was much discussed before and during the conference Renewables 2004 in Bonn; however little action has been taken so far.

References

BBC (2003) BBC News, world edition, 24 February 2003, accessed 28 February 2003

Boyle, S. (2002) 'Major UK energy review opts for renewable energy', *Renewable Energy Report*, vol 35, 10 January 2002, pp37–42

Deffeyes, K. S. (2001) *Hubbert's Peak: The Impending World of Oil Shortage*, Princeton

University Press, Princeton
DLR (Deutsches Zentrum für Luft- und Raumfahrt), Institut für Energie- und Umweltforschung and Wuppertal Institut für Klima, Umwelt und Energie (2004) *Ökologisch optimierter Ausbau der Nutzung erneuerbarer Energien in Deutschland.* Forschungsvorhaben im Auftrag des Bundesministeriums für Umwelt, Naturschutz und Reaktorsicherheit FKZ 901 41 803, Berlin
European Commission (1993) *Towards Sustainability. Fifth Environmental Action Programme*, European Commission, Brussels
European Commission (2000) *Towards a European Strategy for the Security of Energy Supply, Green Paper*, European Commission, Luxembourg
European Commission (2004) *External Costs: Research Results on Socio-environmental Damages due to Electricity and Transport*, Directorate-General for Research, Brussels
European Environment Agency (2002) 'Energy and the environment in the European Union', *Environmental issue report* no 31, European Environment Agency, Brussels
European Environment Agency (2004) 'Energy subsidies in the European Union'*, EEA Technical Report* 1/2004, pp14–17
Garrad Hassan/Greenpeace (2004) *Sea Wind Europe,* www.greenpeace.org.uk/seawindeurope
Gielen, D. and Podkanski, J. (2004) 'CO_2 capture and storage: An energy modeling perspective', *Energy* (Energieverwertungsagentur, Vienna), no 3, pp3–4
Hirshman, W. P. (2004) 'Do you want to know a secret?' *Photon International*, vol 1, pp32–35
IEA (International Energy Agency) (2004) 'IEA side event', International Conference for Renewable Energies 2004, www.iea.org/Textbase/work/2004/bonn/sellers.pdf
Jacobsson, S. and Lauber, V. (forthcoming) 'The politics and policy of energy system transformation – explaining the German diffusion of renewable energy technology', *Energy Policy*
Johannsson, T. B., McCormick, K., Neij, L. and Turkenburg, W. (2004) 'The potentials of renewable energy'. Thematic background paper for the International Conference for Renewable Energies, Bonn
Knight, S. (2004a) 'Shocking claims on reserve requirements', *Windpower Monthly*, vol 20, no 2, pp45–46
Knight, S. (2004b) 'Devious castings of wind as grid villain', *Windpower Monthly*, vol 20, no 11, pp56–57
Krewitt, W. (2002) 'External costs of energy – Do the answers match the questions? Looking back at 10 years of ExternE', *Energy Policy*, vol 30, no 10, pp839–848
Martin, K. (2004) 'Power by the hour', *Photon International*, no 12, pp38–41
Milborrow, D. (2005) 'Goodbye gas and squaring up to coal', *Windpower Monthly*, vol 21, no 1, pp31–35
Milborrow, D. and Harrison, L. (2004) 'The real cost of integrating wind', *Windpower Monthly*, vol 20, no 2, pp35–39
Møller, T. (2004) 'Power sector pushes for green strategy', *Windpower Monthly*, vol 20, no 12, pp19–20
Olson, M. (1965) *The Logic of Collective Action,* Harvard University Press, Cambridge, MA
Pacala, S. and Socolow, R. (2004) 'Stabilization wedges: Solving the climate problem for the next 50 years with current technologies', *Science*, vol 305, no 5686, pp986–972
Photon (2004) 'Thai RPS may call up for up to 2000 of grid-connected PV by 2015', *Photon International*, vol 12, p42
Pigou, A. C. (1912) *Wealth and Welfare*, London
PV TRAC (Photovoltaic Technology Research Advisory Council) (2004) *A Vision for PV Technology for 2030 and Beyond*, Brussels, http://europa.eu.int/comm/research/energy/pdf/vision-report-final.pdf

Rechsteiner, R. (2003) *Grün gewinnt*, Zurich, Orell Füssli
Sawin, J. L. (2004) 'Mainstreaming renewable energy in the 21st century', *Worldwatch Paper* 169, May 2004, Washington, DC
Traube, K. (2004) 'Renaissance der Atomenergie?', *Solarzeitalter*, vol 16, no 4, pp3–6

Part 1

Context

2

Oil Depletion

Werner Zittel and Jörg Schindler

The imminent turning point

The future supply of crude oil is only indirectly governed by the amount of reserves still available; it is much more determined by future production capabilities. An eventual future supply crisis will not be the result of the production of the 'last drop' of oil; long before this ever happens the peaking of world oil production will mark a turning point for our economies. Then a period of time which has seen ever growing production volumes will be followed by a time of steadily shrinking production year after year.

The peak of worldwide oil production will be a unique historic event. It has never occurred before and it will never occur again. Due to the singularity of this event we have not developed an intellectual or emotional sensitivity that could help us in coping with it.

The production peak will not send an advance signal to the markets as to when it will happen (and anyway by no means 10, 20 or even 30 years in advance). In fact there are no long term signals to which the crude oil markets and the consumers can react. The markets react only to short term imbalances between supply and demand; it is therefore futile and dangerous to attempt to assess the future supply situation on the basis of current prices.

The life cycle of oil production

The three phases

The production profile over the life cycle of any oil region can be divided into three phases. The first is a phase of continual production increase ('pre-peak'); in the second phase production is stagnant ('at peak' or 'plateau'); and finally there is a phase of continually declining production ('decline').

These three phases of oil production constitute a general trend. In the following we will explain this production profile by taking Norwegian oil production as an example and then we will sketch the worldwide situation.

Figure 2.1 shows Norwegian oil production where the production profile of each individual field is presented in chronological order.

Phase 1: 'Pre-peak'

No oil company in the world is holding production from its oilfields in reserve to take advantage of improved economic prospects in the future. The largest and most productive oilfields are brought on-stream first. These fields have the highest production rates, and will produce oil for the longest time. In the early phase of production in any basin an expansion of production is quite easily achieved by adding new wells within already producing fields or by developing further fields.

Phase 2: 'At peak'

The longer production goes on, the more the pressure in the oilfield drops while the water level rises. Then at some point in time the production rate begins to decrease. In this situation the addition of further wells within an already producing field leads to a further drop of pressure and therefore succeeds only for a very short time in upholding the rate of production. Modern state-of-the-art production methods (for example increasing the pressure of the field by water or gas injection, or reducing the viscosity of the oil by chemical or thermal treatment), which now have been successfully applied for more than 20 years, can postpone and sometimes soften the decline a little, but they cannot reverse it.

As soon as the large oilfields of a region have passed their production maximum, an ever-increasing number of new smaller fields have to be developed to compensate for the decline of the older, already producing fields. In such a situation it becomes increasingly difficult to extend the production of the whole region. As more and more smaller and less attractive fields enter the production, the 'treadmill' situation of having to develop ever more fields is aggravated by the concurrent faster depletion of these smaller fields. In Norway this phase started in about 1995 or 1996.

Phase 3: 'Decline'

Then, at some point in time, the production decline in the already producing fields gets so steep that it can no longer be compensated by the development of new fields. This is the time when the production of the whole region starts to enter the decline phase. To a certain degree, oil companies and oil producing states can influence this decline pattern by their production strategies. Thus the longer the producers try to uphold the production maximum, the steeper the ensuing decline will be.

Examples

Example 1: The oil production of Norway

Figure 2.1 shows the production profile of Norway. The decline phase started in 2002.

Example 2: The oil production of the UK

Figure 2.2 shows the production profile of the UK. There, production peaked in 1999. The production of all the big fields has been declining for many years,

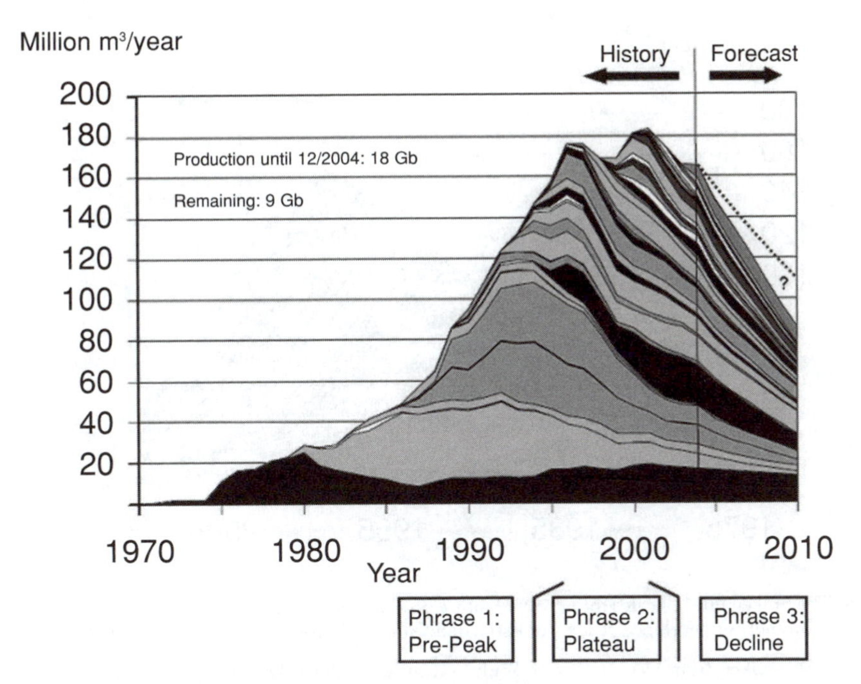

Note: The broken line indicates the expected future production profile – taking into account the already known, but not yet developed fields. An Estimated Ultimate Recovery of 26Gb is assumed.

Source: Data taken from National Petroleum Directorate 2004; Data for 2004 are estimated from January to October production; Analysis and Forecast: LBST (2004).

Figure 2.1 Norwegian oil production profile showing individual fields

but for the first time in 2000, bringing new fields into production could not compensate for this decline. Since then, total production declined until 2004 by a total of 27 per cent. The sharp decrease in production after 1985 shown in the figure was a consequence of the Piper Alpha accident, which led to stricter safety regulations and required some time for the industry to adapt. The subsequent production increase was possible because of the existence of the then current known finds – but this is different from the present situation.

Example 3: The oil production of Alaska

The future oil production of Alaska can be taken from the forecast of the Department of Natural Resources in Alaska, upon which the following figure is based.

The decline of the large fields might be offset for some time by the development of already known smaller fields, but it is more likely that delays in commencing new projects will lead to a continuing overall decline in production. Indeed, the discrepancy between the field data from the

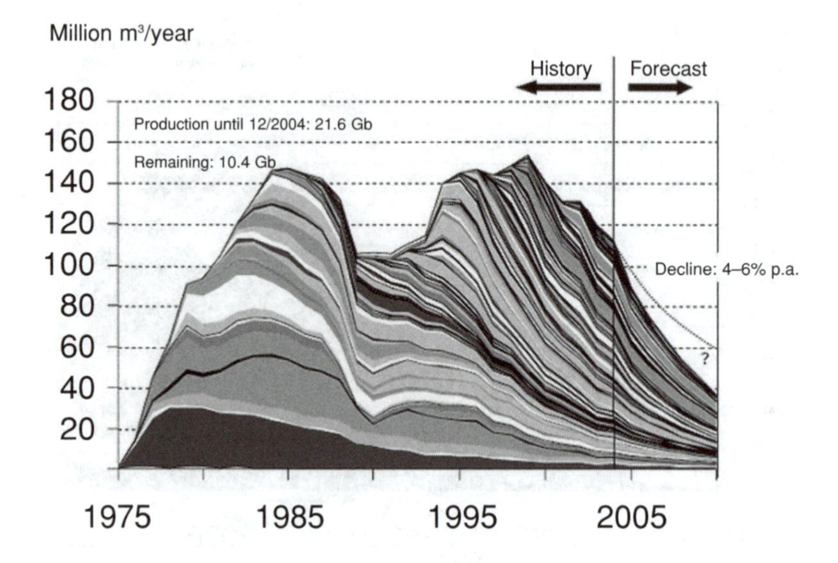

Note: The broken line indicates the expected future production profile – taking into account the already known, but not yet developed fields. An Estimated Ultimate Recovery of 32Gb is assumed.

Source: Data taken from Department of Trade and Industry, 2004; Data for 2004 are estimated from January to August production; Analysis and Forecast: LBST, 2004.

Figure 2.2 UK oil production profile showing individual fields

Department of Natural Resources in Alaska – taken from the 2000 forecast – and the actual production data from the Department of Energy over the last two years supports this argument.

Example 4: US – Decline analysis of regional production profiles

'Excessive activity' can stop the production decline for a short while – but, according to all experience, only at the cost of even higher decline rates in subsequent years. This can be demonstrated by the production profile of the US, shown in Figure 2.4. The US reached its production peak in 1970. Even the growing production in Alaska in the following years could only compensate for the decline of the old fields for a short while. In the early 1980s there was excessive activity in bringing new wells into production. This resulted in a slight increase in the production of the 'Lower 48' states. But within a few years production started to decline again at even higher rates than before – and the long-term decline trend was approaching the values one would have expected from the experience of the 1970s.

Nor could the prevailing trend of a declining domestic oil production and rising imports be reversed in later years – in spite of an economic situation that

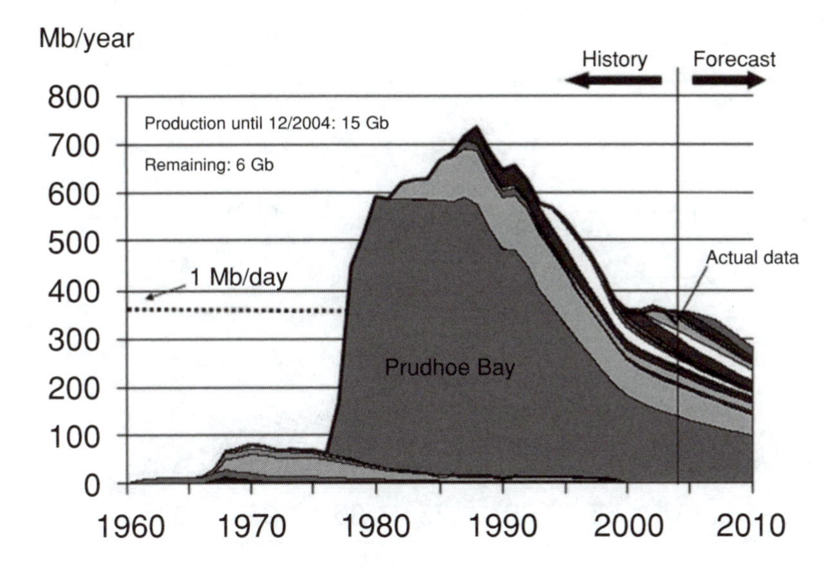

Note: The peaking oil production from Prudhoe Bay in 1989 set the scene for the unavoidable decline in production in the following years. Production from small fields has succeeded in halting the decline at less than half the level of peak production, but this can last for only a few years, and at much higher costs than in the early production phase.

Source: For field by field data and forecast; Department of Natural Resources, Division of Oil and Gas (2000) Annual Report. Source for actual data: Department of Energy – Energy Information Administration (2004); The actual production for 2004 is extrapolated including November data.

Figure 2.3 Field by field analysis of oil production in Alaska

dampened the rise of oil demand and favoured exploration and production of oil, and also in spite of the announcement of a new energy policy by the government aimed at increasing American oil production. Figure 2.4 shows the historical oil production of the US.

Alaska passed its maximum more than ten years ago. Texas passed its maximum 30 years ago. Today, Texas produces no more than it did in the 1930s. The other states passed peak production in the early 1970s and are declining year by year

Average oil production per well has decreased in Texas to about six barrels/day, while in Pennsylvania – where the oil boom started 140 years ago – it is even lower.

These production rates can be compared to the generation of renewable energy: an average oil well in Texas produces about the same amount of energy as might be provided in the form of thermal energy by solar collectors of 1800m^2 in size – that is less than 45m square.

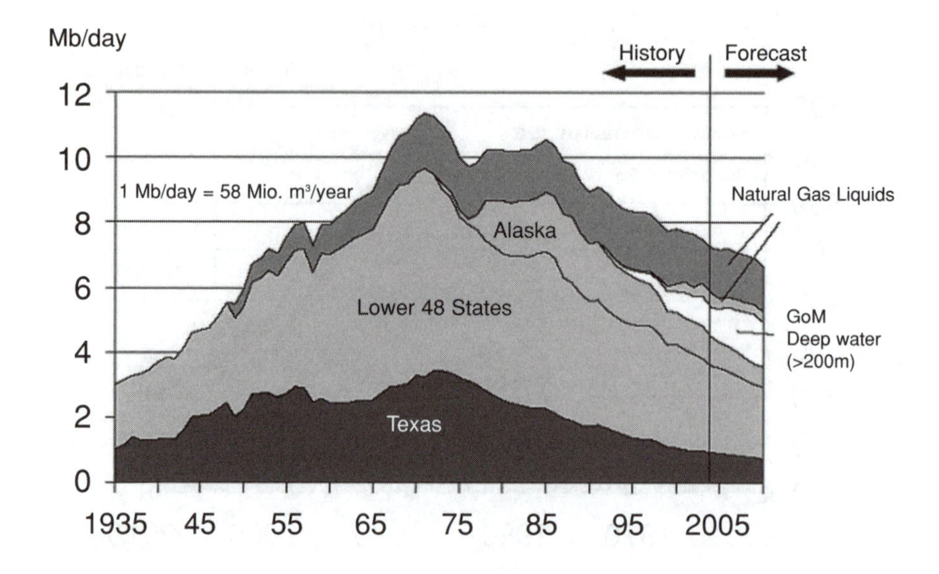

Note: The future oil production profile for the declining oil regions of Texas and 'Rest of the US' is controlled simply by the physics of depletion, allowing a straightforward extrapolation of existing trends.

Source: Data for Texas: Texas Railroad Commission; other data from Department of Energy – EIA; 2004 data are extrapolated from January to October data; 2004 data for deepwater Gulf of Mexico are LBST estimate; Forecast: LBST.

Figure 2.4 The future oil production profile for the declining US oil regions

Only in the deepwater area in the Gulf of Mexico (GoM) can production still be increased, as all other areas have already passed their production maxima. Even production from shallow sites in the GoM is already declining.

Of special importance is the field complex, 'Thunder Horse', which holds about one billion barrels of reserves and is by far the largest discovery in recent years. A study by the US Department of the Interior in May 2001 suggested that the production from all deepwater fields in the GoM might still increase to 1–1.2 megabarrels per day (Mb/d) by 2005 (see Department of Interior, 2001). Unless major new discoveries are made this year and next, of which there is at present little sign, production even in the deepwater areas of the GoM will start to decline soon after the connection of the Thunder Horse fields in around 2006. The GoM production forecast is discussed in more detail below.

Most official forecasts predict increasing natural gas liquids (NGL) production in the near and medium term future, but there is reason for serious doubt. It is much more likely that NGL production will strongly decline in the coming years, in response to the tight natural gas supply, which is even more severe in the US than in Canada.

Enhanced Oil Recovery (EOR) from ageing fields

So-called secondary and tertiary measures are often cited as a viable source of enhanced future oil recovery rates and as a reason to blame forecasters for being too pessimistic regarding future oil production. Indeed, these measures are widely used and have the potential to enhance the oil production rate. When the pressure in the oilfield has decreased to a level where the viscosity of oil dominates the production profile, leading to smaller extraction rates, pressure can be raised artificially by water or gas injection, or by reducing the viscosity with the injection of steam or a chemical surfactant.

However, for various reasons we should not overestimate the influence of these measures:

- EOR measures have already been applied for more than 20 years, and these measures are accounted for in production forecasts. There will not be any sudden jump in the future – continuous progress is and was always part of the production forecasts. There are two major examples for this: (1) the production profile of German oil. After its peak in 1968 the production continuously declined despite efforts to implement EOR techniques, and (2) the production of Prudhoe Bay as already explained in Figure 2.3. This field is at the technological forefront and every possible new measure was exploited to enhance production and to avoid a decline, with almost no success. Today more water is extracted from the wells than oil, water that was injected into the field to increase the pressure
- EOR measures are only applicable beyond peak production when the pressure level is low. These measures cannot reverse a decline into an upward production profile for any substantial period of time
- EOR measures are most effective in certain fields with complex geology, which exhibit a low recovery factor
- These measures are only effective in the sense that more dollars are gained with the extra oil than have to be spent for the measure
- Usually these measures increase the production rate for a short period of time, but enhance the decline in the long term – they are only intended to extract the oil faster, but not to increase the overall oil recovery.

To illustrate this, the influence of EOR measures at one of the largest US fields is investigated in Figure 2.5. The Yates field, which was found in 1926 in Texas, has produced since 1929. Since peak production in 1970 the production rate has declined by more than 75 per cent. In 1993 hot steam (TAGS) and chemicals (WALRUS) were injected to enhance the production rate. This measure was successful for about four years. Afterwards the decline was even steeper, exceeding 25 per cent per year instead of 8.4 per cent as before. Today the production rate is below the level it would be without these measures. To assess the overall influence of this measure, out of the 1.4 billion barrels of oil that have been produced since 1929, only 40 million are due to enhanced oil recovery – an increase of about 3 per cent.

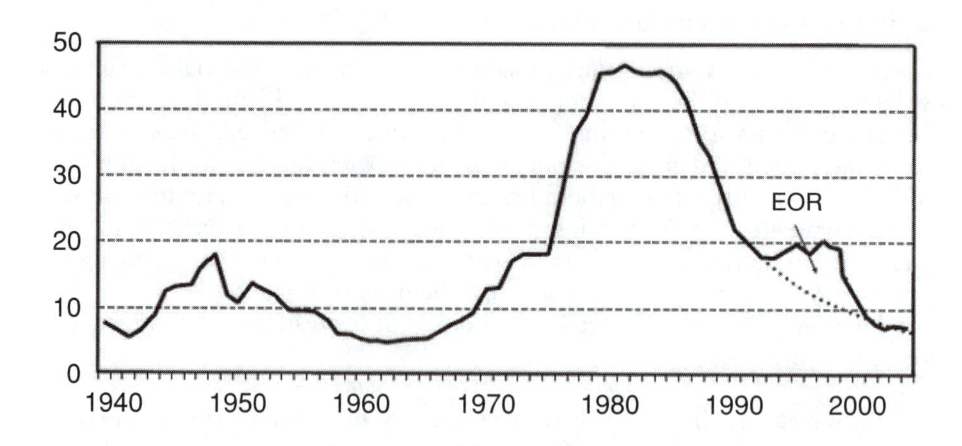

Note: Total output increase was 40 million barrels, with respect to total output since production start of 1.4 billion barrels (an increase of 3 per cent). If the initial recovery rate was 20 per cent, these EOR measures would have shifted the recovery rate to 20.6 per cent.

Source: RRC (2004) (Texas Railroad Commission, various statistics, see www.rrc.state.tx.us).

Figure 2.5 Yates field

The influence of non-conventional oil

Example 1: The oil production of Canada

Very often one hears the argument that depletion of the *conventional* oil supply will be offset by production from *non-conventional* sources, mainly oil from tar sands. It is relatively easy to evaluate the potential of non-conventional oil production in North America. Because of the long lead times and the necessary huge investments, planning has to be done long in advance. These plans, which are documented, allow a reasonable forecast of what is possible over the medium term.

Figure 2.6 shows the history of oil production in Canada. *Conventional* oil production reached a maximum about 30 years ago. The downward trend has been reversed in recent years, due to production from the offshore Hibernia field, several hundred kilometres east of Newfoundland. Hibernia was discovered in 1978 and is believed to contain about 600 million barrels of oil. The field was brought into production only a few years ago, because of difficult operational conditions.

Heavy oil production in Alberta and Saskatchewan has been increasing in recent years, with the production of synthetic crude from tar sands from Alberta starting as early as 1967. Up to now, production has come from only a few localities where the tar sands are close to the surface, allowing a relatively

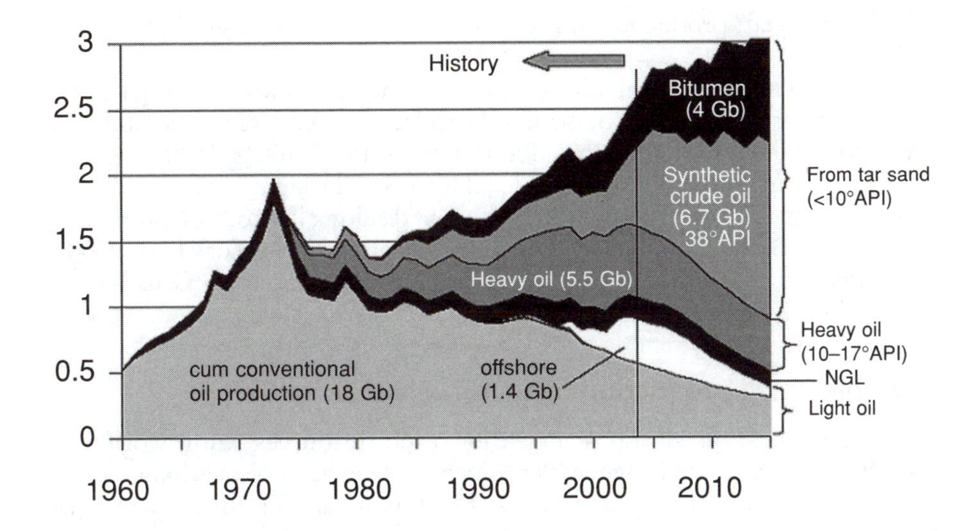

Note: Total oil production including conventional oil and non-conventional oil.

Source: 1960–1974 data US DoE Energy Information Administration; 1975–2004 data from National Energy Board; 2005–2015 forecast LBST based on NEB estimate, May 2004.

Figure 2.6 Total oil production in Canada

cheap extraction. About two-thirds of the recovered bitumen is processed to synthetic crude in Canada.

The onshore crude oil supply from conventional sources will continue to decline over the coming years, following a standard pattern that is already well established. Natural gas liquids (NGLs) are another important element, but production is more likely to decline than increase in the near future as most Canadian gas wells have shown substantial decline rates over recent years.

It looks as if North America as a whole will run into severe gas supply problems. Due to increased gas prices, the operators are likely to market NGL blended with natural gas rather than extract the liquids by processing. This practice occurred during the winter of 2000/2001 when gas prices in the US skyrocketed; NGL production declined by some 30 per cent, since NGLs were left in the natural gas instead of being marketed separately.

The best hope for increasing Canadian conventional oil production rests with offshore areas, mainly east of Newfoundland. Hibernia itself has already reached maximum capacity, but can continue to produce in declining amounts for some years to come. Terra Nova is another large field in the vicinity with reserves of about 400 million barrels. It was brought into production late in 2002 and is expected to reach a maximum production plateau of about 150 kilobarrels per day (kb/d). The third large discovery in that area, White Rose, was found in 1984 with an estimated reserve of about 230 million barrels. It is scheduled to start production in 2006 with an estimated peak production close to 100kb/d.

Most non-conventional oil comes from the tar sands of Alberta. The financial plans of the large companies have already been announced and form the basis for the forecasts by the Alberta Energy and Utilities Board.

The production of synthetic crude oil (SCO) will roughly triple by 2015, as will the production of non-upgraded bitumen, resulting in a total oil production in Alberta of 2.2Mb/day in 2015, an increase of about 1.2Mb/day from 2000.

Adding these sources together gives a forecast for total Canadian oil production as shown in Figure 2.6.

Example 2: Oil production of Venezuela

The other location of non-conventional oil resources with worldwide importance is the extra heavy oil deposits in Venezuela. The development and production from these deposits started 15–20 years ago. In the late 1980s BP invented a method of mixing oil, water and chemicals in a type of synthetic crude called Orimulsion. Although this method facilitates extraction somewhat, the environmental impacts are tremendous, and many countries refuse it. Today several projects have been completed or are under construction, which will increase the total production capacity in 2015 to about 1.2Mb/d, up from 0.4Mb/d in 2002.

Though theoretically a huge resource, tar sands and extra heavy oil in Canada and Venezuela will have only a low impact on the worldwide supply situation for at least the next decade. This is due to the fact that only a small amount of the reserves can be extracted economically. These projects are already planned; further extensions will be much more difficult. The long planning and construction times and the long return of investment times ties up capital for long periods, which hinders fast reactions and fast enlargements. Furthermore, the low extraction rate makes any production increase very slow and subject to long-term planning.

Finally, one should not forget the energy and water resources that are needed to extract the oil from the ground. Just to give an idea of the size of these efforts: today a quarter of all water used in Alberta is directed to the oil and gas industry. Canadian Natural Resources Inc (CNR) is investing about CAN$5 billion to push the production rate of the project horizon to 235kb/d by 2011. Within that project there are plans to divert a river for several miles and to use the water to allow on-site extraction of 3.5 cubic metres per second. Large extensions of these projects are simply not realistic when one considers the needs for other resources and the environmental side effects.

Besides the water consumption there is already competition between tar sands and natural gas extraction: gas deposits close to the tar sands facilitate the lifting of oil. But once these gas fields are emptied, the pressure drops and oil from these deposits becomes inaccessible for today's technologies. In fact, the Alberta Energy and Utilities board has already stopped the development of gas fields close to these areas (AEUB, 2003). Another competition arises from the gas consumption of the upgrader for synthetic crude oil processing. At a time of ageing and declining gas fields, of course, the scarce availability will limit the production and upgrading of additional oil from tar sands and increase future production costs.

Signs of the imminent peak of production

The fact that the oil industry is spending so much money and using all available high technologies to explore for oil in unfavourable areas (such as polar, Caspian, Deep Sea) might be interpreted as the industry's admission of the fact that ever decreasing amounts of oil are being found in other (easier to access) places. Otherwise the industry would look for that oil first.

As one can see in detail from Figures 2.7 and 2.8, ever more oil producing countries are reaching their peak of production in spite of very favourable economic conditions: the high prices of crude oil.

The aggregate production peak of all countries outside OPEC and FSU occurred in 2000 or 2001.

Table 2.1 shows the assignment of all the important oil producing countries to the three production phases defined above. This assignment is done according to our assessment with regard to the production profile in the past and the ratio of reserves/discoveries (that is the share still available for production). Basic data are taken from the industry database. The detailed assignment, country by country follows that given in Zittel (2001). But in contrast to our grouping three years ago we changed the grouping of some countries. The ten largest producers in the group of countries with declining production are the US (peak in 1970), Iran (1974), Norway (2001), Venezuela (1968), the UK (1999), Indonesia (1977), Libya (1970), Oman (2001), Egypt (1973) and Argentina (1999).

The largest producers at or close to peak production are Mexico, China, Canada, Kuwait, Nigeria, Qatar, Malaysia and Neutral Zone. One might argue that Canada has a vast potential of unconventional oil from tar sands. However, due to the declining production from conventional oil sources, extensions from tar sands will be needed to make up for the decline in production. There will be little left to extend total production, at least for the next 10–15 years.

The largest producers in the group of countries that still might increase their production volumes are Saudi Arabia, Russia, Iraq, United Arab Emirates, Algeria, Brazil, Kazakhstan, Angola, Azerbaijan and Thailand.

There might be some points of discussion, for example whether Venezuela will decline in its future production though it has vast unconventional oil deposits, whether Iran will decline in the near future, whether Iraq can really increase its production volumes substantially over the next 10–15 years, whether Saudi Arabia still can increase considerably, for how long Algeria and Angola still can increase their volumes and so on. Most prominently, the reserve estimates of the OPEC countries are in discussion, and whether these countries still have the potential for substantial capacity extensions.

These questions are addressed on p41. For the present purpose a rough classification is sufficient in order to illuminate the basic features of what depletion means.

Table 2.1 *Production in 2003, cumulative discoveries and reserves[a] of all countries grouped by phase of production*

Region	Production 2003 (Mb/d)	Share of world production (%)	Cumulative discoveries (Gb)	Reserves 2002 (Gb)	Still available (%)
Decline	29	38.2	745	293	39
At peak	17.6	23.1	304	128	42
Pre-peak	29.4	38.7	1095	772	70
Total	76	100	2144	1193	56

Note: a Cumulative discoveries minus cumulative production.

A little more than a third of the world's oil production is coming from the few countries still able to increase their production substantially (these are the countries at pre-peak). Those countries own two thirds of the world's remaining oil reserves and have to date used about 30 per cent of their reserves. Nearly half of the world's known oil reserves have already been produced.

Scenarios of future oil production

For the discussion of the question of annual production growth in the future, we group the oil producing countries according to phase of production (see Table 2.1) and then we make assumptions regarding possible and necessary production rates. For this purpose we make projections for 2010 and 2020.

We concentrate our discussion on two scenarios: the first scenario assumes oil demand remaining constant over the next two decades; the second scenario assumes a (moderate) increase in future oil demand. It is our purpose to keep these scenarios very simple in order to make it easy to follow the argument and to enable assessments to be made by the reader.

Scenario 1: Constant demand

This scenario rests on the following assumptions. The world demand remains constant until 2020. The average annual production decrease of the countries in 'Decline' is assumed to be 2.5 per cent. Countries now in the 'At peak' group are expected to start their decline in 2006. The production gap caused by the declining countries has to be compensated for by increased production of the countries in the 'Pre-peak' group and/or by the additional production of non-conventional oil (NC-oil).

Scenario 2: Rising demand

Assumptions for this scenario include the following. The future demand growth will average 1.5 per cent annually (comparable with the average growth

Table 2.2 *Scenario 1: Average daily production (Mb per day)*

Year	2003	2010	2020
Decline	29	25	19.4
At peak	17.6	15.9	12.4
Pre-peak + NC	29.4	35.1	44.2
Total	76	76	76

rate since 1975). The average annual production decrease of the countries in 'Decline' is assumed (again) to be 2.5 per cent, and countries now in the 'At peak' group are expected to start their decline in 2006. The production gap caused by the declining countries has to be offset by the countries in the 'Pre-peak' group and/or by the production of more NC-oil).

Table 2.3 *Scenario 2: Average daily production (Mb/d per day)*

	2003	2010	2020
Decline	29	25	19.4
At peak	17.6	15.9	12.4
Pre-peak + NC	29.4	42.2	64.7
Total	76	83.1	96.4

We think that both scenarios are rather on the optimistic side because the average decline rates of the countries in 'Decline' are assumed to be only 2.5 per cent. This decline rate equals about the average decline rate experienced in the US in individual basins after passing the peak. In offshore basins however, decline rates of 5–10 per cent annually are more likely. As illustrations of this, Alaskan production has halved since 1989, and in the UK production fell by 27 per cent in four consecutive years after passing the peak. Figure 2.7 shows the production profile of the three groups for the two scenarios.

Scenario 1 shows how much the 'Pre-peak' countries would have to increase their production over the next 20 years to ensure constant supply.

In scenario 2 the task for the 'Pre-peak' countries (including increased production of NC-oil) is even tougher: in order to offset the decline in the other regions, their production would have to increase by 2.3Mb/d every year (equivalent to 8 per cent annually) for 20 years!

How realistic are these scenarios?

How probable are the developments described in these scenarios? The following considerations may be helpful in assessing the situation:

- The US EIA estimates that production from the Caspian Sea under favourable conditions could grow by 2.7Mb/d to 4Mb/d by 2010 and could

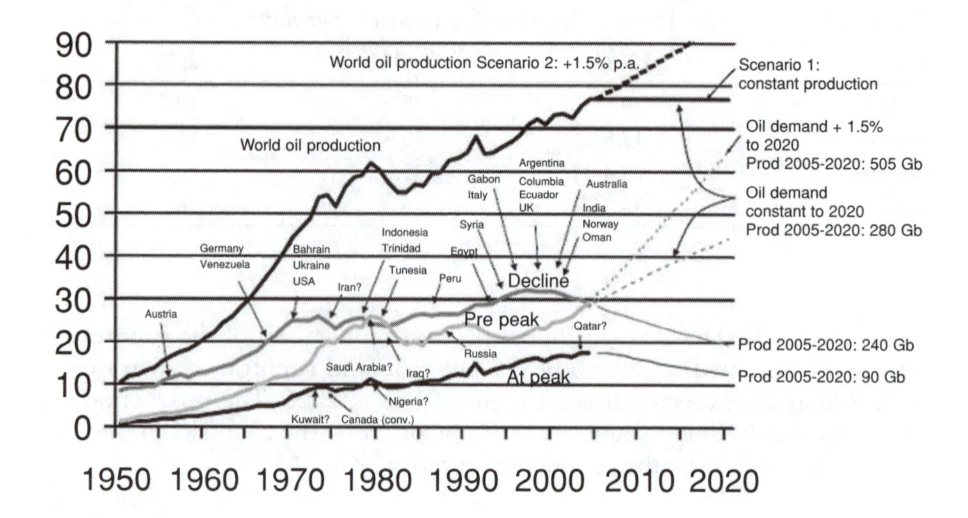

Note: All oil producing countries are assigned to one of the groups 'Decline', 'At peak' or 'Pre-peak', according to the phase in which they are producing.

Source: IHS Energy 2003; Scenario forecast: LBST.

Figure 2.7 World oil production scenarios

possibly reach 6Mb/d in 2020 (EIA, 2001). As discussed below, an increase of about 1–1.5Mb/d is much more likely

- Oil production in the deep sea off Angola until recently was expected by the industry to grow to about 2.5Mb/d by 2015. This would amount to an increase of production of 1.6Mb/d over the next 11 years. Already today these forecasts are deemed to be too optimistic (Luanda, 2000) and an expected increase of about 1–1.5Mb/d seems more realistic
- All known projects up to 2015 dedicated to produce NC-oil from Canadian oil sands add up to an expected production of 2.2Mb/d. This amounts to an expected production growth of 1.2Mb/d within the next nine years
- If we add up the above increases in production regarded as possible by industry we arrive at about 4.2Mb/d. But if demand grows according to scenario 2, there remain more than 12.8Mb/d in 2010 and 35Mb/d in 2020, which have to come from these and other sources
- One should also note that the assumed production increase of 1.5 per cent annually in scenario 2 results in about 96Mb/d for the year 2020. This is still modest with respect to the oil consumption increase assumed by IEA in its World Energy Outlook 2004. There, the IEA forecasts an increase of production by 2020 to about 107Mb/d and to 121Mb/d by 2030 (IEA, 2004)

- The chairman of BP, John Browne, was reported saying alongside of the G8 gathering in Davos in the spring of 2001 that he expected the maximum production capacity of the industry to be somewhere around 90Mb/d.

We very much doubt that even the peak capacity of 90Mb/d as estimated by John Browne can actually be reached and think it much more likely that world peak production will eventually lie somewhere in the region of 80Mb/d.

Worldwide experience shows that the production peak in an oil region happens when about half of the ultimately recoverable oil has been extracted. However, due to the poor available data, these figures have to be seen only as a rough estimate. But with regard to this estimate and together with the above considerations and mainly based on further arguments regarding individual production capacities as listed below, we are very confident that the worldwide production peak is very close.

In contrast to other estimates, the US Geological Survey (USGS) uses different definitions of resources in its assessments (USGS, 2000). The USGS assessment leads to the observation that the US oil production peaked in 1970 when only one third of the USGS-estimated ultimate recoverable oil was extracted. If we translate this in relation to the worldwide situation – where the USGS estimated ultimate recoverable world resources of 3300Gb oil and natural gas liquids in the mean case – then world oil production should peak when about 1100Gb of oil are consumed. This is right now!

The geologist Colin Campbell estimates that about 2200Gb of oil (including crude, NGL and heavy oil) can ultimately be extracted. He forecasts that the world production maximum will occur around 2005, based on the Hubbert model (Campbell et al, 2002).

To develop a better understanding of this assessment, the next chapter focuses on the same topic, but from a slightly different point of view. Also the following sections give further hints and circumstantial evidence regarding the imminent peak of oil production.

Oil production outside OPEC and outside the former Soviet Union

Over the years, the number of countries that have passed their production peak has grown. Figure 2.8 lists the countries outside OPEC and outside the former Soviet Union in the order of their peak production date. Indonesia is the only important representative of OPEC on the list, having passed its peak production 25 years ago. Countries that still might increase their production are given at the right-hand side of the figure. The most important ones are Angola and Brazil. There is a degree of uncertainty whether China and Mexico have already passed their production peak or whether they are still a few years off it. However, this does not change the general pattern much.

The data for North America are taken from Figures 2.4 and 2.6, but exclude bitumen and synthetic crude oil from tar sands as NC-oil. The NC-oil from Canada and Venezuela is added explicitly in Figure 2.17.

The future production pattern of the countries in decline is easily calculated from trend extrapolations. After adding in possible production increases, mainly confined to Angola and Brazil where new deepwater discoveries have been made, we get an overall estimate of decline of some 13–14Mb/d until 2015.

Unless demand is cut, the world will depend on additional supplies from the former Soviet Union (including the Caspian) and the Middle East OPEC countries, and on additional non-conventional sources in Alberta (as outlined above) or Venezuela. Canada might provide about 1.2Mb/d of additional oil until 2015. Another 0.8Mb/d could come from other unconventional oil projects in Venezuela. As already mentioned above the Caspian region could possibly supply 2.7Mb/d additionally in 2010, though 1–1.5Mb/d is much more likely. Let us assume that in 2015 about 2.5Mb/day will be added from the Caspian region.

The missing 8–9Mb/d would have to come from Russia and the Middle East OPEC countries.

It remains to be seen if this is achievable, as the available data for these countries are of poor quality and do not allow reliable forecasts. However, an increasing demand for oil would need even further resources. Another 13Mb/d would be needed if world demand were to rise by 1.5 per cent per year on average up to 2015.

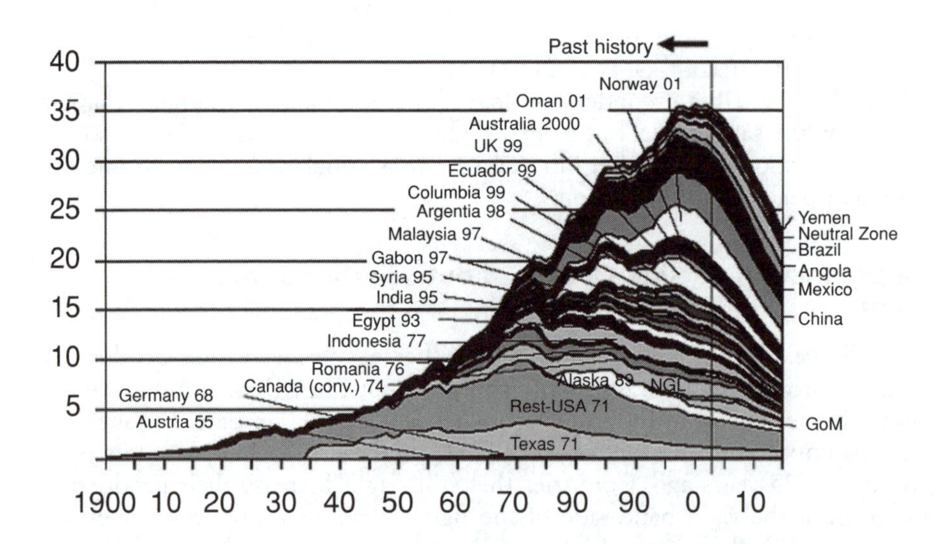

Note: a Only Indonesia is included, having passed its production peak in 1977.

Source: government statistics for US, Canada, UK and Norway; State company statistics in Mexico, China, Brazil and Malaysia; other data from IHS-Energy; forecast LBST, 2004.

Figure 2.8 Oil production summing up all countries outside the former Soviet Union and the OPEC countries

More oil from Russia and the Caspian?

Russian oil production reached its peak in 1987 with about 11.4Mb/d. Some years later in 1996 production fell to 6Mb/d. This decline was partly due to the breakdown of the country's economy, but also due to having passed peak production in important Russian oilfields.

Thus the largest oilfield, Samotlar, with a reported size of 28Gb, had reached peak production in the mid-1980s when it produced 1.3Mb/d. Today its production rate is down to 0.1Mb/d. Recent data indicate that the ultimate recovery of the field is overestimated in official documents by about 8Gb, leaving only 20Gb as ultimately recoverable – an overestimate of about 40 per cent. Even the second largest field, Romashkino, holding 18Gb, has already passed peak production and seems to be overestimated by 10 per cent. Further analysis shows that Russian oil discoveries in general are overestimated by about 20–30 per cent. And most of the large fields have already passed peak production.

In the last five years, oil production has recovered to about 8.6Mb/d. Official statements claim that further production increases will shift the level close to or even above the record levels of the 1980s. However, some scepticism might be allowed. First of all, the reported production volumes cannot be checked. Only exported quantities are measurable for outside observers. It is also possible that oil is removed from the domestic market and exported instead to gain hard currency. Second, once the largest fields are beyond peak production the gradual decline has to be compensated with ever rising new discoveries and field developments. Though there is some potential for further discoveries in Russia, it is very likely that the best opportunities were taken first in those early years after the economic transition. It is very likely that the recent upward trend will be disrupted soon and will turn into a decline as experienced earlier during the 1990s.

Many people believe that the Caspian Sea area holds huge amounts of still undiscovered or undeveloped oil. But actually, only four super giants are worth mentioning: Tengiz, Karachaganak and Kashagan in Kazakhstan, and the complex Chirag-Azeri-Guneshli in Azerbaijan. In total, these fields hold about 20Gb of oil or condensate. But surprisingly, big oil companies that took part in the early development of these discoveries withdrew from operational consortia or postponed their plans: BP and Statoil withdrew from the latest discovery, Kashagan; Chevron postponed plans at the already producing field Tengiz; Exxon after drilling a dry hole in Azerbaijan and having spent about US$1 billion on its exploration; or even the Russian company, Lukoil, which – as a surprise to observers – sold its assets in Azerbaijan. The withdrawal of several billion dollars of investment makes it very doubtful whether the envisaged future production levels are still realistic. Moreover, the remaining consortium for the development of Kashagan had serious discussions with the regional government about future plans. They have now decided to boost production levels even further than previously planned to 450kb/d instead of 100kb/d. But at the same time the earliest flow of the first oil will be in 2007, instead of 2005 as previously planned. This delay is due to serious development problems caused by geology and geographical location (huge depth under shallow water in an extreme climate, high pressure and high sulphur content). This information points more to increasing problems than to increasing oil flows, and makes it very likely that this area is far from being an oil bonanza.

In view of all these facts and considerations we conclude that for the whole former Soviet Union, a production increase of about 4Mb/d is still possible up until 2010; thereafter a decline will set in. Some early extensions of Russian developments in the coming years will most likely turn into decline after 2005, while on the other hand the development of additional Caspian oil will come into production mainly after 2007.

Frontier areas

Some people also hope for huge reserves of undiscovered oil in the deep sea of the oceans. However, looking somewhat closer these people might find themselves disappointed in a few years, along with the oil companies that are also seeking them out.

Only very rare combinations of events in the geological past are responsible for the accumulation of oil in ultra deep waters. These conditions are found in front of Palaeozoic river deltas, for example east of Northern Brazil, west of Angola and Nigeria, in the Gulf of Mexico (GoM) and in the Southern Chinese Sea. Within the last 20 years about 35Gb of oil has been found in these areas, and the discovery rate in most deep sea areas has already slowed down during the last two years. Therefore one can hope to find maybe another 20–30Gb in the future.

Angola and Brazil together produce about 2.4Mb/d at present with some potential to double that rate in 5–8 years for a very short period before the inevitable decline sets in.

Example 1: deep water oil in Angola

Reserves in BP Statistical Review remained at 5Gb for many years and in 2004 they were given as 8.8Gb, completely ignoring the huge discoveries during the last years as shown in Figure 2.9 (p39). Since their development has started it is very likely that over the next few years Angola's 'proved' reserves – as used in BP Statistical Review of World Energy – will increase substantially, as soon as these discoveries are developed and their reserves are categorized as 'proved reserves'. However, despite these reserve increases, future discoveries will be much smaller following the already visible trends. Presumably, total discoveries in Angola will remain below 20Gb, leaving about 12–15Gb for future production. These reserves will not shift the annual output above 1.5–2Mb/year.

Example 2: deep water oil in the Gulf of Mexico

Another frontier area that needs a closer look is the GoM offshore area referred to above. Figure 2.10 disaggregates the cumulative discoveries in that region since 1945 into various depth classes. The early period saw the development of the continental shelf. To date, about 11Gb has been found there without any hope for new large discoveries. During the 1980s the frontier

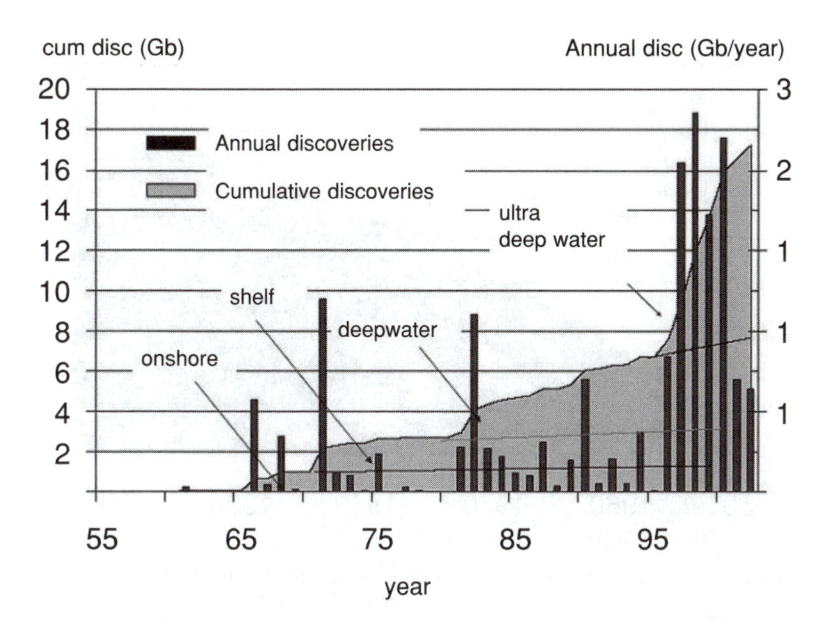

Note: In each basin the large discoveries were made first, followed by small additions; it is very likely that discoveries in the ultra deep water peaked in 1998 and future prospects are getting worse.

Source: Industry database.

Figure 2.9 Annual (bars) and cumulative (area) discoveries in Angola

was shifted to deeper basins, up to 1000 feet, 2000 feet, 3000 feet and even below. Over the last 20 years, less than 2Gb was found in areas between 500 and 2000 feet, with no sign of hope for the future. Over the last ten years attention was drawn to even deeper water levels, with another 2Gb found in these new frontier areas. Just over the last few years remarkable discoveries were made in ultra deep waters below 4000 and even 5000 feet of water depth, as already indicated above with the exceptional Thunderhorse field (~1Gb).

But due to its maturity, production from the continental shelf is already in decline. Even many new developments of deepwater fields are turning into decline, sometimes only a few months after the commencement of production. But nevertheless, several large ultra deepwater fields are not yet developed. Therefore production from this area will still rise in the coming years.

A detailed field-by-field assessment based on published monthly production data indicates that the total production from all connected fields in the deepwater GoM is declining (Figure 2.11). The huge bubble in the second half of this decade is due to connections of most of the known, but not yet developed fields. For this figure an optimistic initial reserve of about 6.5Gb was assumed, more than double the proven discoveries as given by the MMS

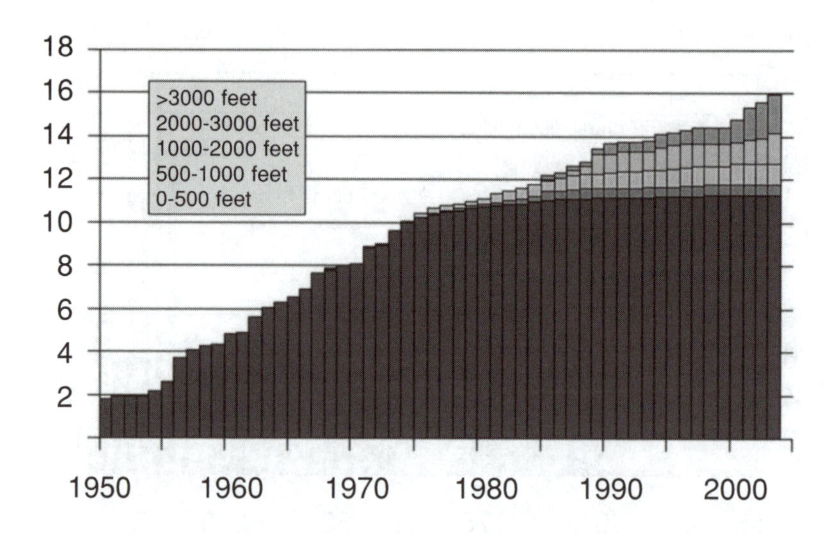

Note: New discoveries are still made but only at locations with more than 3000 feet water depth.

Source: MMS, 2003.

Figure 2.10 Oil discoveries in the Gulf of Mexico split into different depth classes

(fields below 1000 feet water depth in Figure 2.10).

A total of 6–7Gb in the deepwater area sums up to 17–18Gb of total GoM oil fields including shallow water shelf oil developments (see Figure 2.10, above).

But even there new discoveries are getting scarce. Actually, most of the deepwater fields were found in the late 1980s and in a second wave in the late 1990s. The latest and most up-to-date largest discovery was the Thunderhorse complex in 2000/2001. Thunderhorse is hoped to come on-stream around 2006–2007 with a designed capacity of about 250kb/d or 90Mb/year. At present, oil companies are devoting their attention mainly to the development of these discoveries. The number of active rigs in Figure 2.12 shows a strongly declining interest in new exploration, suggesting that presumably most of the oil in the deepwater GoM has already been found. The diminishing exploration interest of oil companies is also reflected in declining monthly lease sales. Figure 2.12 also includes the average monthly oil price. This shows that there is almost no correlation between high oil prices and enhanced exploration activities, as is often claimed by some experts.

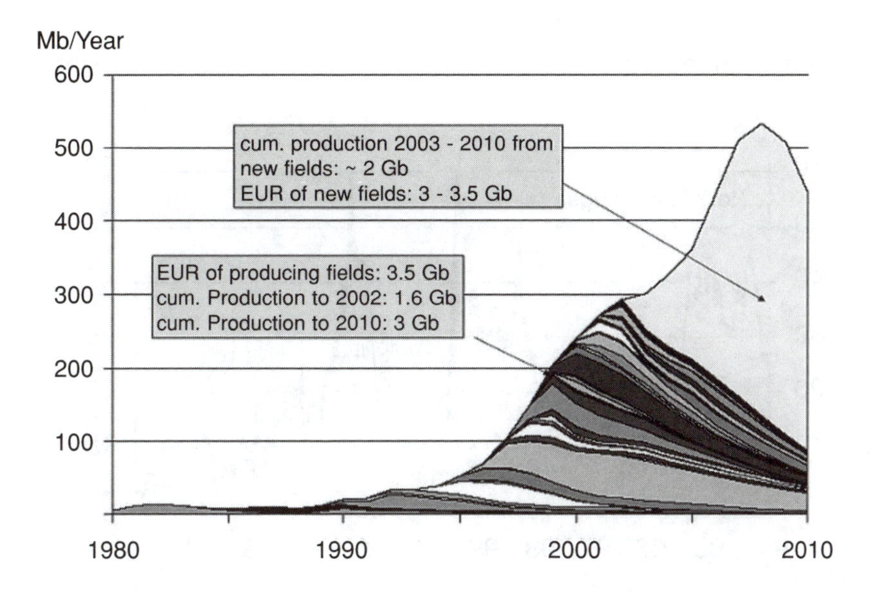

Note: More than 1000 feet water depth; fields already producing in 2002 are listed individually; optimistically expected field developments from known fields are summarized under the shaded area.

Source: MMS, 2003; analysis LBST.

Figure 2.11 Production profile of producing fields in the deep water Gulf of Mexico

A 'profit warning' for future oil from OPEC countries

The Middle East countries are the 'wild card' in future oil production. Indeed, many people believe that their reserves will feed the rising world oil demand for many decades to come. But on looking a little closer we note that Saudi Arabia's large oilfields have already been producing for many decades and are close to or beyond peak production. More expensive offshore oilfields already contribute 25 per cent to the country's oil output. One should not be surprised to find that production in Saudi Arabia cannot be increased any more.

It turns out that eventually only one country still has the potential to step up its production capacity by several million barrels per day – Iraq. But even there production growth is only possible if large investment takes place soon. However, there is as yet no sign that this is happening. This makes it very doubtful whether Middle East oil producers will be able to offset a future oil scarcity resulting from the decline in other countries. This task becomes even harder because within this decade the production decline in other regions will accelerate.

Actual production data show that the large production rates of the late 1970s were never seen again in OPEC countries. Even in 2004 when all

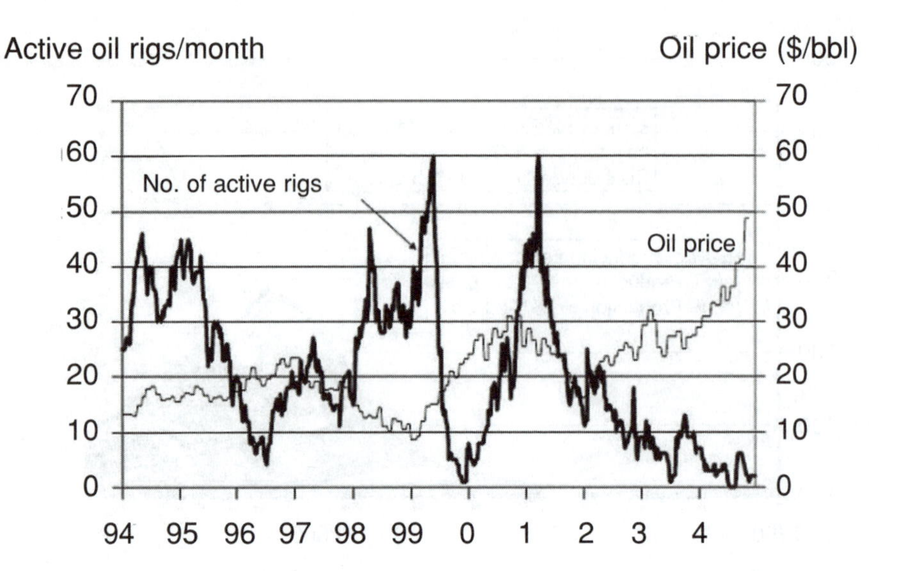

Note: Though oil prices are considerably higher than in 1999, the number of active rigs fell to very low levels, indicating that most of the oil in the Gulf of Mexico is already believed to be found.

Source: Baker Hughes.

Figure 2.12 Number of active oil rigs in the Gulf of Mexico

countries were producing at maximum capacity the production did not exceed the rate of 20 years ago and was even below the 1998 output. It seems that OPEC countries as a whole have already been in decline since 1998, and will remain so unless substantial new capacity is added very soon. At least it is remarkable that for the first time in history, in 1998–2002, total OPEC oil production (excluding Iraq) declined while world oil consumption still increased, though at a very moderate rate. It might also be noted that the Iraq oil embargo was lifted just at a time when these countries did not (or could not?) increase their production any more. Only with the help of Iraqi wells could the 2000 production surpass that of 1998.

Example 1: Saudi Arabia

Figure 2.14 shows the ranking of the 30 largest oilfields in Saudi Arabia. The four largest fields have produced for at least 40 years. Their initial reserve contains more than half of Saudi Arabia's discoveries. The largest fields (Ghawar) alone contain between 85 and 120Gb, depending on different sources. However, a field's actual size is of secondary importance as soon as the production maximum is reached. And the known facts indicate that this has happened already or is about to happen. Today more than 7 million barrels of

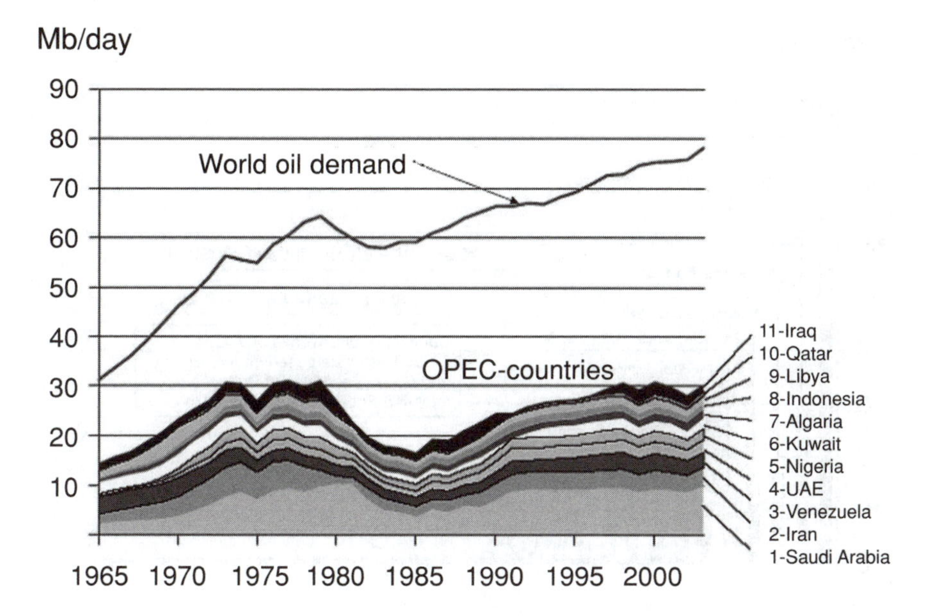

Source: BP, 2004.

Figure 2.13 OPEC production according to BP statistics

water are pumped into the field daily to keep the pressure at the present level for the extraction of 4–4.5 million barrels of oil per day, making this field responsible for more than half of Saudi Arabia's oil output. The huge water demand is a clear indication that its production rate cannot be extended any more. Within a few years this field might begin to decline, if it has not already started to do so. As Figure 2.15 shows, total oil production from onshore fields has been in decline since at least 1996.

The second largest field is the offshore field Safaniyah, which has also been producing since the late 1950s. Even that field is already half emptied. Though not definitely known, the production profile of Saudi Arabia over recent years indicates that this field is close to or already has passed peak production. At least the production from offshore fields peaked in 1991 and has declined since then. Nearly all large fields have been producing for many years. Additional capacity can only be added from smaller, not yet producing fields. But since their size is small compared with producing fields, fast and expensive field developments are needed to counteract present production profiles. But these investments have not been observed so far. These indications make it probable that the present decline (since 1996) cannot be reversed, at least not unless much more intensive development activities are begun.

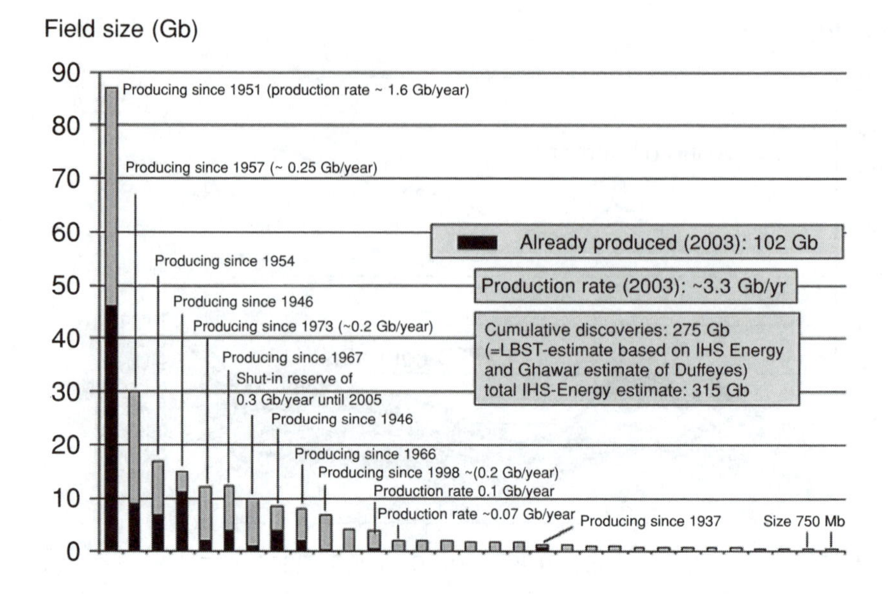

Source: Laherrère et al (1996), except Ghawhar (Deffeyes); depletion estimate based on Saudi Aramco publications and LBST estimate, 2004.

Figure 2.14 Field analysis of Saudi Arabia oilfields

Example 2: Oman

As a surprise to many observers, Oman passed its peak production in 2000. This sudden decline was triggered by the decline of its largest field, Yibal. Yibal contributed 20 per cent of total oil production in Oman. It was depleted very fast as almost from the beginning the pressure was increased by water injection and the oil was extracted with many horizontal wells. The water level in the field rose unprecedentedly quickly and reached the horizontal wells very fast, sometimes bypassing oil which remained isolated in the pores of the limestone. Yibal gives an extreme example of how new production technologies might increase the production rate for some years, but result in the extraction of smaller quantities of oil over the long term unless applied cautiously. There are some indications that the largest field in Saudi Arabia (Ghawhar) is being exploited with a similar technique. At least, water injection began 40 years ago and 500 out of the 3000 wells are horizontally drilled. It might be that the oil production of Oman gives an example of what could happen with the oil production in Saudi Arabia very soon.

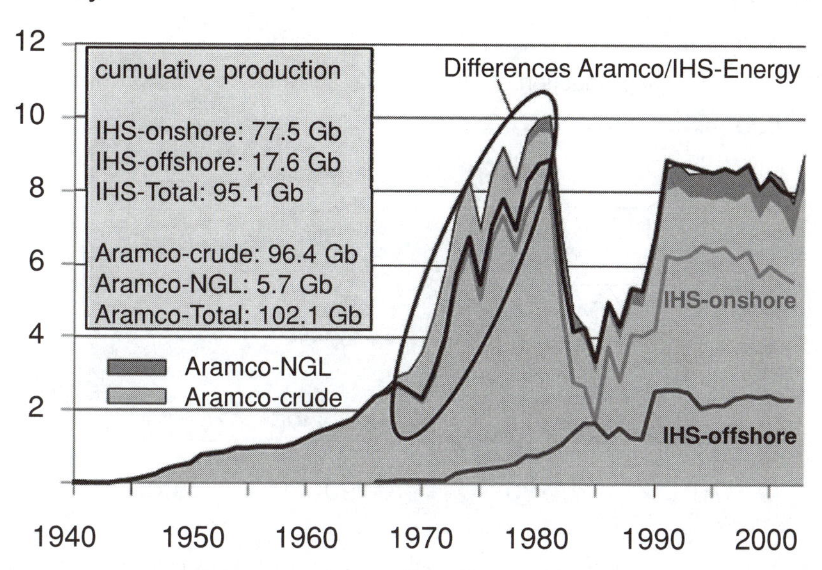

Note: As published by Aramco and the Industry database; it is remarkable that oil production from onshore fields has been in steady decline since 1996; offshore production contributes a quarter, while in the late 1970s all oil came from a few onshore fields.

Source: Crude and NGL production: Aramco; onshore and offshore production: Industry database.

Figure 2.15 Oil production of Saudi Arabia

World oil production

Figure 2.17 summarizes the history of world production distributed among the different producing regions: OPEC is counted without Indonesia; FSU includes the Caspian Region; *non-conventional* production from Canada and Venezuela is shown explicitly; and 'Rest of world' (ROW) is the same as given in Figure 2.8 (p36).

Forecasts to 2015 are made in line with the data and figures discussed above. Since future production data for the ROW countries are probably quite accurate, any further demand increase above today's levels would have to be supplied by OPEC or Russia. In spite of the above discussion, however, it might be interesting to see where future OPEC production extensions could come from, if at all. If we assume for the moment that the production from Russia and Caspian countries can increase by about 4Mb/day to 15Mb/day in 2010 and will decline thereafter by about 3 per cent per year, we can estimate the production extension needed from OPEC countries, just to compensate for these declines. This shows from another perspective, as above, that extending production is becoming more and more difficult as ever fewer

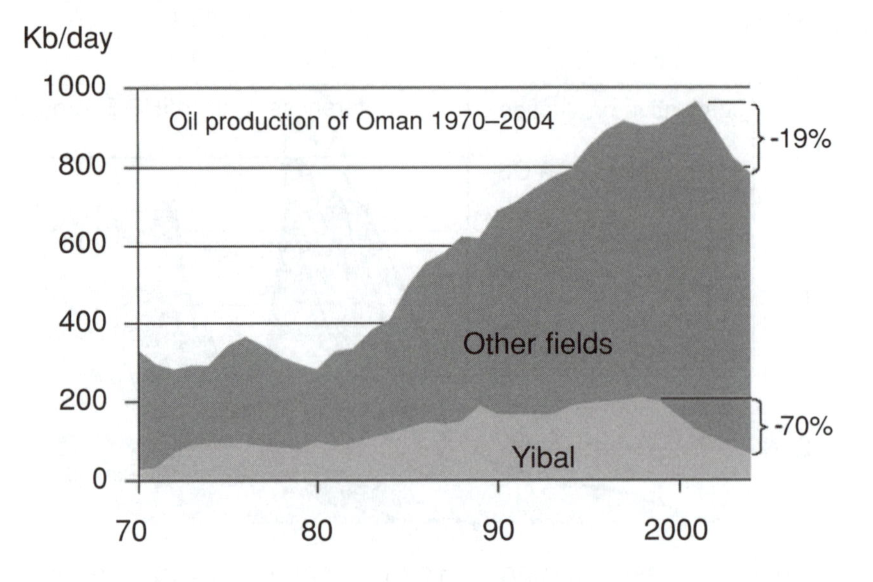

Source: BP Statistical Review of World Energy (Online Statistics with annual updates at www.bp.com), IHS-Energy and Laherrère, 2004; data for 2004 are estimated by LBST.

Figure 2.16 Production analysis of Oman

regions are able to do so, and first have to compensate for the decline in other regions. And in the face of the above discussed facts, such further increases are more and more doubtful.

The economic situation of oil companies indicates imminent problems

Future oil production rates are limited by geological and technical constraints, but economic factors also influence future spending on exploration and production. And vice versa – once a growing industry arrives at a turning point due to diminishing assets, this will influence the economic situation of individual companies and of the industry as a whole. This chapter draws together economic indicators that further support the evidence of an imminent peak of world oil production.

Example 1: BP and Amoco

Amoco spent large amounts of money on unsuccessful exploration efforts, as can be seen from Figure 2.18. Entering the business as a late starter, the good oil prospects were explored very fast with the first 100 wildcats. With the next 300 wildcats only another 4Gb were discovered. Five to six hundred further explorative drillings could only add about 1Gb, all discoveries adding up to

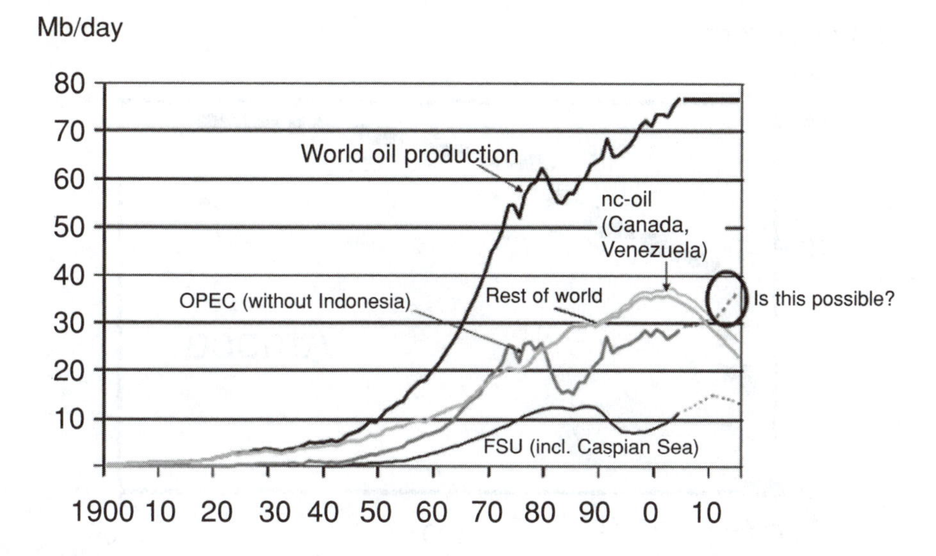

Note: Explicitly shown is the rising contribution from tar sands and heavy oil from production in North America and Venezuela, not included in 'ROW' production. The forecast is based on the assumption that total world oil production might stay flat. To offset the decline in ROW countries, an increase of oil production by 10 Mb/d has to come from the other regions: OPEC and FSU. If oil demand were to rise above the given level, oil production in these countries would have to rise further – which would need investments which are not yet visible.

Source: IHS-Energy, NEB Canada (2004), Analysis and forecast: LBST.

Figure 2.17 Forecast of world oil production up to 2015 split into OPEC (without Indonesia), former Soviet Union (FSU) and Rest of World (ROW)

16Gb. This poor result was the main reason for the financial problems of the company and its eventual merger with BP. Later on the same story happened to Atlantic Richfield Corporation (ARCO). Both companies merged with BP to form the joint company, BP-Amoco-ARCO, which has now downsized again to BP, both in terms of its name and of its staff. In other words these mergers were nothing more than a downscaling of the industry hidden to the public.

The whole industry comes under severe financial pressure as soon as investment in new exploration fails to maintain the same return of investment (ROI) as in former years. The only remaining possibility for companies to increase their production volumes is via mergers and acquisitions.

This situation is completely different from five years ago and is already reflected in the market and illustrated by recent profit warnings. Thus BP downscaled its production goals three times within the last year. The record profits of the recent years are solely based on increasing oil and gas prices while production volume was flat or slightly declining.

Looking at BP's present assets, about 70 per cent of the oil production in 2000 came from the North Sea and the US (Alaska). In both areas production

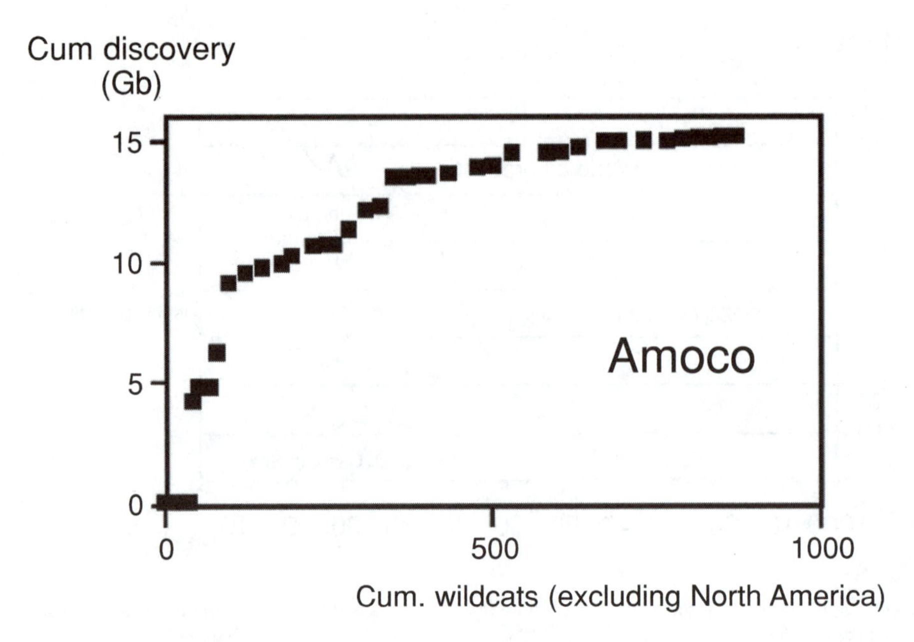

Source: Colin Campbell.

Figure 2.18 The disappointing oil exploration of AMOCO exhibited in the cumulative discoveries resulting from new wildcats

is declining year by year. Therefore it would be consistent to sell these assets, as has been announced very recently. The company argued that the new liquidity would be used to reinvest in more profitable projects. The most important large-scale projects of recent years are the involvement in Chirag-Azeri-Guneshli (offshore Azerbaijan) and Thunderhorse (Gulf of Mexico). But these require huge development costs. Other attempts to establish new projects were less successful. In 2000 the company withdrew its shares from the largest discovery in the Caspian Sea (Kashagan) for being unprofitable. More and more, BP has to turn to less profitable assets (compared with its past performance).

In recent years, total production of the company started to decline. However, with the investment of about US$7 billion this could be offset for a short while by the acquisition of a 50 per cent stake in the Russian producer TNK. This acquisition might be interpreted as a sign that it is getting cheaper for the companies to acquire assets from others instead of looking for new discoveries. The production share of the individual region can be seen in Figure 2.19. This behaviour is just what Colin Campbell stated five years ago when he said 'there is nothing left (for the oil companies) than to eat each other'.

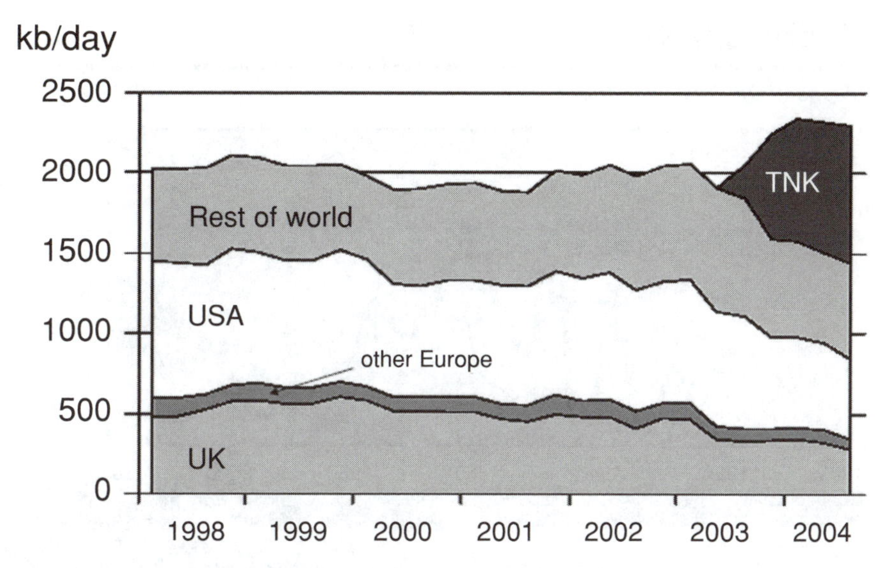

Note: Only the acquisition of TNK shares helped to avoid a dramatic decline. It will be interesting to see whether the new field developments in the Gulf of Mexico (Thunderhorse) or in Azerbaijan (Azeri-Chirag-Guneshli) can help to stop this decline for some time.

Source: Quarterly company reports.

Figure 2.19 Oil production of BP in individual regions

Example 2: ExxonMobil

ExxonMobil is the leading privately owned oil (and gas) company, which until recently completely ignored the problems with which oil companies are confronted: climate change was seen as an issue where the oil industry must not take action; depletion didn't seem to affect their profits and behaviour; renewable energies were considered to be decades away from now, thus prohibiting any investment in this area.

However, in the period covering 2003–2005 there have been signs of change. For the first time high ranking managers made several speeches in which they communicated to the public that the whole industry is facing new and fundamental problems:

- ExxonMobil has published a graph that shows the historical pattern of oil discovery, which was first compiled and published by Colin Campbell and Jean Laherrère. This graph is based on the industry database and makes use of backdating reserve reassessments of known fields to the date of the original field discovery. It shows the historical fact that peak discovery took place in the 1960s
- ExxonMobil admits the fact that discoveries fell in the last two decades despite strongly increased oil prices. There is no empirical correlation between oil prices and discoveries

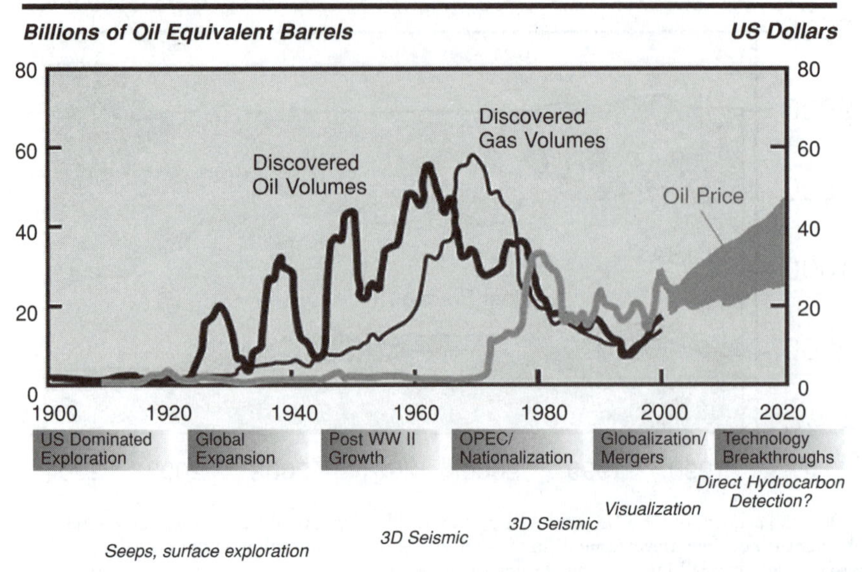

Note: This figure is well in line with data from the industry database and in striking contrast to public domain information sites. It is interesting to see ExxonMobil expecting a rising oil price in coming years.

Source: Drawing on Longwell, 2002.

Figure 2.20 Smoothed annual discoveries of oil and gas as seen by ExxonMobil

- ExxonMobil states that depletion in producing fields is a fact. Therefore in order to maintain today's production level, about 40Mb/d of new capacity will have to be developed within this decade
- ExxonMobil states that the oil industry is facing a new situation that is completely different from the past. To increase production by 2010 to those levels to which oil demand will rise (according to IEA estimates) requires expenses for exploration and development in the order of $100 billion per year, a total of $1000 billion by 2010. This is substantially more than in the past.

It is also interesting to see worldwide gas discoveries peaking only a few years later than oil did, though worldwide interest in gas assets increased tremendously over the last two decades. Can this be read as a signal that in the long run ExxonMobil also does not believe in natural gas as a substitute for oil?

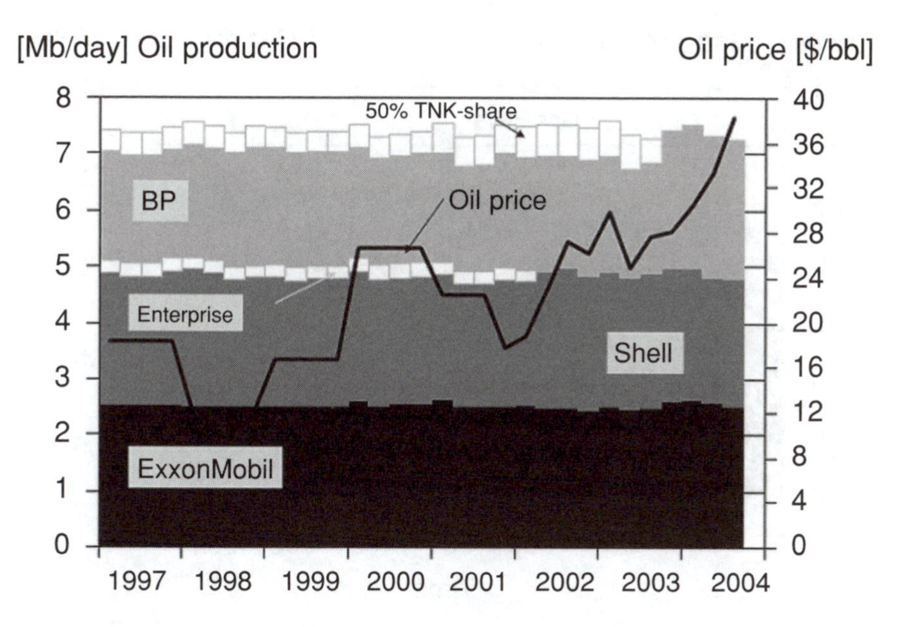

Note: ExxonMobil, Shell (which in 2002 bought Enterprise) and BP.

Production rates and average sales prices are taken from quarterly and annual company reports. Total production since first quarter 1997 is declining by a few per cent while sales prices are increasing.

Source: quarterly company reports.

Figure 2.21 Oil production rate (Mb/day) of the three largest private western oil companies

Example 3: The big three – ExxonMobil, BP and Shell

Figure 2.21 shows the oil production of the three largest western private oil companies over the last six years. Actually it hit its highest level in 1998 when the oil price was at its lowest level. Since that time the declining (or steady) production could only be compensated by higher oil prices and an increased share of gas production (see Figure 2.22, p52]). The importance of gas is also underlined by the exorbitantly high gas prices in recent years, which rallied when gas in North America became a scarce commodity. Also, since 1999 the expenditure on oil and gas exploration, and production of these companies has increased by more than 50 per cent (see Figure 2.24, p55). However, total production is already below the 1997 production level when taking account of acquisitions.

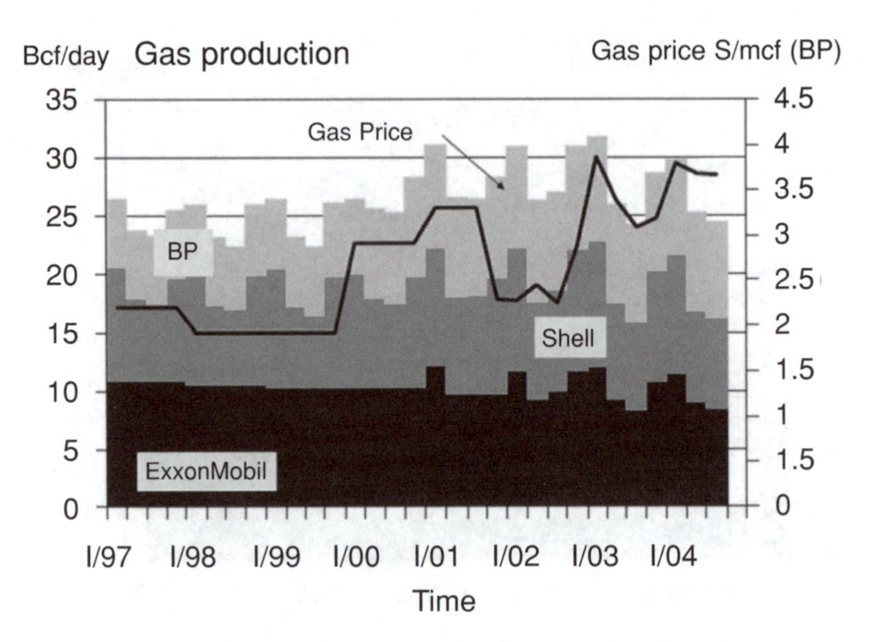

Note: Production rates and average sales prices are taken from quarterly and annual company reports. Total production and sales prices since first quarter 1997 were increasing, outweighing the volume losses of crude oil production.

Source: quarterly company reports.

Figure 2.22 Gas production rate of the three largest private western oil companies

Difference between reserves in public domain statistics and industry databases

Diminishing oil assets are in striking contrast to public communications stating that oil reserves are growing, and have been doing so for decades without any sign of a reverse trend. What sources are these different views based on?

The 'Economist' point of view

This view is based on annual reserve statistics, which are published each year by oil companies (for example BP Statistical Review of World Energy). Those statistics are based on two completely different facts:

1 new discoveries that are made during the course of the year are added to reserves;
2 the estimated recoverable amounts of oil in ageing fields are re-evaluated and systematically increased year by year. These field revisions are due to

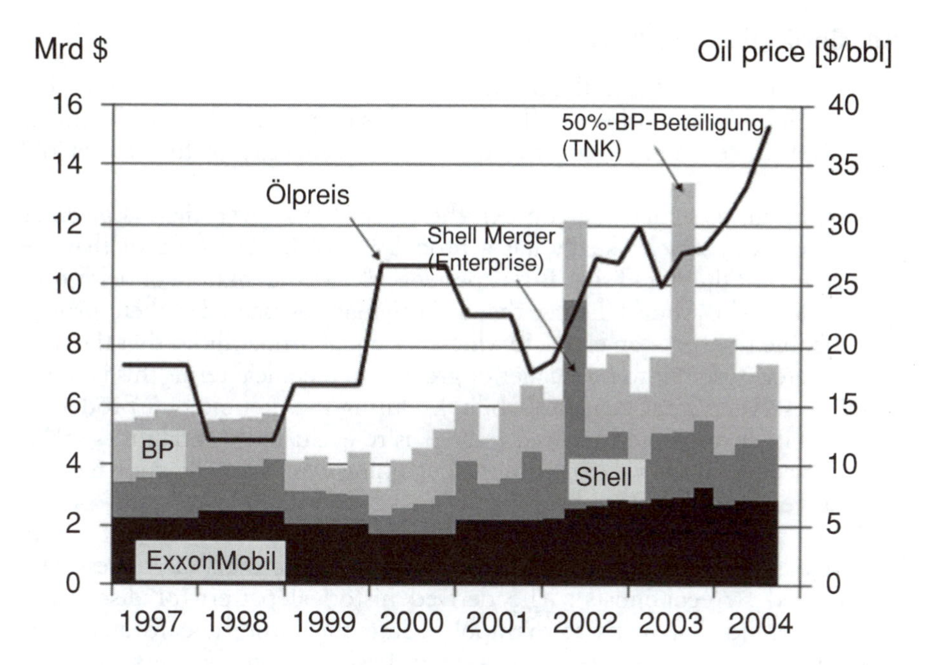

Note: In 2002 Shell bought the medium scale company Enterprise for about 5 billion dollars; BP acquired a 50 per cent share of the Russian company TNK in 2003.

Source: quarterly company reports.

Figure 2.23 Quarterly expenditure for exploration and production of the three largest private western oil companies

> an early underestimate in annual reports. This guarantees that year by year proven reserves are increasing, thus hiding the real situation regarding new discoveries.

This is common practice for the reporting of reserves by private oil companies. With the help of these systematic upward revisions, years with disappointing exploration success can be hidden, and the produced quantities smoothly replaced in the company statistics. This accounts for the fact that oil reserves have almost continuously increased for more than 40 years, though each year about 25Gb or 2 per cent of known reserves are removed by production.

The reserve figures used in financial contexts and shareholder meetings are completely different from those that address the question of how much oil has already been found and how much will still be found.

The main reason, however, for the apparently unchanged world reserves year after year is the reporting practice of state owned companies. More than 70 countries have reported unchanged reserves for many years, despite substantial production.

The 'Geologist' point of view

This view is concerned with the question of how much oil will still be found in a certain oil basin or worldwide. To address this question a clear distinction is necessary between new discoveries and the re-evaluation of the size of old fields.

First of all, in a financial context, the reserve of a given field is given as 'proved reserve', which usually has a probability of 80–90 per cent that the field has at least the stated size. But experience shows that mature oilfields turn out in reality to be around the size that was originally estimated as their 'proved and probable reserve' – an estimate which has equal probabilities that the field will ultimately turn out to be smaller or greater. In technical terms this is called the P50 reserve (50 per cent probability). During the lifetime of a producing field the initially estimated 'proved reserve' is re-evaluated several times and is finally very close to the value that in the beginning was internally known as the P50 reserve.

To get a correct historical pattern of discoveries, reassessments of an old field are based on the year of discovery and not on the years of reassessment (as favoured by economists). The derived historical pattern of discoveries displays a trend that helps to extrapolate into the future and to assess the prospects for future discoveries in a given basin in coming years. Such an analysis is essential for the geologists' decision as to where it is still worth looking for oil and where not. In nearly all oil provinces, the same pattern can be observed: large discoveries are made early and with minimal effort. In later years the size of individual and annual discoveries gets smaller and smaller. The cumulative discoveries over the years saturate and approach an asymptotic value, which might be seen as the estimated ultimate potential for the oil recovery of a region.

An important reason why reassessments of ageing oilfields do not influence the production pattern and the production peak is that they are mostly done for old fields after their production maximum. This leads to the paradoxical situation that the oil production in an individual field can decline even though the ultimate reserve in the field increases. In these cases the reassessment has no influence on the future production profile. This can be demonstrated with many examples. The reassessment of the largest UK oilfield (Forties) is shown in Table 2.4.

Table 2.4 *Production rate of Forties, the largest UK oilfield, and estimated ultimate recovery for the field as stated by the operator*

Year	1980	1984	1987	1997	1999
Production (kb/d)	182	156	128	41	28
Estimated Ultimate Recovery (Gb)	1.8	2.0	2.4	2.5	2.7

Source: Brown Book, 2002 and Laherrère, 2001

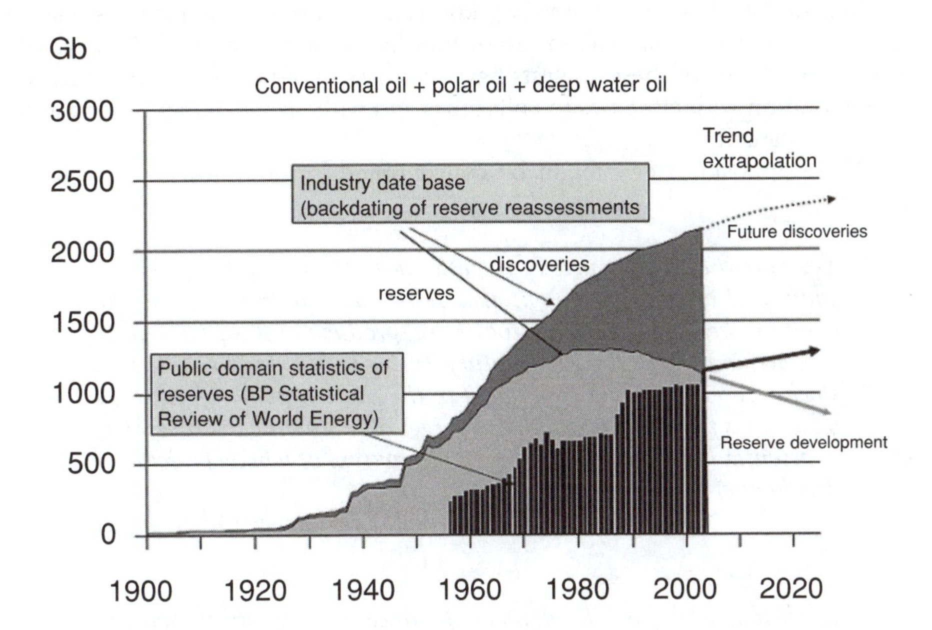

Note: Though their absolute size is similar, the most striking difference is the backdating of reserve revisions to the discovery year. This gives a completely different historical pattern, which changes the forecast for future discoveries and reserves considerably.

Source: IHS-Energy and BP Statistical Review of World Energy.

Figure 2.24 The difference between industry data and public domain statistics of reserves

Another example is the case of the reserve estimates for the US, which are reassessed each year resulting in almost constant oil reserves over many years, though each year oil is removed by production. Despite these reassessments, the US oil production has been in decline for 30 years (as shown in Figure 2.4, p26).

Some comments on the current debate

Discussion of future oil and gas supply is dominated by two opposing positions. Economically oriented observers base their knowledge mainly on information derived from the financial departments of their companies. They argue that market mechanisms are sufficient to guarantee future discoveries and production rates. Many therefore believe that even in the long term, supply shortages can be avoided. Price increases will happen only for short times until supply and demand reach equilibrium again, driven by market forces and the panacea 'technology'.

In contrast, those who base their knowledge more on the 'geologist view' argue that the historical peak in discoveries has to be followed by the peak of production. Many of these analysts have many years of personal experience in oil exploration with close ties to colleagues and with access to exploration and discovery statistics.

These opposing views might be concentrated in the following two statements:

> *The potential significance of the alternative theory of the origin of additional oil and gas potential is self-evident with respect to the issues of the longevity of hydrocarbons' production prospects and to production costs in the 21st century. Instead of having to consider a stock reserve already accumulated in the finite number of so-called oil and gas plays, the possibility emerges of evaluating hydrocarbons as essentially renewable sources in the context of whatever demand developments may emerge.*
>
> Peter Odell, Economist (Odell, 2001)

> *Geologists look for oil, engineers produce oil, economists sell oil. Beware of economists who tell you how much is there.*
>
> Colin J. Campbell, Geologist

To give some orientation in this ongoing discussion it is also helpful to have a closer look at the type of arguments used by different groups and the motives and interests behind them.

Oil industry

Oil companies frequently publish reassuring press releases, speeches and 'studies' on the future availability of oil. The oil industry has a natural financial interest in doing this:

- The signal of diminishing resources could induce the consumers to reduce their oil consumption even faster than necessary. In such a case the oil industry would be left with more oil available on the market than consumers would want to buy. Certainly, this would be bad for business
- The signal of diminishing resources and hints at 'diminishing assets' could lead shareholders and investors to redirect their investments into new business opportunities with more growth potential. This also would be bad for business
- It would be best if consumers and shareholders would stay loyal to oil even at declining production rates and rising prices. This would bring about the highest earnings at minimum costs. Therefore, the best communication is to convince the consumer that any current problems are only temporary – and thus to keep them dependent on oil even in worsening times.

Because of these motives the industry are unlikely to admit that the future availability of oil might be a problem and rather tends to communicate growing reserves. All this however does not necessarily prevent the oil industry from investing in new business areas.

Consumers and the public

The consumer is interested in using cheap energy in everyday life to gain as much comfort as possible. Therefore disturbing messages are seen as endangering their present way of life and questioning their own behaviour, and it is therefore very likely that they will be discarded. On the other hand, messages that confirm their present lifestyle find open ears and are readily believed. So, in a way, the interests of all participants in the market point in the same direction: it is more comfortable not to be aware of the imminent problems caused by peak oil production.

Economic research institutes

Economists concentrate on the market mechanism, which tends to balance demand and supply. The optimum price is reached when supply and demand are in balance. Imbalances between supply and demand are reflected in changing prices, which in turn trigger enhanced production or demand until both are in balance again.

In this reasoning oil scarcity will lead to rising prices, which in turn is an incentive for the oil industry to invest more into exploration and field development, and consequently the oil supply will rise. The argument that every new exploration tends to be less successful because the amount of the remaining undiscovered oil is diminishing is countered with the argument that because of this, efforts have to be intensified by spending still more on exploration. And once the prices for oil are high enough, the market mechanism will ensure that alternatives to oil are smoothly introduced into the market.

In short, the market perfectly regulates the economy and there is no need for political action with regard to energy supply.

Though this reasoning seems plausible at first sight, we criticize it as never being able to predict a critical supply situation; it does not take care of – among other things – the changing geological, geographical and physical conditions (resource depletion) and the long lead times for reactions which govern the real world. This 'naive optimistic' economic view always results in a sufficient supply situation and therefore is 'unfalsifiable'. Within that thinking it is left to the engineers and technicians to provide the preconditions necessary for the proper working of the economic model. Whether and how this can be done is not considered, but rather the fact that it can be done is deemed as self-evident.

Ecologically oriented institutes

The threat of global warming caused by the burning of fossil fuels dominates the present discussion on the necessity to change our energy economy by reducing the use of fossil fuels and by introducing renewables.

One can observe that ecologically oriented scientists and policy makers frequently give the impression that they think the whole debate on the possibility of imminent oil and gas supply crises is absurd. It even looks as if the topic of resource depletion is a cause of nuisance for them. These people seem to believe that the debate on resource depletion diverts the attention away from the predominant topic of climate change and thus reduces the urgency of action. They seem to believe that in a few years oil prices will return to the low levels we have seen before and that the 'wolf!' cries will again turn out to be without substance. They therefore think that this discussion might discredit the efforts to change the energy system and to reduce emissions.

But between the topics of climate policy and resource depletion of fossil fuels there is no antagonism. Both problems can be addressed by exactly the same measures. Therefore, both topics together may strengthen the pressure and also the willingness to reconsider our energy policy.

Early investigations of peak oil production

Very often one hears the argument that diminishing oil supplies have been predicted several times before, but each time the claims turn out to be false. The report *Limits of Growth* by the Club of Rome, which was published in 1972, is cited in this context. Also it is pointed out that proved oil reserves remain at constant or even increasing levels, that the production to reserves ratio (P/R ratio) indicates that reserves will last for more than 40 years at present consumption levels, and that this value has not changed in the last few decades despite rising consumption.

The discussion of P/R ratios based on proven reserves is nonsense for various reasons. As explained above, proven reserves (or P90 reserves) for systematic reasons always grow and therefore are the wrong measure when trying to assess future availability of oil. Instead, 'proved and probable' or P50 reserves are the better measure (the value to which proved reserves will converge ultimately). On the other hand the question 'for how many more years will the oil last at present consumption rates?' is irrelevant since this R/P ratio is based on constant supply over time while actually the supply is growing in early phases and falling once peak production is passed. The cause of economic disruption is not the date at which the 'last drop of oil' will be consumed, but the point in time at which rising supplies are replaced by continually declining supplies.

Incidentally, the report *Limits of Growth* used the figure of the then known proven reserves in its calculations five times. This resulted in a reserve estimate of about 2280Gb, which even today is well in the range of best estimates (even if only by chance).

But the only detailed historical analysis of the subject that was known to a wider public was the *Global 2000* report to the President of the US, published in 1980 by US government agencies. The then estimated amount for ultimate oil recovery was well in line with today's knowledge. Even the estimated date of peak production is roughly accurate when one uses the factual growth rates of demand in the past two decades. In the report this date was estimated for different levels of demand growth. The annual growth rates of 2, 3 or even 5 per cent calculated in this report would have led to a peak of production between 1985 and 1995. If we insert the average rate of 1.3 per cent per year since 1975 into the graph, then peak production will take place around 2002 or 2003. This is well in line with today's (and also our) estimates that peak production will happen sometime between 2000 and 2010.

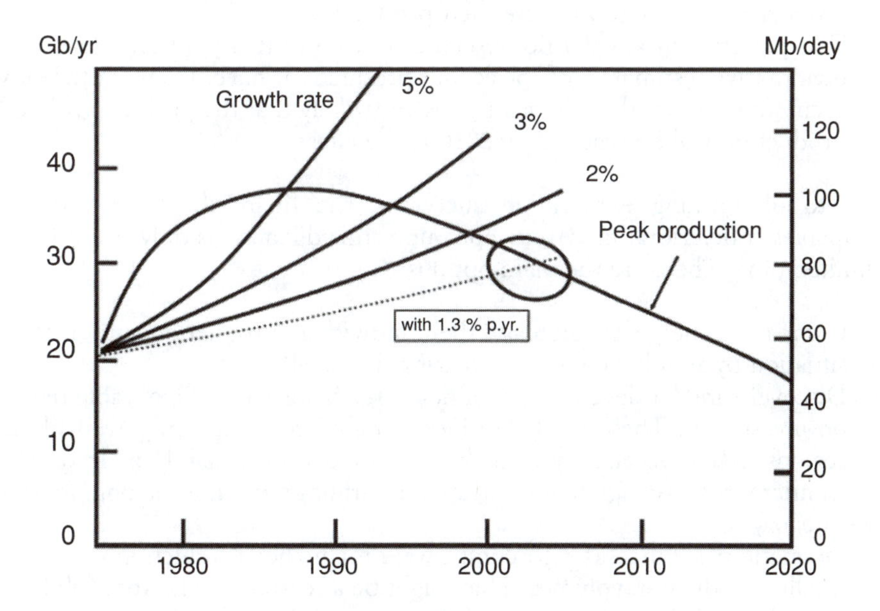

Note: The forecasts from 1979 assumed 2, 3 or even 5 per cent annual growth rate, given expected dates of peak production between 1985 and 1995 (Drawing on 'Global 2000', published in 1980). The dotted line however shows the real production growth between 1975 and 2000 with 1.3 per cent annually, indicating peak production at around 2002.

Source: Global 2000; actual development: LBST 2002.

Figure 2.25 Early forecasts of the date of peak oil production, assuming various annual production growth rates

Summary

The main facts, theses and conclusions are:

- The peak of oil discoveries was reached in the 1960s. This is a historical fact

- This peak in discoveries has to be followed by a peak in production, since we can only produce what has been found before
- The production peak of individual fields is a historical fact; almost all large oilfields have already passed their production maximum and are in decline
- The aggregation of the production profiles of individual fields (with their individual peaks) sums up to a production peak of individual oil regions. Historically peak production was reached, for example, in Austria in 1955, in Germany in 1968, in the US in 1971 and in Indonesia in 1977. Recent regions joining the list of countries with declining production rates are Gabon (1997), UK (1999), Australia (2000), Oman (2000) and Norway (2001)
- The aggregate decline of mature regions is getting steeper with every new 'member of the club'. In order just to keep overall production flat, ever fewer regions have to increase their production
- This pattern can already be observed over more than 30 years. Even the quantitative estimates of peak oil production have been sufficiently accurate for more than 20 years. It is very likely that the peak of world oil production will be reached by 2010 at the latest.

Due to diminishing exploration successes, the financial situation of oil companies is deteriorating. An escape out of this dilemma is only possible for a limited time. These are the main options:

- Companies merge to increase their individual production and reserve situation by simultaneously downsizing their staff
- Discovery and/or development of new large fields with a favourable return on investment. These fields are increasingly scarce and only available in regions that have already been known for decades, mainly in Iraq. The frontier areas (Caspian and deep sea drilling) do not belong to this category
- Dramatic price increases to compensate for higher production costs and smaller production volumes. That might be a reason why ExxonMobil has started to point out the fact of oil depletion: perhaps they want to prepare the public for higher prices in the future.

References

AEUB (2003) 'Proposed conservation policy affecting gas production in oil sands areas', *Alberta Energy and Utilities Board News*, 22 July 2003, www.eub.gov.ab.ca/BBS/new/Projects/GasBitumenPolicy.htm

Brown Book (2002) Various editions of the Brown Book, London, UK Department of Trade and Industry

Campbell, J., Liesenborghs, F., Schindler, J. and Zittel, W. (2002) *Ölwechsel!*, München, Deutscher Taschenbuch Verlag

Department of Interior (2001) *Gulf of Mexico Outer Continental Shelf. Daily Oil and Gas Production Rate Projections from 2001 to 2005*, OCS-Report MMS 2001-044, Department of Interior, Minerals Management Service, May 2001

EIA (2001) *'Country Analysis', The Caspian Sea*, US-Department of Energy, Energy Information Administration

Global 2000 (1980) *The Global 2000 Report to the President*, Washington, DC, edited by the Council on the Environmental Quality and the US Foreign Office

IEA (2004) *World Energy Outlook 2004*, Paris, International Energy Agency

IHS (2002) *Petroleum Exploration and Production Statistics (PEPS)*, Geneva, IHS-Energy

Laherrère, J. (2001) 'Estimates of oil reserves', Paper presented at *EMF/IEA/IEW Meeting*, Laxenburg, Vienna

Laherrère, J. H., Perrodon, A. and Campbell, C. J. (ed) (1996) *The World's Gas Potential*, Geneva, Petroconsultants

Longwell, H. J. (2002) 'The future of the oil and gas industry: Past approaches, new challenges', *World Energy*, vol 5, no 3, pp102–105

Luanda (2000) 'Angola to restrict oil output', *Financial Times*, 3 November

NEB (National Energy Board) Canada (2004) 'Canada's oil sands: Opportunities and challenges to 2015', 15 May 2004, www.neb-one.gc.ca/energy/EnergyReports/EMAOilSandsOpportunitiesChallenges2015/EMAOilSandsOpportunities2015QA_e.htm

Odell, P. (2001) 'The long-term future supply potential of oil', International Energy Workshop, 19–21 June 2001, IIASA, Laxemburg, Vienna, Austria, 1st plenary presentation by Odell, available at www.iiasa.ac.at/Research/ECS/IEW2001/pdffiles/Papers/Odell.pdf accessed 10 February 2005

L-B-Systemtechnik GmbH, München

Texas Railroad Commission (2003) Internet information at www.rrc.state.tx.us

USGS (2000) 'US geological survey world petroleum assessment 2000', *Geological Survey*, Washington, US Department of Interior, US

Zittel, W. (2001) 'Analysis of UK oil production', *A contribution to the Association for the Study of Peak Oil*, L-B-Systemtechnik GmbH, München

3

Global Climate Change and Renewable Energy: Exploring the Links

Ian H. Rowlands

Introduction

The purpose of this chapter is to explore the links between global climate change and renewable energy. To do this, the chapter is divided into six sections. After this brief introduction, the extent to which renewable energy could conceivably address the challenge of global climate change is explored in the second section. (This also provides the reader with a brief introduction to the challenge of global climate change). Third, the degree to which renewable energy has been offered as a 'solution' during international negotiations on global climate change is explored. Fourth, given that much of the focus regarding climate change is now upon the implementation of global agreements, the extent to which renewable energy is playing a role in national discussions is then examined. To give further substance to this investigation, three case studies are presented – namely, Germany, the United Kingdom (UK) and the United States (US). Potential reasons for the differences that have emerged between European countries and the US – regarding climate change responses, generally, at the national level – are also studied. In the fifth section, ongoing and future issues regarding the links between global climate change and renewable energy are identified. Finally, a summary is offered and the main conclusions are presented in the sixth section.

Links between global climate change and renewable energy: The extent to which the latter could serve to address the former

Global climate change is resulting from the increased build-up of anthropogenic greenhouse gases (primarily carbon dioxide) in the atmosphere. The average concentration of carbon dioxide in the global atmosphere has risen from a pre-industrial level of 280 parts per million (ppm) in 1750 to today's level of approximately 370 ppm, an increase of more than 30 per cent (IPCC, 2001a). In turn, average global temperature has risen by about 0.6°C during the 20th century. With 'business-as-usual', the concentration of carbon dioxide and other greenhouse gases will continue to rise, along with average global temperatures. Experts anticipate an increase in the latter of somewhere between 1.4°C and 5.8°C over the period 1990–2100 (IPCC, 2001a). Follow-on environmental changes would include shifting climatic zones and rising sea levels, perhaps as much as 88 centimetres by 2100 (IPCC, 2001a). The economic, social and political consequences of such changes are potentially dramatic.

Most anthropogenic sources of greenhouse gases are the result of the extraction and combustion of fossil fuels (that is, coal, oil and natural gas). Additional causes of human-induced climate change include land-use activities and emissions of certain kinds of industrial chemicals. Tables 3.1, 3.2 and 3.3 provide more information about the origins of greenhouse gases. The first provides information about the relative contribution of different molecules, for the European Union (EU) countries[1] collectively, and the US. (The focus is upon these countries, because they will be examined more closely in the fourth section of this chapter.) Data from Table 3.2 allow for the study of climate change causes in another way – more specifically, considering the kinds of human activities that most contribute to the phenomenon. (Note that global figures are also included here.) Finally, Table 3.3 focuses upon energy-related greenhouse gas emissions, identifying the respective share of global climate change that is caused by different fuels. Given that two different areas – namely, the EU and the US – are compared, it is important to recognize not only the different 'make-up' of each area's emissions profile (as presented in these three tables), but also the differences in each area's absolute and relative contributions to global climate change. In brief, the US is responsible for approximately 2 per cent of the world's energy-related carbon dioxide emissions, and the EU for 13 per cent. In terms of per capita emissions, the US figure is 5.4 tonnes carbon per person and the corresponding EU figure is 2.3 tonnes. The global average, meanwhile, is 1.1 tonnes per person. (All calculations are for 2000 and are following Marland et al, 2003.)

Table 3.1 *Contributions, by substance, to global climate change (European Union and United States), not including land-use change and forestry activities*

	European Union (2002) % share	United States (2002) % share
Carbon dioxide	82	83
Methane	8	9
Nitrous oxide	8	6
Industrial chemicals	2	2

Sources: EEA (2004) and EPA (2004).

Table 3.2 *Contributions, by human activity, to greenhouse gas emissions (globally, European Union and United States)*

Globally (1990) % share	European Union (2000) % share	United States (2002) % share
Agriculture: 29	Agriculture: 10	Agriculture: 8
Buildings: 21	Buildings: 17	Electricity generation: 33
Industry: 32	Energy industries: 31	Industry: 19
Transportation: 14	Industry: 17	Residential and commercial: 13
Waste: 3	Transportation: 22	Transportation: 27
	Waste disposal: 3	

Note: figures may not add up to 100, because of rounding.

Sources: IPCC (2001b), IEA (2002b), EEA (2002) and EPA (2004).

Table 3.3 *Contributions, by fuel, to energy-related greenhouse gas emissions (globally, European Union and United States)*

	Globally (2000) % share	European Union (1999) % share	United States (2001) % share
Coal	37	27	36
Natural gas	20	25	22
Oil/petroleum	42	47	42

Note: figures may not add up to 100, because of rounding.

Sources: EIA (2003), Marland et al (2003) and EPA (2003).

Table 3.4 *Range of estimates of greenhouse gas-intensity of selected 'conventional' and 'renewable' fuels (for electricity generation), in grams of carbon dioxide equivalent per kWh electricity generated*

Conventional fuels	Renewable fuels
Coal: 790–1182	Biomass: 15–101
Large-scale hydro: 2–48	Geothermal: 15–97
Natural gas: 362–653	Solar PV: 13–731
Nuclear: 2–59	Wind: 6–124
Oil: 733–935	

Sources: IEA (1998), Nuclear Energy Institute (2003), Paul Scherrer Institut (2002) and Serchuk (2000).

Renewable energies have great potential to reduce energy-related greenhouse gas emissions. (Note that energy-related greenhouse gas emissions still make up the vast majority of the global warming challenge, especially in northern countries, even when land-use activities are included.) The extent to which the displacement of non-renewable energy sources by renewable energy sources will reduce greenhouse gas emissions will be a function, of course, of the relative greenhouse gas-intensity of the two 'fuels' involved. Table 3.4 presents some data about the relative greenhouse gas-intensity of both 'conventional' and 'renewable' energy sources, from a life-cycle perspective. From this, it is clear that different displacement choices will yield different 'gains' in terms of greenhouse gas emission reductions. But given that the average greenhouse gas-intensity of fossil fuel use in the world is just over 500 grams carbon dioxide equivalent[2] per kWh (for electricity, at least),[3] and that the average greenhouse gas intensity of all renewables is something of the order of 50–100 grams carbon dioxide equivalent per kWh, there is clearly scope for significant gains. The extent to which these gains are realizable is, of course, another issue.

Renewable energy in global climate change negotiations

The purpose of this section is to examine the role that 'renewable energy' has played in international negotiations on global climate change to date. Space limitations preclude an exhaustive investigation. Instead, attention will be focused upon key documents produced by two of the most important international bodies considering the global climate change challenge. The first body is the Intergovernmental Panel on Climate Change (IPCC).

The IPCC was established in 1988 by the World Meteorological Organization (WMO) and the United Nations Environment Programme (UNEP). Made up of climate change experts nominated by national governments, the IPCC is intended to present the world's 'state-of-the-art' knowledge with respect to scientific, technological and socioeconomic aspects of climate change. To date, it has published three assessment reports.

The second body considered here is the negotiating group taking the lead in working out international agreements on climate change. Initially referred to as the Intergovernmental Negotiating Committee for a Framework Convention on Climate Change, these representatives, in May 1992, agreed the Framework Convention on Climate Change (FCCC). This document was 'opened for signature' at the Earth Summit in June of that year and entered into force in March 1994. (As of 24 May 2004, 189 countries had ratified the FCCC.) International negotiators, convened under the terms of the FCCC – with their most high-profile work occurring at annual 'Conferences of the Parties (to the FCCC)' – agreed, at the Third Conference of the Parties in Kyoto, Japan, in 1997, the Kyoto Protocol to the FCCC. Although ratified by 127 countries (as of 2 November 2004), this international agreement has yet to enter into force. Russia, however, had indicated that it was set to complete the process of ratification, so many anticipated that the Kyoto Protocol would soon enter into force.[4] In any case, it still remains an important expression of international sentiment on global climate change.

IPCC's First Assessment Report

In 1991, the IPCC's First Assessment Report was published. The IPCC's Working Group III investigated 'response strategies'. The Energy and Industry subgroup argued that 'to reduce greenhouse gas emissions from energy systems', the second of four main sets of options is 'fuel substitution by energy sources that have lower or no greenhouse gas emissions' (IPCC, 1991, p60). Much of the focus in the rest of the executive summary, however, is upon the short term (10 or 15 years) gains that could be made from increased energy efficiency and conservation (the subgroup's first main set of options). However, even in the medium and long term options laid out, the amount of attention paid to new and renewable resources is modest. Under 'non-fossil and low emission energy sources', nuclear power plants are listed first and solar power is the only renewable resource mentioned by name in terms of electricity generation (IPCC, 1991, p55).

United Nations Framework Convention on Climate Change

Agreed in May 1992, the Framework Convention on Climate Change (FCCC) has often been called, at best, a 'first step' in limiting greenhouse gas emissions. Largely to accommodate US resistance, the agreement simply encouraged industrialized countries to bring their greenhouse gas emissions in 2000 back to what they were in 1990 (Agrawala and Andresen, 1999, p461). Given the presence of such a 'non-target', perhaps it is not surprising that there was relatively little direction as to what countries should do to restrict their greenhouse gas emissions. Indeed, there are only limited references to 'energy' in the entire text of the FCCC. In Article 4(1)(c), for example, 'energy' is identified as one (indeed, the first) of six 'relevant' sectors for the control, reduction or prevention of anthropogenic emissions of greenhouse gases. Another reference is in the Preamble, where it is noted that developing countries' greenhouse gas emissions will need to grow, but that while this is happening there should be efforts to achieve 'greater energy efficiency'. There is no explicit mention of 'renewable energy' anywhere in the FCCC.

IPCC's Second Assessment Report

The IPCC's Second Assessment Report was released in 1996. Working Group II of the IPCC examined the scientific and technical aspects of impacts, adaptation and mitigation. For our purposes, we consider those sections that focus on options for mitigation (that is, emission reduction and sink enhancement).

The role of energy is 'front-and-centre' in this work, although the emphasis is upon 'efficiency' rather than 'substitution'. Acknowledging that it is a crude measure, we nevertheless note that the 'Summary for Policymakers' devotes 42.5 column centimetres to 'improving energy efficiency' options, and only 8 column centimetres to 'renewable energy' options (IPCC, 1996a, pp13–15). Worth noting is not only that the relative attention accorded to renewable energy is much smaller, but also that its appearance follows references to fuel switching to lower-carbon sources, decarbonization of flue gases and carbon dioxide storage and increased use of nuclear energy. Similar levels of attention are given in the 'technical summary'. However, more space is devoted to renewable sources of energy in the actual text of the report itself, reviewing different sources in more depth.

Meanwhile, IPCC Working Group III – which focused upon the economic and social dimensions of climate change – considered the costs of different mitigation options. With its requirement to focus upon 'cost-effective' measures,[5] it is perhaps not surprising that 'a portfolio of possible actions that policy-makers could consider' includes, at the top of the list, 'implementing energy efficiency measures...'. Reference to 'conducting technological research aimed at minimizing emissions of greenhouse gases from continued use of fossil fuels and developing commercial non-fossil energy sources' can be found 12th (of 13) on the list (IPCC, 1996b, p6).

Kyoto Protocol to the FCCC

The Kyoto Protocol, agreed in December 1997, obliges industrialized countries to limit their greenhouse gas emissions by 2008–2012, with reference to a base year of 1990. Given the specific obligations, perhaps it is not surprising to find greater direction (as compared to the FCCC) with regard to how the limitations shall be achieved. For instance, Article 2(1), when laying out example 'policies and measures' identifies 'enhancement of energy efficiency' first (without, however, explicitly saying that the list is in any kind of preferential ordering). Fourth, meanwhile, is 'Research on, and promotion, development and increased use of, new and renewable forms of energy...'. In total, eight sets of policies and measures are presented. Moreover, while six sectors are identified as being particularly important in Article 10(b)(i) of the Kyoto Protocol, three of them appear to be 'privileged', with 'energy' being the first of these three.

IPCC's Third Assessment Report

According to the IPCC's Third Assessment Report, released in 2001, renewable energy is expected to play a relatively modest role in climate change mitigation activities during the coming decade, though its role will increase somewhat during the subsequent ten years.

Between 2000 and 2010, the substitution of wind for coal or gas and hydropower for coal (IPCC, 2001c, p32) might generate reductions of something of the order of 100 MtC_{eq}[6] per year (estimate following IPCC, 2001c, p264). Although the use of biofuels in transportation – for example, ethanol – is also highlighted as a mitigation option (IPCC, 2001c, p197), the relative contribution of renewable energy to mitigation activities within the transportation sector as a whole is not clearly spelled out in the IPCC's reports. Given that the emphasis is upon 'efficiency improvements' (IPCC, 2001c, p30), perhaps it can be assumed to be relatively minor. Even without such a pessimistic assumption, however, renewable energy sources would appear to contribute reductions of, at most, about 200 MtC_{eq} per year – perhaps 10 per cent of the reductions that would be anticipated if a modestly ambitious plan for restricting the growth in global emissions were put in place.[7]

Turning to 2020, it is expected that renewable energy will increase its share, with respect to total mitigation potential. This reflects, it is anticipated, continued progress along the so-called 'technological learning curve' (IEA, 2000), which serves to reduce costs, and the fact that – at least in the electricity supply industry – the capital stock of existing generation plants will need to be replaced (IPCC, 2001c, p40). The result is that the contribution could be 800 MtC_{eq} per year – almost 20 per cent of the reductions anticipated with a 'modest cap'.[8]

Activity at the international level continues. The IPCC is preparing its fourth assessment report, which is currently scheduled to be completed in 2007. The 'Conferences of the Parties to the FCCC' continue as well. The tenth meeting was scheduled to be held in Buenos Aires, Argentina in December 2004. Additionally, other international fora have recently examined the links between global climate change and renewable energy as one of their agenda items. The Commission on Sustainable Development and the World Summit on Sustainable Development are but a couple of such examples (Rowlands, forthcoming). Moreover, an international conference – held in Bonn, Germany in June 2004 – explicitly considered renewable energy.

Additionally, more detailed consideration of assessment options at the international level (as reflected in the IPCC's Third Assessment Report) is being mirrored at the national level. This is not particularly surprising, as the world moves from international agreements on climate change to national implementation of the same (which thus requires national-level policies). The next section of this chapter looks at this – trying to draw out, in particular, the role of renewable energy in emerging national climate change plans and actions. We focus, in particular, upon three countries – Germany, the UK and the US.

Renewable energy in national-level implementation

Germany

Of all the industrialized countries, Germany has reduced its relative greenhouse gas emissions by the greatest extent. In 1990, total emissions amounted to 1249 $MtCO_{2eq}$ By 2002, they had fallen by 19 per cent, to 1016 $MtCO_{2eq}$ (EEA, 2004). Given this, it makes sense to reflect further upon Germany's climate change policy, and to consider the role of renewable energy therein.

Of course, Germany has unique circumstances. With reunification in 1990, the country now included – as part of its 'base year' calculations for the purposes of the climate change regime (namely, 1990) – the emissions from the former East Germany, which contained many energy-intensive activities. These were subsequently shut down, or transformed with new capital so that they reached higher levels of energy- and carbon-efficiency. The result was that emissions from the former East Germany fell by 27 per cent between 1990 and 1991 alone (Beuermann and Jäger, 1996, p191). Indeed, between 1990 and 1995, energy-related (temperature-corrected) carbon dioxide emissions fell by approximately 43 per cent (Schleich et al, 2001, p367). Therefore, this helped the reunified Germany meet its climate change goals. Indeed, Schleich et al (2001) argue that reunification – and the associated acts which were not primarily motivated by climate change aspirations – accounted for approximately 50 per cent of Germany's greenhouse gas emission reductions (from what they would otherwise have been) during the 1990s. These have become known as the 'fall-wall effect' (IEA, 2002a, p125) or, alternatively, 'wall fall profits' (Schleich et al, 2001, p364; Michaelowa, 2003, p31).

But that still leaves the other half of emission reductions to explain. To achieve these reductions, the German government introduced a series of policies and plans – what one critic has called a 'hodgepodge of ... instruments' (Michaelowa, 2003, p34). Some of these were primarily intended to address climate change ambitions, while others had greenhouse gas mitigation as a secondary aim at best.

Interestingly, it has been suggested that the most significant single policy for climate change involved renewable energy. The Feed-in Law of 1990 and its successor, the Renewable Energy Sources Act of 2000, are credited with emission reductions of 12.6Mt. Indeed, the former – on its own – is credited with almost one-fifth of Germany's energy-related carbon dioxide reductions that were induced by policies during the 1990s (Schleich et al, 2001).

Other energy-related measures included fuel taxes. 'The two increases of mineral oil taxes in 1991 (12.5 Euro cents) and 1994 (8 Euro cents), almost doubled the mineral oil taxes in force prior to 1991' (Schleich et al, 2001, p372). During the latter part of the 1990s, and into the next decade, fuel taxes were part of the broader Ecological Tax Reform programme introduced by the new coalition government in 1999 and implemented in several steps until 2003.

Schleich et al (2001, p371) argue that the 'most prominent policy measure in the industrial and services sectors is the voluntary agreements between the majority of German industrial associations, organized under the Federation of German Industries (BDI) and the federal government in 1995–1996 and 2000, which include commitments by industry to specific energy reduction'. These were quite detailed. Critics suggest, however, that many of the 'goals' agreed to by businesses had, in fact, already been achieved (for example Michaelowa, 2003, p35).

Activities outside the energy sector were also important. Deep cuts in methane and nitrous oxide emissions (41 per cent and 34 per cent, respectively, during the 1990s) helped to improve the overall figures. The former was largely catalysed by a 1993 regulation that obliged landfills to recover methane, and the latter by the chemical industry's voluntary agreements to reduce emissions from adipic acid production (Michaelowa, 2003, p37).

Germany's Kyoto target is a 21 per cent reduction (within the EU target of an 8 per cent reduction). This appears within the country's reach. Whether, however, Germany will achieve its own self-imposed target of a 25 per cent reduction in carbon dioxide emissions (from 1990 levels) by 2005 is much more in doubt. Indeed, emission reductions during the past few years have not been as dramatic as they were during the early 1990s. Moreover, with a commitment to phase out nuclear power, and the backdrop of a near-recession and the process of structural economic reform, ongoing reductions may be a challenge. Nevertheless, given Germany's size and influence – in both global economic affairs generally, and environmental issues in particular – its journey will be followed with much interest.

United Kingdom

Like Germany, the UK is in a relatively good position with respect to its greenhouse gas emissions. In 2002, its emissions were more than 14 per cent below 1990 levels (EEA, 2004), which meant that it had already effectively achieved its share of the EU's 8 per cent Kyoto target – namely, a 12.5 per cent reduction. However, also like Germany, much of that achievement can be attributed to activities that were not inspired by climate change ambitions. In the case of the UK, it resulted from the increased use of natural gas in power generation and the associated decline of the coal industry following restructuring of the country's electricity supply industry. In 1990, coal-fired power stations in the UK accounted for 69 per cent of the country's electricity supply; the role of natural gas was virtually negligible at that time. By 2000, coal's share had fallen to 30 per cent, while natural gas had become the single most important resource, with gas-fired power stations in the UK meeting 39 per cent of the country's electricity demand. This so-called 'dash-for-gas' is largely responsible for the 26 per cent reduction in carbon dioxide emissions from power stations in the UK. And it was the reduction in emissions from power stations that was responsible for virtually all of the UK's greenhouse gas emission reductions during the 1990s (DTI, 2000).

But the Kyoto Protocol's 'target year' is still almost a decade away. Therefore, much attention in the UK is now focused upon how these reductions can be consolidated and further ones achieved. This has prompted a recent upsurge in interest in climate change policy in the UK – a distinct change from most of the past decade.

There was relatively little political interest in climate change policy (at least at the national level) in the UK before the mid-1990s (O'Riordan and Rowbotham, 1996). Up until that time, declining carbon dioxide emissions meant that the UK could claim to be one of the few countries that would meet the FCCC's 'intention' – that is, to reduce year 2000 emissions to what they were in 1990. Some commentators (for example Eyre, 2001, p310) identify an increase in road transport fuel tax as the only regulatory measure; little else – apart from 'voluntary measures' – was put in place.

But agreement to the Kyoto Protocol, and the subsequent EU burden-sharing agreement (the aforementioned 12.5 per cent reduction target) helped to catalyse additional consideration. In October 1998, the government launched its main consultation on a comprehensive greenhouse gas emissions reduction strategy, which ultimately resulted in the Climate Change Programme. In addition to the measures laid out in this document, two other measures have attracted considerable attention. One is the 'Climate Change Levy' and the other is the 'Renewables Obligation'. For information about the former, see Varma (2003), while for the latter, see Connor (this publication).

Renewable energy plays an important role in the UK's climate change strategy. Indeed, Wordsworth and Grubb's (2003) recent estimates of the cost of the national climate change mitigation programme suggest that almost 30 per cent of the country's £1.3 billion annual expenditure (in 2002 and 2003) was spent directly on supporting renewable energy in some way. The vast majority (£234 million) was for the renewable obligation certificates, but other significant amounts were available for capital grants for renewable energy (£76 million), climate change levy payment exemptions for renewable energy (estimated at £50 million) and direct research and development funding (by government) for renewables (£18 million). It is not anticipated that these investments will deliver an associated 30 per cent of carbon dioxide reductions (indeed, the renewables obligation, at 18 per cent of expenditure, is expected to provide only 11 per cent of the total reductions in the first implementation period, that is until 2008–2012 (DETR, 2000)), though, together, they are still significant parts of the UK programme. Indeed, for suggestions that they are too high, see Wordsworth and Grubb (2003, p83).

In addition to the Kyoto Protocol obligations, the UK has a stated aim to reduce carbon dioxide emissions to 20 per cent below 1990 levels by 2010. Given the variety of tools being used in the UK to try to reach this goal, that country's experience may offer lessons for others to follow.

United States

The United States – under either the first Bush Administration or the two Clinton Administrations – was rarely identified as a 'leader' on the global climate change issue. In fact, it had made considerable – and successful – efforts to reduce the scope of both the Framework Convention and the Kyoto Protocol during negotiations that preceded the conclusion of each (Agrawala and Andresen, 1999). There was nevertheless a widespread assumption that the country would continue to be engaged in the development of the climate change regime – if not, least of all, to protect its own interests. Therefore, the dramatic shift of position enacted by the new Bush Administration in 2001 was all the more startling.[9]

In March 2001, President Bush announced that the US would withdraw from the Kyoto Protocol process, for he declared that he would 'not accept a plan that will harm our economy and hurt American workers' (quoted in Rosencranz, 2002, p227). Just under a year later, an alternative plan was unveiled by the Bush Administration. It proposed to reduce greenhouse gas emission intensity by 18 per cent by 2012. Given that economic output was predicted to continue to increase during this period, this translated into an anticipated 12 per cent increase in greenhouse gas emissions during the same period. Indeed, the Pew Center on Global Climate Change suggests that this would put US emissions at 30 per cent above their 1990 levels – a far cry from the Kyoto Protocol's 7 per cent reduction commitment.[10]

The President's strategy for limiting greenhouse gas emissions largely places an emphasis upon 'business as usual'. Because there was a 16 per cent reduction in emission intensity during the 1990s (Pew Center, 2002), it will not take much initiative to achieve the goal laid out. Indeed, many say that the US$4.5 billion in funding announced at the same time was largely a repackaging of commitments previously made. Thus, given that there is little 'climate change strategy' at the national level,[11] there is little to say about the content of 'renewable energy' therein.

More action, however, is taking place at the sub-national level (Rabe, 2002). Indeed, there have been numerous documented activities on climate change taken by individual states. Interestingly, however, virtually every one of them has an associated 'co-benefit' – that is, an accompanying local improvement (often economic development) that allows the climate change policy to be packaged as a 'win-win'. In the case of Texas, for example, one in-depth analysis suggests that local air quality concerns and rural economic development were two of the most important motivators; there is no mention of global climate change (CRS, 2000).

Explaining European and American differences

Although the German and British 'performances' on global climate change are not representative of the entire EU experience – for more, see, for example, EEA (2004) – it is still the case that Europeans are taking a much more 'proactive' (and 'serious') approach to the challenge of global climate change than are the Americans. Given this, a question that immediately arises is: 'Why, given their similar economic and political situations, and given the global impacts of climate change, are the two areas adopting different tacks?' In response, a few ideas can be advanced.

First, it may be because the respective costs and benefits of action (versus inaction) are different in these two parts of the world. Significant action to reduce net greenhouse gas emissions would mean that the impacts arising from global climate change would not be as large as they otherwise would have been. The scale of such impacts may be of greater consequence to Europeans than Americans: with a much-higher coastline-to-land area ratio (18.5 versus 2.1[12]), for example, more Europeans would be immediately 'in the front line' of climate change. Similarly, with a larger land mass, in absolute terms (9.6 million square kilometres versus 3.2 million square kilometres) and a smaller population (290 million versus 380 million), those in the US may feel better placed to absorb any climate change impacts (all data taken from CIA, 2003). Indeed, publications that highlight the relatively small economic consequences of any climate change that may occur with 'business-as-usual' (for example Mendelsohn, 2001), let alone those that herald the benefits of a carbon-enriched atmosphere (for example Greening Earth Society, 2004), appear to have higher profile in the US as compared to Europe.[13]

But when attention is turned to the costs of action to reduce net greenhouse gas emissions, the US appears to be the country in which more people believe that these costs would be significant. This may be because Americans perceive their existing energy system to be extremely attractive in terms of costs, and the alternative system to be one that would not only be much more expensive, but which would also place them at a competitive disadvantage, internationally. Let us consider each of these elements in turn.

The US is well known for being the home of cheap energy. Consumers in that country pay much less for their energy needs than do those in Europe. A litre of petrol (gasoline) at the pumps, for example, cost US$0.435 in the US in February 2004. In both Germany and the UK, the price for the same litre of petrol was more than three times higher (US$1.358 and US$1.402, respectively (CEC, 2004)). As a result, when prescriptions to meet the challenge of climate change involve higher prices for energy (to reduce demand), opposition is usually quickly mounted and eventually effective in the US. Consider, for example, the 1993 proposal (from the Clinton Administration) to introduce a 'BTU tax' – a tax based on the heat content of the fuel. 'This tax was expected to raise about US$72 billion in tax revenues over five years to help cut the federal deficit while leading to reduced greenhouse gas emissions by stimulating more efficient consumption of energy' (Agrawala and Andresen, 1999, p461). It, however, never materialized, for 'concerted lobbying from diverse interest groups' (ibid, p462) effectively defeated it. In the end, an extremely modest gasoline tax (amounting to only 4.3 US cents a gallon) was all that went forward. President Jimmy Carter's failed efforts to mount a 'moral equivalent of war', in the face of the energy crisis of the late 1970s, show that Americans' defence of their right to cheap energy predates the rise of climate change up the political agenda. Although Europeans have also occasionally shown their own resistance to energy price rises (the failed efforts of EC Commissioner Carlo Ripa di Mena to introduce a 'carbon tax' in 1992 provides the best such example (Zito, 2002)), there does not appear to be the same deep-rooted opposition to energy price rises. Indeed, Europe's lower carbon intensity levels (almost 50 per cent lower for the largest EU countries[14]) suggest that they are not as dependent upon a cheap supply of carbon-intensive energy sources for their wellbeing.

Additionally, it appears that Europe may have an advantage in terms of the 'alternative energy system' often being proposed as a response to the challenge of global climate change. Most significantly, Europe is the global leader in wind power technology. Not only is almost three-quarters of the world's windpower capacity presently in the countries of the EU (with only 15 per cent in the US) (AWEA, 2003), but '90 per cent of the world's wind turbine manufacturers are based in Europe' (Asmus, 2002). Economic advantage, therefore, might be forthcoming for European companies (Rowlands, 2003). Thus, different trans-Atlantic actions may be explained in terms of respective interests: while the US does not feel particularly threatened by global climate change, it does feel that it would be badly hurt by a change in the fundamental structure of its energy system; alternatively, Europe feels that it would be significantly affected by climate change and it also foresees many benefits (in addition to environmental ones) arising from increased use of new energy systems.

A second possible explanation – building upon parts of the first – takes perceived costs and benefits as its starting point, but emphasizes only parts of that broader argument. More specifically, there are those who argue that because the current Bush Administration has so many links with the oil industry, it is motivated to protect those interests, narrowly defined. Thus, this builds upon the broader hypothesis that business has a particularly strong influence in US decision making (for example Gonzalez, 2001) by directing attention to the fact that many key personnel in the US Administration have past associations with oil companies – for example, Vice-President Dick Cheney was chief executive officer of Haliburton oil company, and National Security Adviser Condoleeza Rice was a director of Chevron (Kay, 2001). Indeed, some critics have called the Bush team an 'oil and gas administration' (Kay, 2001). Given that many US oil companies have not been as active as their European counterparts in diversifying their business activities into less carbon-intensive activities (for example renewable fuels), they may perceive their wellbeing to be inversely correlated with the success of the Kyoto Protocol (for example Rowlands, 2000; Levy and Egan, 2003).

As a third potential explanation for trans-Atlantic differences, recognize that during the past few years, the US has tended to base its risk assessments on 'scientific factors', while Europe has been more apt to use the 'precautionary principle' and to include 'social factors' in its deliberations (compare with Kramer, 2002, pp16–17; Pollack and Shaffer, 2001). Vogel (2001, p29), for example, argues that the 'spread of the precautionary principle within Europe ... serves to both promote and legitimize the politicization of regulatory decision-making, privileging the responsiveness of policy-makers to public opinion'. Given the nature of the global climate change challenge – long lead times, 'lags' in the response of the system, and so on – it may well be that Europeans are more willing to exercise precaution, while those in the US search for unequivocal 'facts'.

Fourth, others may argue that the US does not have the luxury of entering into commitments lightly. If it ratifies an international treaty, the terms of its legal system mean that it is held accountable to reach the goals laid out. If it does not fulfil its commitments, then it may be sued by its own constituencies (see, for example, Lempert, 2001, p115). Some in the US argue that other countries are not subject to the same level of scrutiny.

Finally, Robert Kagan might argue that there exist differences in the policy approach to global climate change because Europeans are from Venus and Americans are from Mars (Kagan, 2003). In other words, global climate change is the kind of issue that the Europeans are ready to address – and, indeed, like to address – because their 'soft' power can be applied to it. By contrast, the Americans have 'higher security concerns' on their mind, so thus bring their 'hard power' to bear on more traditional 'high politics' issues. Indeed, recent polling might lend support to this. Although large portions of both Europe's and the US' population do see global climate change as an extremely important, or critical, threat to their own nation's interest,[15] what might be more indicative is the ranking (compared to other problems) that global climate change receives. When looking just at the proportion of respondents to one poll who claimed that it was an 'extremely important' or 'critical' issue, global climate change ranked third highest ('most important') of 11 threats for Europeans, but only eighth highest ('most critical') of 11 threats for Americans. More strikingly, when we look at the respective share of the respondents who said that global climate change was 'not important' (7 per cent in Europe and 17 per cent in the US), the issue ranks second of 11 for Europeans (that is, there was only one other issue that received a lower percentage of 'not important' responses) and tenth of 11 for Americans. Thus, one might conclude that global climate change is seen as important by both sets of peoples, but that other issues (see Table 3.5) push it off the radar screen in the US (CCFR and GMF (2002)).

Table 3.5 *Possible threats to the vital interest of respondents' countries, as queried in a series of national polls*

- Political turmoil in Russia
- Economic competition from the US ('... from Europe')
- The development of China as a world power
- Islamic fundamentalism
- International terrorism
- Large numbers of immigrants and refugees coming into (own country)
- Global warming
- Globalization
- Military conflict between Israel and its Arab neighbours
- Iraq developing weapons of mass destruction
- Tensions between India and Pakistan

Source: CCFR and GMF (2002).

Thus, five different explanations are presented here, in light of the varied approaches to the challenge of global climate change being taken on each side of the Atlantic Ocean. Our purpose is not to determine, unequivocally, which is 'right'. Instead, the various explanations are meant to give the reader a flavour of the kinds of issues being discussed with respect to transatlantic issues and global climate change. For more about this set of questions, see, for example, Bang (2004), Bodansky (2003), Busby (2003), Schreurs (2004) and Thorning (2000).

Future challenges

Notwithstanding exceptions in individual countries, renewable energy appears to occupy a relatively modest role in the global strategy to address the challenge of global climate change. Within international analyses and negotiations, efforts to promote energy efficiency occupy the most prominent positions. This is certainly understandable, and to some extent desirable. But the proverbial story should not end there.

With the Kyoto Protocol's only target dates being the period 2008–2012, the focus during much of the climate change discussions has been upon the coming decade. Additionally, with the need – for 'political reasons' – to adopt the most 'cost-effective strategies', the focus has been upon solutions that can deliver limited greenhouse gas emission reductions (of the degree demanded by the Kyoto Protocol) over the next few years at the least cost. To a great extent, energy efficiency improvements best fit the bill.

To some degree, this is a positive development. Low-cost greenhouse gas emission reductions (indeed, they are often 'no-cost' or 'low-gain', because they have associated 'co-benefits' (such as improved local air quality)) could help to develop popular and political support for climate change-related activities. The problem, however, is that energy efficiency gains will only get us 'so far'.

Singular focus upon the Kyoto Protocol's reduction targets (approximately 5 per cent reductions from 1990 levels, for industrialized countries as a whole, and no limits upon global emissions) fails to consider the fact that much greater emission reductions are actually required if we want to achieve the ambition of the FCCC, as articulated in Article 2 – that is, 'stabilization of greenhouse gas concentrations in the atmosphere at a level that would prevent dangerous anthropogenic interference with the climate system'. In fact, scientists estimate that emission reductions of the order of 60–80 per cent are required if the aim is to be met. (And it should be remembered that such a reduction in emissions would still mean dramatic increases in concentrations of greenhouse gases in the atmosphere. The impacts of this would continue to have to be dealt with.)

Energy efficiency, however, means that achievements are being made upon the base of a 'carbon-based economy' (and society). Therefore, even with dramatic increases in efficiency, growing output, predicated upon carbon-based technologies, means that greenhouse gas emissions will still be considerable. Consider the following simple example.

Imagine that a firm, a country, or whatever unit you prefer were 'growing' in real terms at 3 per cent a year. It was also, however, improving its 'carbon efficiency' (in terms of output per unit of carbon dioxide released) by 4 per cent a year. Given that the latter is greater than the former, its emissions were in fact decreasing. (For the sake of comparison, the EU's gross domestic product (GDP) during the 1990s was growing at approximately 2 per cent a year and its carbon efficiency was improving by 2 per cent a year. The result was that its emissions remained approximately the same (IEA, 2002b, p75). (If we take our example from the year 2000, we see that its carbon emissions will have declined by 9.2 per cent by 2010. This is, of course, to be applauded. Even though its economy has grown by 34 per cent during the decade, an increase in its carbon efficiency of 48 per cent during the same period means that emissions will have fallen. Indeed, such a decline would probably be deemed a 'success' in light of the Kyoto Protocol targets.

A different assessment arises, however, when a target of 80 per cent is identified. Given the same trajectory, one might think that an almost 10 per cent reduction in ten years would bode well for an 80 per cent reduction. The fact is, however, that this would not be achieved until 2167. Indeed, it takes a 700-fold increase in carbon efficiency to achieve this reduction (because of an associated 139-fold increase in output during the same period). If reductions of this order are required in the medium term, efficiency gains themselves will not achieve the goal. Indeed, not only will they not achieve the goal, but they may also make eventual achievement of larger reductions that much more difficult. This may be because improvements in the efficiency of the existing system will serve to further 'lock-in' the structure of that same system. Moreover, that same system will be supported by what Unrah (2000) calls a 'technico-industrial complex' – elite-level actors interested in continuing down the path of efficiency. Supported in such a way, change may be that much more difficult to achieve.

Indeed, many argue that incentives need to be given now to encourage greater investment in renewable energies (in spite of their apparent higher costs, at least using present financial measures). Grubb et al (2002), for example, maintain that 'induced technical change' (that is, increasing technological choices that are available through mechanisms that encourage private sector involvement; this is as opposed to 'autonomous technical change', which is not driven by energy market conditions or expectations) can result in numerous net benefits for society – in particular, the long-run costs of climatic stabilization may be quite low (Grubb et al, 2002, p304). So-called 'learning by doing' can reduce the price of new technology quickly and dramatically, so that what were perceived to be expensive options in economic models that rely solely upon autonomous technical change benefit greatly from widespread application and increased economies of scale. Any climate change policy that relies, therefore, upon 'inventors, university departments or government research and development' (the 'main potential sources' of autonomous technical change) will probably be more expensive. Such observations would encourage a significant role for renewable energy in response strategies.[16]

Conclusions

The purpose of this chapter has been to explore the links between global climate change and renewable energy. After demonstrating that renewable energy has the potential to make a significant contribution to meeting the challenge of global climate change, a review of selected international and national activities revealed that its role, to date, has been relatively modest. While there have been isolated incidents where renewable energy has played a significant part in reducing greenhouse gas emissions (for example, in the case of Germany's 'feed-in law') or is planning to do so in the near future (for instance, in the UK's climate change strategy), renewable energy has usually played a relatively small role in the overall portfolio of responses. The chapter concluded, however, by noting that 'energy efficiency' – which is currently dominating policy discussions – will only get us 'so far': while it may be the most cost-effective route towards fulfilling the requirements of the Kyoto Protocol, additional action will surely be needed. To meet the ultimate goals associated with the challenges of global climate change, renewable energy will have to take on a much larger role than currently envisaged. Given this, action to support increased development and use of renewable energy must be forthcoming in the short term. For it is the short term experience – both successes and failures – that will lead to effective implementation in the longer term.

Notes

1. Note that 'European Union' figures refer to the 15 member states that were part of the Union as of 1 January 2004.
2. 'Carbon dioxide equivalent' is used to refer to the emissions from various greenhouse gases based upon their global warming potential. In other words, it represents the amount of carbon dioxide that would generate the same climatic changes as the combination of greenhouse gases actually present.
3. Author's calculation given that the share of electricity, worldwide, is reported as follows: coal 34 per cent, renewables (primarily large-scale hydropower) 21 per cent, natural gas 19 per cent, nuclear power 19 per cent and oil 7 per cent (EIA, 2003). The average of the emission ranges presented in Table 3.4 was then used for the calculation.
4. To enter into force, not only must 55 countries ratify the Kyoto Protocol, but together the ratifying countries must also represent at least 55 per cent of Annex I (industrialized) countries' 1990 greenhouse gas emissions. With the United States (36 per cent of Annex I countries' emissions) refusing to ratify the Protocol, Russia's (17 per cent) participation was critical. As of 2 November 2004, the 35 Annex I countries that had ratified the Kyoto Protocol together accounted for 44 per cent of Annex I countries' 1990 greenhouse gas emissions.
5. Article 3(3) of the FCCC notes that '...policies and measures to deal with climate change should be cost-effective so as to ensure global benefits at the lowest possible cost'.
6. C_{eq} stands for 'carbon equivalent'. See Note 2 for an explanation.
7. This 'modest ambition' reflects a scenario in which a suite of emission reductions implemented by 2010 causes energy-related sources of greenhouse gases to rise from 7650 MtC_{eq} (average value of the range presented) in 1990 to 10,500 MtC_{eq} in 2010 (IPCC, 2001c, p264). This would represent an annual increase of 1.6 per cent per year. For the sake of comparison, the average annual growth rate of carbon dioxide from 1990 to 2000 was recently reported as 0.8 per cent per year (Marland et al, 2003).
8. Again, this modest path suggests that emissions go from 7650 MtC_{eq} in 1990 to 9675 MtC_{eq} per year in 2020. This represents an average annual increase of 0.8 per cent, between 1990 and 2020 (IPCC, 2001c, p264).
9. For more on the development of the US position, see, for example, Anderson (2002), Pew Center (2002) and Rosencranz (2002),
10. For the Pew Center's analysis, see Pew Center (2002).
11. This is not to say that other initiatives have not been taken. They have. Prominent ones included the Byrd-Stevens strategy on climate change (the 'US Climate Change Strategy and Technology Innovation Act' in 2001) and the Lieberman-McCain cap and trade proposal (the 'Climate Stewardship Act' in 2003).
12. Figures are the result of dividing total coastline (in metres) by total land area (in square kilometres). Figures for the calculations are taken from CIA (2003).
13. For an argument that major conservative think-tanks in the United States mobilized to challenge the legitimacy of global climate change as a social problem, see McCright and Dunlap (2003).
14. In terms of carbon dioxide per unit GDP, the 2001 figure for the United States was 0.63 kilogrammes carbon dioxide per 1995 US$ GDP, in terms of purchasing power parity. The equivalent figures for the largest EU countries were: France (0.28), Germany (0.44), Italy (0.33) and the UK (0.42) (IEA, 2003).
15. A recent poll, for example, found that a similar proportion of Europeans and Americans considered global warming to be an 'extremely important (or critical) threat to the vital interest of their country in the next ten years' (49 per cent of

respondents in Europe, and 46 per cent in the US) (CCFR and GMF (2002). For more polling results, see Brechin (2003).

16 Grubb et al (2002) also recognize the importance of 'spillover effects' – more specifically, the diffusion of technology from the industrialized countries (those that presently have emission limitation obligations under the Kyoto Protocol) to the developing countries (which do not). This highlights the importance of mechanisms like the Clean Development Mechanism and the Global Environment Facility as key links between global climate change and renewable energy.

References

Agrawala, S. and Andresen, S. (1999) 'Indispensability and indefensibility? The United States in the Climate Treaty Negotiations', *Global Governance*, vol 5, no 4, pp457–482

Anderson, K. (2002) 'The Climate Policy Debate in the U.S. Congress', in Schneider, S., Rosencranz, A. and Niles, J. (eds) *Climate Change Policy: A Survey*, Washington, DC, Island Press, pp235–250

Asmus, P. (2002) 'Gone with the wind: Is California losing its clean power edge to Texas?', *Faultline*, 16 August, www.faultline.org/news/2002/08/wind1.html accessed in April 2003

AWEA (2003) 'record growth for global wind power in 2002', Washington, DC, American Wind Energy Association, News Release, 3 March, www.awea.org/news/news030303gbl.html accessed in May 2003

Bang, G. (2004) *Sources of Influence in Climate Change Policymaking: A Comparative Analysis of Norway, Germany, and the United States*, Oslo, Department of Political Science, University of Oslo

Beuermann. C. and Jäger, J. (1996) 'Climate change politics in Germany: How long will any double dividend last?', in O'Riordan, T. and Jäger, J. (eds) *Politics of Climate Change: A European Perspective*, London, Routledge, pp186–227

Bodansky, D. (2003) 'Transatlantic environmental relations', in Peterson, J. and Pollack, M. (eds), *Europe, America, Bush: Transatlantic Relations in the Twenty-first Century*, London, Routledge, pp59–68

Brechin, S. (2003) 'Comparative public opinion and knowledge on global climatic change and the Kyoto Protocol: The US versus the World?', *International Journal of Sociology and Social Policy*, vol 23, no 10, pp106–134

Busby, J. (2003) 'Climate change blues: Why the United States and Europe just can't get along', *Current History*, March, pp113–118

CCFR and GMF (2002) '*Worldviews 2002*', Chicago Council on Foreign Relations and German Marshall Fund of the United States, www.worldviews.org accessed in April 2003

CEC (2004) 'Selected world gasoline prices', Sacramento, CA: California Energy Commission, www.energy.ca.gov/gasoline/statistics/world_gasoline_prices. html accessed in November 2004

CIA (2003) *The World Factbook 2003*, Washington, DC, Central Intelligence Agency

CRS (2000) '*RPS Case Study: Texas*', San Francisco, CA, Center for Resource Solutions, May, www.resource-solutions.org/Library/librarypdfs/IntPolicy-USRPSCASE-STUDIES.pdf accessed in July 2003

DETR (2000) *Climate Change: The UK Programme, Summary*, London, Department of the Environment, Transport and the Regions

DTI (2000) *Energy Paper 68: Energy Projections for the UK*, London, Department of Trade and Industry, November

EEA (2004) *Annual European Community Greenhouse Gas Inventory 1990–2002 and Inventory Report 2004: Submission to the UNFCCC Secretariat*, Copenhagen, European Environment Agency, Technical Report No 2

EEA (2002) *Greenhouse Gas Emission Trends in Europe, 1990–2000*, Copenhagen, European Environment Agency, Topic Report 7/2002

EIA (2003) *International Energy Outlook 2003*, Washington, DC, Energy Information Administration, 1 May, Report # DOE/EIA-0484(2003)

EPA (2004) *Inventory of U.S. Greenhouse Gas Emissions and Sinks: 1990–2002, Final Version*, Washington, DC, United States Environmental Protection Agency, EPA 430-R-03-004, April

EPA (2003) *Inventory of U.S. Greenhouse Gas Emissions and Sinks: 1990–2001, Final Version,* Washington, DC, United States Environmental Protection Agency, EPA 430-R-03-004, April

Eyre, N. (2001) 'Carbon reduction in the real world: How the UK will surpass its Kyoto obligations', *Climate Policy*, vol 1, pp309–326

Gonzalez, G. (2001) *Corporate Power and the Environment: The Political Economy of U.S. Environmental Policy*, Lanham, MD, Rowman & Littlefield Publishers, Inc

Greening Earth Society (2004) 'Climate change FAQs', www.greeningearthsociety.org/faqs_climate.html accessed in March 2004

Grubb, M., Köhler, J. and Anderson, D. (2002) 'Induced technical change in energy and environmental modeling: Analytical approaches and policy implications', *Annual Review of Energy and the Environment*, vol 27, pp271–308

IEA (2003) *Key World Energy Statistics*, Paris, International Energy Agency

IEA (2002a) *Energy Policies of IEA Countries, 2002 Review*, Paris, International Energy Agency

IEA (2002b) *Dealing with Climate Change: Policies and Measures in IEA Member Countries, 2002 Edition*, Paris, International Energy Agency

IEA (2000) *Experience Curves for Energy Technology Policy*, Paris, International Energy Agency

IEA (1998) *Benign Energy? The Environmental Implications of Renewables*, Paris, International Energy Agency

IPCC (2001a) *Climate Change 2001: The Scientific Basis*, Cambridge, Cambridge University Press

IPCC (2001b) *Climate Change 2001: Synthesis Report*, Cambridge, Cambridge University Press

IPCC (2001c) *Climate Change 2001: Mitigation*, Cambridge, Cambridge University Press

IPCC (1996a) *Climate Change 1995: Impacts, Adaptation and Mitigation of Climate Change: Scientific-Technical Analyses*, Cambridge, Cambridge University Press

IPCC (1996b) *Climate Change 1995: Economic and Social Dimensions of Climate Change*, Cambridge, Cambridge University Press

IPCC (1991) *Climate Change: The IPCC Response Strategies*, Washington, DC, Island Press

Kagan, R. (2003) *Of Paradise and Power: America and Europe in the New World Order*, New York, Knopf

Kay, K. (2001) 'Analysis: Oil and the Bush Cabinet', London, BBC News; http://news.bbc.co.uk/2/hi/americas/1138009.stm, 29 January; accessed in March 2004

Kramer, L. (2002) 'Development of environmental policies in the United States and Europe: Convergence or divergence?', San Domenico di Fiesole, Robert Schuman Centre for Advanced Studies, Working Paper No 2002/33, May

Lempert, R. (2001) 'Finding transatlantic common ground on climate change?', *The International Spectator*, vol 36, no 2, pp109–121

Levy, D. and Egan, D. (2003) 'A neo-Gramscian approach to corporate political strategy: conflict and accommodation in the climate change negotiations', *Journal of Management Studies*, vol 40, no 4, pp803–829

Marland, G., Boden, T. and Andres, R. (2003) 'Global, regional, and national fossil fuel CO_2 emissions', in *Trends: A Compendium of Data on Global Change*, Oak Ridge, TN, Carbon Dioxide Information Analysis Center, Oak Ridge National Laboratory, US Department of Energy

McCright, A. and Dunlap, R. (2003) 'Defeating Kyoto: The Conservative movement's impact on US climate change policy' *Social Problems*, vol 50, no 3, pp348–373

Mendelsohn, R. (2001) *Global Warming and the American Economy: A Regional Assessment of Climate Change Impacts*, Cheltenham, Edward Elgar

Michaelowa, A. (2003) 'Germany – A pioneer on earthen feet?', *Climate Policy*, vol 3, pp31–43

Nuclear Energy Institute (2003) 'Life-cycle emissions analysis', Washington, DC, Nuclear Energy Institute, www.nei.org/doc.asp?docid=1038 accessed in July 2003

O'Riordan, T. and Rowbotham, E. (1996) 'Struggling for credibility: The United Kingdom's response', in O'Riordan, T. and Jäger, J. (eds) *Politics of Climate Change: A European Perspective*, London, Routledge, pp227–267

Paul Scherrer Institut (2002) 'Life cycle assessment (LCA)', Villigen PSI, Switzerland, Paul Scherrer Institut

Pew Center (2002) *Climate Change Activities in the United States*, Washington, DC, Pew Center on Global Climate Change, June

Pew Center (no date) *Pew Center Analysis of President Bush's February 14th Climate Change Plan*, Washington, DC, Pew Center on Global Climate Change, www.pewclimate.org/policy/response_bushpolicy.cfm accessed in July 2003

Pollack, M. and Shaffer, G. (2001) 'The challenge of reconciling regulatory differences: Food safety and GMOs in the transatlantic relationship', in Pollack, M. and Shaffer, G. (eds) *Transatlantic Governance in the Global Economy*, Lanham, Rowman & Littlefield Publishers Inc, pp153–178

Rabe, B. (2002) *Greenhouse & Statehouse: The Evolving State Government Role in Climate Change*, Washington, DC, Pew Center on Global Climate Change, November

Rosencranz, A. (2002) 'U.S. climate change policy', in Schneider, S. H., Rosencranz, A. and Niles, J. (eds), *Climate Change Policy: A Survey*, Washington, DC, Island Press, pp221–234

Rowlands, I. (forthcoming) 'Renewable energy and international politics', in Dauvergne, P. (ed), *International Handbook of Environmental Politics*, Cheltenham, Edward Elgar

Rowlands, I. (2003) 'Renewable electricity and transatlantic relations: Exploring the issues', San Domenico di Fiesole, Robert Schuman Centre for Advanced Studies, Working Paper No 2003/17

Rowlands, I. (2000) 'Beauty and the beast? BP's and Exxon's positions on global climate change', *Environment and Planning C: Government and Policy,* vol 18, pp39–54

Schleich, J. et al (2001) 'Greenhouse gas reductions in Germany – Lucky strike or hard work?', *Climate Policy*, vol 1, pp363–380

Schreurs, M. (2004) 'The climate change divide: The European Union, the United States, and the future of the Kyoto Protocol', in Vig, N. and Faure, M. (eds), *Green Giants? Environmental Policies of the United States and the European Union*, Cambridge, MA, MIT Press, pp207–230

Serchuk, A. (2000) 'The environmental imperative for renewable energy: An update', Washington, DC, Renewable Energy Policy Project

Thorning, M. (2000) 'Climate change policy: Contrasting the US and the European Union approaches', Washington, DC, American Council for Capital Formation

Unrah, G. (2000) 'Understanding carbon lock-in', *Energy Policy*, vol 28, pp817–830

Varma, A. (2003) 'UK's climate change levy: Cost effectiveness, competitiveness and environmental impacts', *Energy Policy*, vol 31, pp51–61

Vogel, D. (2001) 'Ships passing in the night: The changing politics of risk regulation in Europe and the United States', San Domenico di Fiesole, Robert Schuman Centre for Advanced Studies, Working Paper No 2001/16, June

Wordsworth, A. and Grubb, M. (2003) 'Quantifying the UK's incentives for Low Carbon Investment', *Climate Policy*, vol 3, pp77–88

Zito, A. (2002) 'Integrating the environment into the European Union: The history of the controversial carbon tax', in Jordan, A. (ed) *Environmental Policy in the European Union: Actors, Institutions & Processes*, London, Earthscan, pp241–255

4

Renewable Energy for Developing Countries: Challenges and Opportunities

Giulio Volpi

The problem: Access to clean energy

Access to affordable, modern energy services is a prerequisite for sustainable development and poverty alleviation and, more specifically, for achieving each of the Millennium Development Goals (MDGs).[1] Yet there are currently two billion people worldwide, mainly in rural areas of developing countries, who lack access to modern forms of energy. This figure is roughly the same as in 1970, as population growth has in these areas offset the achievement of bringing electricity and other modern energies to over a billion people during this period. In the coming decades, the world will also face the challenge of meeting the energy needs of those living in urban areas – where the most population growth is expected to occur. Together this has been called the 2+2 billion problem, that is meeting the energy needs of those presently outside the modern economic system, and providing for the added population to be expected – virtually all of whom will be in the South and in the cities (UNDP, 2000; ICS, 2002).

The challenge of climate change

If the unmet energy needs of the South are met in the same way as those of the North have been, through uncontrolled use of fossil fuels, this will result in more local and global environmental pollution. Global climate change is already occurring and will increase flooding, hurricanes, droughts and rising sea levels. Southern countries will suffer more than the rich countries because they are most vulnerable, yet least able to cope. Analysis strongly suggests that large reductions of global greenhouse gas emissions are needed if the global climate is to be stabilized (IPCC, 2001). However, under its 'Business as usual scenario' the International Energy Agency predicts a 70 per cent increase in carbon emissions by 2020, one-third of which would result from electricity generation in developing countries (IEA, 2000). Therefore, unless poorer nations are given support to develop renewable and efficient energy systems, they could become significant emitters of GHG (greenhouse gas) emissions.

Sustainable energy: A priority for development

Access to sustainable energy services is an essential element of sustainable development, as identified by the Commission on Sustainable Development (CSD):

> *To implement the goal accepted by the international community to halve the proportion of people living on less than US$1 per day by 2015, access to affordable energy services is a prerequisite.*
>
> CSD, 2001

However, energy has so far largely fallen betw,een the environment and development agendas.[2] The lack of clean and affordable energy has serious consequences for health, quality of life and the economy. Reliance on traditional fuels such as wood and animal and crop waste for cooking and heating, results in widespread respiratory illnesses. Prolonged exposure to smoke dramatically increases the likelihood of very young children and women developing severe health problems such as acute respiratory infection and chronic lung disease, as well as raising the risk of stillbirths, and is now being linked to blindness and immune system changes (WB, 2000).[3]

The need to carry fuel manually over ever greater distances is time-consuming, leads to musculo-skeletal problems and diverts labour from more productive uses. It accentuates the related problem of deforestation and soil erosion, particularly in arid and semi-arid areas. It also helps perpetuate gender inequality as the burden falls mainly on women and girls. Studies in Nepal, for instance, have shown that many women spend as much as two and a half hours every day searching for firewood and animal dung for cooking. This daily grind traps them in poverty by preventing them from putting their time to more productive uses. The unavailability of energy in rural areas in the quality and forms needed for processing raw materials and other industrial applications limits these regions' potential for economic growth. The resulting shortage of productive employment opportunities, which drives the surplus population to cities, thus aggravates problems of urbanization (IT Power, 2001).

The role of renewables

All scenarios for reaching sustainability in energy use envisage a significant increase in the share of primary energy from renewable sources (UNDP, 2000; G8 Renewables Task Force, 2001; IPCC, 2001). 'Renewable energy' commonly refers both to traditional biomass (firewood, animal wastes and crop residues burned in stoves) and modern technologies based on solar, wind, biomass, geothermal, small hydropower, wave and sea flow. Our definition, also called 'new renewables', excludes large hydropower systems because, besides being an already mature technology, large hydropower poses many threats to the environment and the communities involved, as has been documented by the World Commission on Dams (WCD, 2000).

Renewable energy potential

As shown in Figure 4.1, while traditional biomass provides about 7–11 per cent of global primary energy supply, new renewable energy provides about 2 per cent. This may seem small, but ought to be compared with the 2.2 per cent contribution made by large hydropower, which is so important around the world. Besides traditional biomass, small hydropower in China and transport ethanol in Brazil are among the largest supplied in developing countries. Globally, contributions from wind power and solar photovoltaics (PV) are still small, although these technologies have witnessed an annual growth of 30 per cent in recent years. Analysis suggests that with the necessary support measures and political will, 'new' renewable energy could reach a share of 10 per cent by 2010 (Goldemberg et al, 2002).

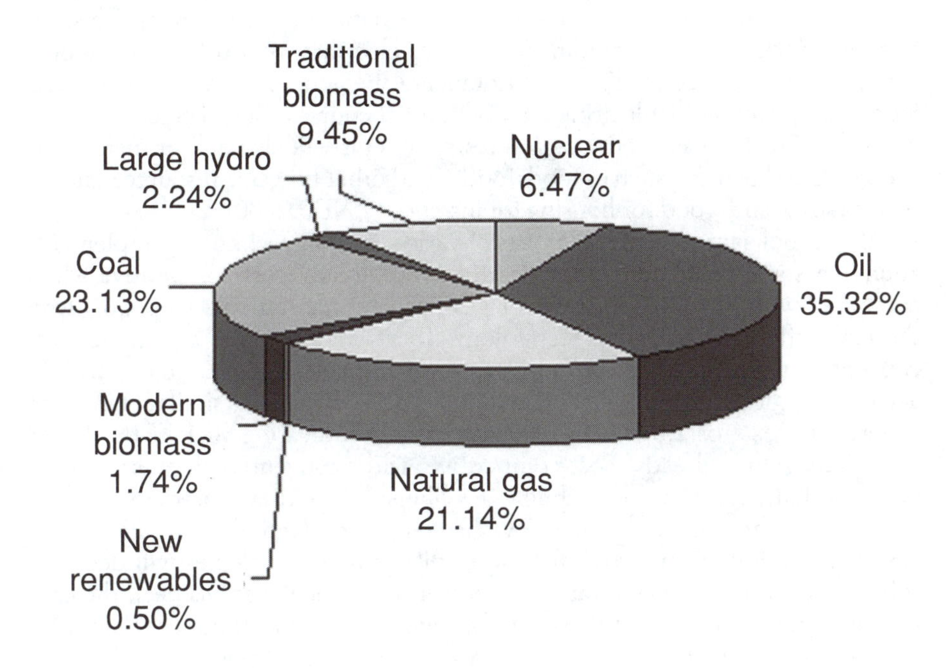

Source: UNDP, 2004

Figure 4.1 World energy consumption in 1998 (TPES)

Presently the world pays more than US$1 billion in official energy bills per year and invests between $290 and $430 billion annually in new energy infrastructure (United Nations, 2001). Redirecting a fraction of these funds would help to reach the 10 per cent target by 2010. This target would require increasing the share of clean renewables by one percentage point per year from now onwards. Renewable energy sources have the potential to power the earth several times. Indeed, the oil multinational Shell stated in a recent report that even under the assumption that global per capita energy use triples to 200 gigajoules (GJ) in a few decades, primarily geothermal and solar sources would be enough to fuel the world (Shell International, 2001).

Benefits and costs

Developing countries are increasingly recognizing the various advantages of renewable energy. The latter adds to self-reliance in energy production, a goal increasingly high on the international and national policy agenda of many countries. Renewables can also promote energy security by decentralizing energy supplies with small, modular and rapidly deployable energy projects that are particularly suited to the electrification of rural communities in developing countries (UNDP, 2000).

In those countries, renewables are often the most economical choice in rural areas – where it would be too costly to bring in the grid – while simultaneously offering a clean 'leap' over fossil fuels. Their modular nature and distribution mean they can be built (and paid for) as the demand for energy grows. This can also reduce the need for upgrading the existing electricity network, if there is one. Finally, renewables offer direct environmental benefits, in terms of improved indoor air quality, available drinking water and of course reduced greenhouse gas emissions. For biomass, sound agroforestry systems will also deliver many non-energy related benefits such as food, fodder and other biogenic resource material such as straw and wood for housing for instance (UNEP, 2000).

If one looks into the present costs of various energy technology implementations, in some cases new renewables can be already cost-competitive when compared with new coal, nuclear or gas-fired power stations. Dependent on thermal efficiency, specific technology, location and flue gas treatment, conventional plants display power generation costs of 3.4–8.0 Euro cents/kWh across the globe. Modern biomass and wind display a similar range of generation costs of 2.8–8.0 Euro cents/kWh – indicating that in the lower ranges they are already cost-competitive with conventional energy technologies. Larger renewable technologies like solar thermal electricity, already profoundly cheaper than solar PV, have not received the attention they deserve. In addition, the costs for these still young technologies will decrease substantially if their growth rates continue as dynamically as has been the case for wind and solar. Both of these have grown globally by 30 per cent annually over the last seven years. India for instance grew its wind power base in a few years from 0 to 1700MW capacity and is prepared to increase its new renewable energy, mainly wind and decentralized biomass for rural off-grid application by 10,000MW per year until 2012 (Martinot et al, 2002).

Applications in developing countries

Renewable energy is already in use across the world, providing reliable and clean power to people in rural areas on low incomes and contributing to grid power. Different renewable energies are available for supplying a number of energy services: on the one hand, rural residential and community uses including lighting, entertainment and other conveniences (e.g. wireless telephone); on the other, productive uses such as agriculture, small industry, commercial and social services (e.g. drinking water, education, healthcare). Below is an outline of these uses, the relevant renewable technologies and recent market development (Martinot et al, 2002; IT Power, 2001):

- *Off-grid electricity in the home.* Solar home systems, small-scale wind and family-sized hydropower schemes can provide electricity for lighting and low power appliances such as radios or cooling fans. Solar home systems consist of a photovoltaic solar panel, battery, charge controller and end uses such as fluorescent lamps. There are currently 1.1 million solar home systems and solar lanterns in rural areas of developing countries. The largest markets for solar home systems are India (with a total of 450,000 systems), China (150,000) and Kenya (120,000)
- *Electricity for health centres and schools.* Health centres use solar PV or wind power to generate electricity for the refrigeration of medical supplies, lighting, equipment sterilization and telecommunications. Around 150,000 PV and wind power systems are being used at health clinics, schools and other communal buildings in developing countries
- *Community water pumping.* Many communities use solar PV or wind powered equipment to pump water for drinking, livestock and in some cases irrigation. For instance, between 500,000 and one million wind-powered water pumps are in use in Argentina, with other markets including South Africa (100,000) and Namibia (30,000). Renewable energy can also power purification systems to make clean drinking water. Few examples of this renewable energy application exist however, mainly in the Dominican Republic
- *Cooking.* Biomass (such as wood, straw and animal dung) and biogas (principally methane from composting) provide the cheapest option for cooking in many rural areas, and some rural communities are successfully using solar cookers. Since 1980, donor programmes have disseminated about 220 million improved stoves throughout the South, mostly in China (180 million) and India (30 million)
- *Agriculture and commercial.* Small-scale hydropower provides direct mechanical power for processing agricultural crops such as grains. Water mills are still an important part of many rural economies. Solar drying plays an important role in rural areas. Solar water distillation is of growing importance, particularly in places where climate change is exacerbating the problem of seawater contaminating fresh water supplies
- *Small industry.* Mini-grid or stand-alone systems can power small industries and provide substantial income and create jobs at the local level.

An example is a wind-solar-diesel hybrid that provides 24-hour power for seaweed drying, woodworking and sewing. Solar-photovoltaic systems can support income-generating applications such as evening lighting in shops and charging batteries for local people

- *Grid-connected supplies.* Electricity consumption in developing countries continues to grow rapidly, especially in the industrial sector. Wind farms, small-scale hydropower, biomass, geothermal and other renewable sources are all competitive and viable technologies for grid-based power generation. Half of the global small hydropower capacity is installed in China, while Brazil and the Philippines are the leading producers of biomass power in the South. As for wind, India leads the developing world with 2800MW, followed by China with over 700MW (2004 figures).

As shown in Table 4.1, providing life-sustaining energy to poor communities does not require much energy in developing countries (G8 Renewable Energy Task Force, 2001). Using the most efficient appliances for these life-sustaining services requires as little as 60kWh per person per year. And in cases where renewable energy is used, the emission of pollutants and greenhouse gases is close to zero. In comparison, Europeans on average consume more than 3000kWh of electricity per year just at home.

Table 4.1 *Basic development needs and energy requirements*

Basic needs per day	**Energy requirements per person per year (kWh)**
An electric pump for 5 litres of potable water per capita	2
5 hours basic lighting (50W, CFL) per household	18
Domestic refrigeration (0.5kWh) per household	36
Hospitals – lighting and refrigeration of medical supplies for 100 households	2
Schools – energy for lighting/heating/information technologies for 100 households	2

Source: Adapted from G8 Renewables Taskforce, 2001

The barriers to renewables

As in all new technologies, renewable energies face a number of barriers. After more than 20 years of research and development, it is clear that by now most technical barriers have been overcome. As recognized by the Renewables Energy Task Force (see below) in a report for the Group of Eight leading industrialized nations: 'the barriers to renewable energy are not technological, but rather political, financial, educational and related infrastructure' (G8 Renewables Taskforce, 2001).

Indeed, the first 'barrier' to greater renewable energy penetration is the lack of enabling policy and regulatory frameworks, which usually favour traditional energy sources. This has been the case with the liberalization and privatization of energy markets imposed by international financial institutions such as the World Bank (WB) and the International Monetary Fund (IMF) on most of the developing countries over the last 15 years. Such reforms have been primarily aimed at achieving economic and financial objectives, with little or no care for their environmental and social impacts. They have often reduced the incentives for energy efficiency and renewable energies. On the contrary, experience has shown that aggressive policy targets and specific support measures for renewables are needed to guide and frame the global opening of energy markets in a way that favours rather than undermines the cleanest energy sources.

Another barrier is the massive subsidies benefiting conventional energy systems. While clean energy technologies typically have a higher investment cost offset by lower or near zero operating costs, costs comparisons are distorted by previous and current subsidies for conventional fuels. 'Direct' subsidies include tax reductions and research and development funding. Although subsidies are being cut in many countries, they are still believed to account for more than US$150 billion in public spending (excluding transport) (UNDP, 2000). In Europe, annual subsidies to fossil fuels have been estimated to amount to US$10 billion, plus US$5 billion to nuclear fuels (Free University of Amsterdam, 1999).

An important but often hidden subsidy is the environmental costs related to conventional fuels use – such as damage to biodiversity, air pollution and of course climate change – which are not accounted for in the final energy prices. In the European Union (EU), for instance, it has been estimated that the cost of producing electricity from gas would increase by 30 per cent, if external costs were to be taken into account. This would account for 1–2 per cent of the EU gross domestic product (GDP), or 85–170 billion, not including the costs of global warming (European Commission, 1995). Other hidden subsidies include obligations to purchase a certain form of energy, reduced electricity rates for large (usually industrial) users, infrastructure support and exemptions from risks or liabilities (for example nuclear power plants).

Some developing countries have recently started reducing subsidies, with clear environmental benefits. For instance, over 1996–2000 China reduced subsidies to its coal industries, which resulted in carbon emission cuts of 7.3 per cent, while its GDP increased by 39 per cent. Additional benefits include a 32 per cent reduction in particulate matter, which has significant health effects and contributes to reducing the impact of climate change (Reid et al, 1997). When phasing out environmentally harmful energy subsidies, government should nevertheless carefully consider the potential social impacts. An interim successful strategy is to provide the equivalent to reward the renewable energy technology for its avoided social and environmental costs.

A third barrier is the lack of available start-up finance for renewables, as current financial arrangements in developing countries do not properly address the characteristics of this sector. Funding mechanisms such as development agencies and Export Credit Agencies (ECAs) often lack specific lending portfolios for clean energies and energy efficiency projects. This policy inertia seems to affect even those governments that have adopted ambitious renewable energy targets domestically, such as Germany or the United Kingdom (UK), but are failing to extend this approach in a consistent fashion to their Official Development Aid and export credit support programmes (Maurer and Bhandari, 2000).

International policies and actions

The analysis above clearly suggests the need for policy action at national and international levels to advance renewable and efficient energy systems. Since the United Nations Conference on Environment and Development in 1992, the attention given to this field has certainly increased, although policies and measures to overcome the regulatory and financing barriers to renewables have being implemented too slowly at best. The key recent developments and their relevance for renewable markets are briefly reviewed below.

The Clean Development Mechanism (CDM)

The Clean Development Mechanism (CDM) is one of the two project-based flexible mechanisms of the 1997 Kyoto Protocol. The CDM allows industrialized countries to meet their emissions reductions targets in part through 'carbon credits' bought by investment in low-carbon projects in developing countries. CDM projects are also supposed to result in benefiting developing countries by helping them to achieve sustainable development (Ellis et al, 2004).

Many observers from the UN and other development organizations viewed the CDM as a potential tool for promoting renewables in developing countries and thus assisting in the transition away from fossil fuels. The evidence to date, however, is that most CDM projects merely shift the location at which emissions reductions are made through the Kyoto Protocol without delivering additional sustainable development benefits to host countries, and do not help to catalyse funding for renewable and energy efficiency by industrialized country governments. The most important CDM projects are those that capture or destroy gases with high global warming potentials like methane, nitrous oxide and hydrofluorocarbons (such as HFC-23) at existing facilities. For example, out of the 240 million credits being claimed up to 2012 by 111 projects in December 2004, 40 million come from two HFC-23 projects and another 70 million from one nitrous oxide project. About 46 per cent of all credits are from these three projects alone (Pearson, 2004).

In comparison, renewables projects adopted thus far are estimated to generate about 25 million carbon credits up to 2012, roughly 10 per cent of the total. This is less than the credits resulting from a single HFC-23 project in India and about a third of the credits that will be generated by a South Korean nitrous oxide project. While renewables account for 41 per cent of all projects, comparing credit volumes is a better way to judge how successfully the CDM is profiting renewables. In some cases renewables projects may attract incrementally higher prices – the Dutch government, for example, offers slightly more for credits from renewables projects. However, it is still the case that they will receive a small amount of the total spent on carbon credit purchases by industrialized countries (Haya et al, 2002; Pearson, 2004).

If such trends continue, it is clear that the CDM will result in a rather small, if not insignificant, support for renewable and energy efficient systems in the South, as well as undermine the credibility of the Kyoto Protocol. According to one estimate, the CDM is likely to represent less than 0.5 per cent of the annual renewables market in developing countries, becoming considerably less influential in financial terms than existing funding mechanisms such as the Global Environmental Facility (Salter, 2004).

The Group of Eight (G8) Renewable Energy Task Force

The G8 Renewable Energy Task Force was established at the Okinawa Group of Eight (G8) summit in Okinawa (Japan) in 2000, with the objective of producing concrete policy recommendations to increase the use of renewable energy in developing countries (G8 2000). The Task Force recognized that, over the medium term, increasing the share of renewables within the global energy supply would prove less expensive than a 'business as usual approach'. It also proposed a global indicative target of serving an additional one billion people by 2010, with a major focus on increasing energy access in developing countries.

The main policy recommendation made by the Task Force was to reduce the costs of renewable technologies by expanding markets. To this end, it called on G8 countries to, amongst others (G8 Renewable Energy Task Force):

- remove incentives and other support for environmentally harmful energy technologies;
- implement market-based mechanisms that address externalities, enabling renewable energy technologies to compete in the market on a more equal and fairer basis;
- adopt common environmental guidelines among the G8 Export Credit Agencies (ECAs), which include minimum standards of energy efficiency or carbon intensity for ECA-financed projects.

While these recommendations are not new in themselves, for the first time they were put forward by a high-level group of governmental and mainstream business representatives rather than environmental activists. They were finally not adopted by the G8 governments, specifically due to the US opposition to the one billion people target and, jointly with Canada, its unwillingness to underwrite the phase-out of subsidies to fossil fuel sources.

The World Summit on Sustainable Development

Renewable energy was intensely debated at the World Summit on Sustainable Development (WSSD), held in Johannesburg in 2002. Compared with the Earth Summit in 1992, the issue of access to sustainable energy moved from the periphery to centre stage; however, the WSSD somehow failed to deliver meaningful results. The debate raged over whether to raise targets for use of renewable energy. Some members of the EU and many developing countries, such as Brazil, wanted to set a global goal to increase the world's use of renewables by 2010. Finally, the EU tabled a proposal for a world target of 15 per cent renewable energy by 2010, up from about 14 per cent in 2000 – including large hydro and traditional unsustainable biomass.

On the other hand, the US and OPEC countries, with Venezuela and Saudi Arabia at the fore and joined by Japan and Canada, strongly opposed the whole concept of a global renewables goal. The US made clear its opposition to introducing concrete new targets in the UN document, arguing that these would damage US economic flexibility. The insistence on the omission of timetables was rather due to the Bush administration's close relationship with the oil and coal industry in the US – which has consistently lobbied against renewable energy targets at domestic level. At the same time the G77 and China succumbed to OPEC and made very clear that support for any target on renewables would lead to a split in the G77 and China group. And as developing countries needed a functioning G77 and China group for matters other than energy at WSSD, the expansion of renewables became a victim of this arrangement (La Vina, 2002).

The final text on energy within the WSSD's Plan of Implementation mainly reiterated commitments already negotiated under the Commission for Sustainable Development (CSD), with no time-bound targets forcing countries to switch away from fossil fuels to clean renewables. To overcome the failure to agree upon a global renewable energy target, the EU then launched the Johannesburg Renewable Energy Coalition (JREC). As shown in Table 4.2, the JREC is a group of like-minded nations committed to adopt new and time-bound targets and effective measures for renewable energy uptake (European Commission, 2002).

The 'Renewables 2004' Conference

Following the failure of the WSSD, the International Renewable Energy Conference (Renewables 2004), held in Bonn in July 2004, injected a new political momentum to the promotion of renewables. While the final outcome was the usual result of a series of diplomatic compromises, ministers from over 130 countries underwrote that renewables have a key role to play in the future energy system and committed to a number of national/regional time-bound targets and actions.

The Bonn Political Declaration – endorsed by all countries including the US and Saudi Arabia – recognized that renewables, 'combined with energy efficiency, will become a most important and widely available source of energy and will offer new opportunities for co-operation among all countries' (German Government, 2004). As in the case of the WSSD, governments could not agree on a global target for renewable energy but rather took note 'of countries who have adopted, and others that will adopt targets for enhancing the share of renewables in their national energy mix...' The fact that the EU itself was internally divided on its own longer term renewables commitments did not help in getting political support to new global or regional targets.

The Bonn Declaration also included language urging international financial institutions and regional development banks 'to significantly expand

Table 4.2 JREC Member Countries with Renewable Energy Targets

Country	Renewable Energy Targets
Austria	78.1% of electricity output by 2010
Belgium	6% of electricity output by 2010
Brazil	Additional 3300MW from wind, small hydro, biomass by 2016
Cyprus	6% of electricity output by 2010
Czech Republic	5–6% of TPES by 2010 8–10% of TPES by 2020 8% of electricity output by 2010
Denmark	29% of electricity output by 2010
Estonia	5.1% of electricity output by 2010
Finland	35% of electricity output by 2010
France	21% of electricity output by 2010
Germany	12.5% of electricity output by 2010
Greece	20.1% of electricity output by 2010
Hungary	3.6% of electricity output by 2010
Ireland	13.2% of electricity output by 2010
Italy	25% of electricity output by 2010
Latvia	6% of TPES (excluding large hydro) by 2010 49.3% of electricity output by 2010
Lithuania	12% of TPES by 2010 7% of electricity output by 2010
Luxembourg	5.7% of electricity output by 2010
Mali	15% of TPES by 2020
Malta	5% of electricity output by 2010
Netherlands	12% of electricity output by 2010
New Zealand	30PJ of new capacity (including heat and transport fuels) by 2012
Norway	7TWh from heat and wind by 2010
Poland	7.5% of TPES by 2010 (Development Strategy of Renewable Energy Sector) 14% of TPES by 2020 (Development Strategy of Renewable Energy Sector) 7.5% of electricity output by 2010 (As per Directive 2001/77/EC)
Portugal	45.6% of electricity output by 2010
Singapore	Installation of 50,000m^2 of solar thermal systems by 2012 Complete recovery of energy from municipal waste
Slovak Republic	31% of electricity output by 2010
Slovenia	33.6% of electricity output by 2010
Spain	29.4% of electricity output by 2010
Sweden	60% of electricity output by 2010
Switzerland	3.5TWh from electricity and heat by 2010
Turkey	2% of electricity from wind by 2010
United Kingdom	10% of electricity output by 2010

IEA (2004)

their investments in renewables and energy efficiency and...establish clear objectives for renewable energies in their portfolios.' As a direct response, during the conference several development banks launched special renewable energy funds. For example, the World Bank Group announced a pledge to increase by 20 per cent its lending to renewable energy and energy efficiency projects. However, according to independent analysis, the World Bank's announcement is not more than a public relations exercise, given the fact that the proposed increase will be marginal at best and do little to address the Bank's ongoing bias towards fossil fuels. For instance, over the past decade (1994–2003), the World Bank Group approved over $24.8 billion in financing for fossil fuel extractive and power projects. At the same time, the World Bank Group approved just $1.06 billion in renewable energy projects. The Washington-based institution preferred fossil fuels to renewables by a 23:1 ratio (Vallette and Kretzmann, 2004).

The most significant outcome of the Bonn Conference was perhaps the International Action Programme (IAP), in which developing countries in particular showed a readiness to move forward on promoting renewable energy. Amongst the most ambitious commitments were those from China, the Philippines, the Dominican Republic and Egypt. More specifically, the Philippines committed to doubling their share of renewable power to 40 per cent by 2013. The Philippines is rich in renewable energy sources, such as wind, biomass and geothermal energy. Currently, only the US produces more geothermal energy than the Philippines. Similarly, China committed to increasing its share of renewable power, including hydropower up to 50MW, from the current 5 per cent to 10 per cent by 2010 and 12 per cent by 2020.

An NGO analysis of the IAP, however, shows that only about 20 of the 180 reviewed proposals were truly outstanding (Singer et al, 2004). Among the commitments of industrialized countries, only a few are noteworthy – such as the commitments to increase wind power and renewable energy from Germany, Spain, Denmark and New Zealand. Nevertheless, the German government estimated the implementation of the Action Programme could result in carbon dioxide savings of about 1.2 billion tonnes per year by 2015.

Conclusion

Since the Earth Summit, it has been increasingly recognized that efficient and renewable energy systems can play a key role in meeting both economic development and climate protection goals. The basic conditions for wider deployment of renewable energies both in the developed and the developing worlds – for example reduced costs, fair and stable regulatory frameworks and investor confidence – are progressively materializing, although in a patchy and scattered manner. Nevertheless, the road to a sustainable energy future is still long and full of blocking stones, including opposition from conventional energy lobbies and lack of policy-makers' awareness. International co-operation thus far, whether under the CDM or at the WSSD, has resulted in marginal support for renewable and energy efficient systems in the South.

The political momentum created by the Renewables 2004 conference clearly showed that only strong political will can overcome the numerous barriers, whether regulatory or financial, to jump start renewable energy globally. Two factors will be key to build on and continue this momentum. First, the EU must keep its global policy leadership on renewables. The next political test will be the willingness to adopt an ambitious renewable target for 2020, as requested by the European Parliament, environmental NGOs and clean energy groups. Second, increased multilateral or bilateral funding must be available to support those developing countries seriously embarking on a transition to clean energy sources. Successes in the field of renewable energy development in the South can play a crucial role in facilitating an agreement on future climate change regimes beyond 2010.

Notes

1 The Millennium Development Goals (MDGs) are to: eradicate extreme poverty and hunger, achieve universal primary education, promote gender equality and empower women, reduce child mortality, improve maternal health, combat HIV/AIDS, malaria and other diseases, ensure environmental sustainability, and develop a global partnership for development. Notably there is no MDG on increased access to energy services, though energy-related indicators are used to measure progress made on environmental sustainability..

2 The failure to make this linkage was explicitly identified by the United Nations Development Programme: *'Poverty has received scant attention from an energy perspective. This is remarkable given that energy is central to the satisfaction of basic nutrition and health needs, and that energy services constitute a sizeable share of total household expenditure in developing countries' (UNDP, 2000).*

3 According to the World Health Organization (WHO), *'Acute respiratory infections in children under five years of age are the largest single category of deaths [and disease] from indoor air pollution, apparently being responsible for about 1.2 million premature deaths annually in the early 1990s'* (Smith and Mehta, 2000).

References

Commission for Sustainable Development (2001) 'Decision on energy for sustainable development', Ninth session, Agenda Item 4

Ellis, J. et al (2004) 'Taking stock of progress under the CDM', *OECD Paper*, Paris, OECD

European Commission (1995) *Externalities of Energy,* Brussels

European Commission (2002) *The Way Forward on Renewable Energy*, Johannesburg Renewable Energy Coalition

Free University of Amsterdam (1999) 'Energy subsidies in western Europe', *Report for Greenpeace*, Amsterdam

G8 Renewable Energy Task Force (2001) *Renewable Energy: Development that Lasts – Chairmen's Report*, Rome, Italian Ministry of Environment

German Government (2004) 'Conference report, outcomes and documentation. Renewables 2004 – International Renewable Energy Conference', www.renewable2004.org

Goldemberg, J. et al (2002) 'Energy since Rio: achievements and promising strategies',

Discussion paper, New York, GEF Energy Panel in Preparation for the WSSD
Haya, B. et al (2002) 'Damming the CDM: why big hydro is ruining the Clean Development Mechanism', *International Rivers Network* and *CDM Watch Paper*, Sydney
IEA (2000) *World Energy Outlook*, Paris, International Energy Agency
IEA (2004) 'Johannesburg renewable energy coalition database', JREC, www.iea.org/dbtw-wpd/textbase/pamsdb/jr.aspx
IEA Bioenergy Task (2001) 'Answer to ten frequently asked questions about bioenergy, carbon sinks and their role in global climate change', www.joanneum.at/iea-bioenergy-task38
International Council for Science (ICS) (2002) 'Energy and Transport', *ICSU Series on Science for Sustainable Development*, no 2, Geneva
IPCC (2001) 'Climate Change 2001 – Mitigation', *Contribution of Working Group III to the Third Assessment the IPCC*, Geneva
IT Power (2001) 'Power to tackle poverty, getting renewable energy to the world's poor', *Greenpeace Report*, London
La Vina, A. (2002) 'The success and failures of Johannesburg: A story of many summits', Issue Brief, *World Resources Institute*, http://governance.wri.org/pubs_description.cfm?PubID=3797
Martinot, E. et al (2002) 'Renewable energy markets in developing countries', *Annual Review of Energy and the Environment*, vol 27, pp309–348
Maurer, C. and Bhandari, R. (2000) 'Climate of export credit agencies', issue brief, *World Resources Institute*
Pearson, B. (2004) 'Market failure: Why the CDM won't promote sustainable development', *CDM Watch*, www.cdmwatch.org
Reid, W. et al (1997) 'Are developing countries already doing as much as industrialized countries to slow climate change?' *World Resources Institute*
Salter, L. (2004) 'A clean energy future? The role of the CDM in promoting renewable energy in developing countries', *WWF International*, Manila
Shell International (2001) *Energy needs, choices and possibilities – scenarios to 2050*, London
Singer, S. et al (2004) 'A NGO assessment of the international action plan of the renewables 2004 conference', *WWF International*, unpublished document
Smith, K. and Mehta, S. (2000) *The Burden of Disease from Indoor Air Pollution in Developing Countries: Comparison of Estimates*, Geneva, World Health Organization
UNDP (2000) *World Energy Assessment – energy and the challenge of sustainability*, New York, UNDP, World Energy Council, and UNDES
UNDP (2004) *World Energy Assessment Overview 2004 Update*, New York, UNDP, World Energy Council, and UNDES
UNEP (2000) *Natural Selection: Evolving Choices for Renewable Energy Technology and Policy*, Paris, UNEP
United Nations (2001) report on 'Energy and Transport', *Economic and Social Council*, E/CN.17/2001/PC/20, New York, UN
Vallette, J. and Kretzmann, S. (2004) 'The energy tug-of-war: The winners and losers of world bank fossil fuel finance', *Sustainable Energy and Economy Network*, www.seen.org/PDFs/Tug_of_war.pdf
World Bank (2000) 'Indoor air pollution, energy and health for the poor', *World Bank Newsletter*, no 1, http://Inweb18.worldbank.org
World Commission on Dams (2000) *Dams and Development – A New Framework for Decision Making*, London, World Commission on Dams

Part 2

Policies to Develop Renewable Electricity and its Generation Technologies

Danish Wind Power Policies from 1976 to 2004: A Survey of Policy Making and Techno-economic Innovation

Kristian Hvidtfelt Nielsen[1]

Introduction

This chapter addresses the governmental promotion and regulation of wind power in Denmark from 1976 onwards. Many different policies with different objectives, scopes and approaches have been implemented and consequently the overall picture is one of great complexity and differentiation. I will be surveying the use of the following seven policy mechanisms (adapted from Redlinger et al, 2002):

1 national programmes of research, development, demonstration and information (RDD&I);
2 direct investment subsidies;
3 renewables set-asides (quotas);
4 government-assisted business development;
5 physical planning;
6 power purchase agreements (including recent attempts to establish a market for renewable energy, recently postponed until further notice);
7 environmental taxation.

The guiding question for this survey is: How do wind power policies translate into techno-economic innovations and vice versa? Considering the many policy varieties and their even more numerous possible interactions, one may expect to find that the answers differ greatly from one policy mechanism to another. One would also expect each of these answers to complicate some of the more simple views on the relations between policy making and innovation.

Table 5.1 provides a brief summary of the history of these policy mechanisms with special emphasis on techno-economic innovations in the field of wind power. From the point of view of the current government in office (as of December 2004), Danish wind power policy mechanisms and Danish techno-economic innovations combined have been very successful. Already in 2002 Denmark had met its wind power policy goals for 2005 (1500MW installed), and it is well known that Denmark has a thriving and leading wind industry. In consequence, the Danish government has decided to stall some of the policy mechanisms in operation (the development programme for renewable energy and the wind power quotas, for example).

Such mechanisms have all played an important role in the transition of the Danish energy system from an almost exclusively fossil fuel-based system in 1976 to the present-day penetration by renewable energy of approximately 15 per cent. Clearly, in this transitionary process, all of these different policy mechanisms are interrelated in various ways. In some respect, it may not make sense to treat them individually as they are all historically situated in close relation to each other. Even though we may be able to say something about the specific techno-economic innovations that each of the seven policy mechanisms gives rise to, and I will be doing so in what follows, techno-economic innovations in the field of wind power in Denmark have partly resulted from the combined effect of several of such mechanisms. Policy-making is therefore not only about enacting such mechanisms with specific innovations in mind, but also, and perhaps more significantly, about orchestrating and managing interactions between policy mechanisms. Even so, in the following I will be expanding on each of the seven policy mechanisms separately and examining their implications for techno-economic networks of wind power in Denmark (Callon, 1992; Nielsen, 2001).

Research, development, demonstration and information

Historically, Denmark has had a long tradition of promoting wind power with government grants for research, development, demonstration and information. Around the turn of the 19th century the Danish government supported the wind power experiments of physicist and folk high school teacher Poul la Cour (Hansen, 1985). La Cour gave inspiration to engineer Johannes Juul, who also benefited from government support for the design and construction of the famous Gedser Wind Turbine in the late 1950s (Thorndahl, 1996, 2000).

Since the mid–1970s, various Danish governments have supported wind energy RDD&I through many different channels such as:

- the Energy Information Council (1974–1975);
- the national energy research programme (1976–);
- the development programme for renewable energy, including basis grants for the Folkecenter for Renewable Energy in North Jutland (1982–2002/2003);
- the Ministry of Energy's support scheme for new energy technologies (1981–1985/1986);
- the Information Secretariat for Renewable Energy (1982–1996), which

Table 5.1 *Summary of the seven wind energy policy mechanisms in Denmark*

1 RDD&I	In 1976, two national programmes, the energy research programme and the development programme on renewable energy, were launched with wind energy being one of the main areas. The development programme was closed down in 2002/2003, and the energy research programme goes on. To some extent, the RDD&I of the national energy research programme has been planned and carried out in close relation to the administration of the investment subsidies (see investment subsidies below), and also to the enactment of the new type approval scheme (see business development, below)
2 Investment subsidies	From 1979 to 1989, Danish governments have paid direct investment subsidies as a percentage of total investment costs to wind turbine buyers (actual percentages presented in Table 5.2). These subsidies were intended to be a direct market pull complementing the technology push included in the research programmes. In fact, the pulls and pushes have been closely related to each other through the wind turbine approval scheme that was enabled with some funding from the government's subsidy scheme, but also from the research programme described above
3 Quotas	From 1985 onwards, various agreements between the government and the utilities have been enforced with respect to certain wind energy quotas to be fulfilled by the utilities. Before 2002 these quotas were filled exclusively with latest wind turbine models from Danish manufacturers in order to push the technological development. At present (2004), an international bidding procedure is being carried out for three 150 MW offshore wind power plants (formerly part of the most recent 750 MW offshore quota for the utilities).
4. Business development	In order to replace the approval scheme that fell away with the end of the direct subsidies in 1989, a type approval scheme was enabled in 1990–1992 (including codes of practice and certification schemes). The new scheme of approval was directed towards a general quality lift in the wind industry. Also, in 1992–1993, supplementing the expected boost of quality, an export guarantee of around DKK 50 million for wind turbine manufacturers was established
5 Planning	In 1991, the government established a committee to identify wind turbine sites for the utility quotas. Various planning tools were being designed and put to use on a national scale when, from 1994 onwards, wind power siting became part of longer term municipal planning.
6 Power purchase agreements	Up until 1992 the utilities, wind turbine manufacturers and wind turbine owners agreed on payment for wind electricity and on costs of grid connection. In 1992 this voluntary agreement broke down and, instead, the government set a fixed tariff and divided costs related to grid connection between utilities and wind turbine owners
7 Environmental taxation	Since 1984, wind turbine owners have been exempted from general per-kWh electricity tax on electricity consumption. Also, in 1984, the Danish government enacted a per-kWh tax subsidy for producers of renewable energy. Later, in 1991, the government introduced a CO_2 subsidy for all producers of renewable energy and an additional subsidy for producers of wind power, waterpower and biogas electricity (additional subsidy was removed again in 2000)

continued under the auspices of the Council for Sustainable Development and Renewable Energy (previously known as the Energy Environment Council) (abolished 2002).

Generally, such policy channels are used to enhance the knowledge base with respect to both shorter and longer term stimulation of renewable energy technologies. They give rise to knowledge production, distribution and circulation, directed towards the development of renewable energy technologies and/or markets. Also, they help enable coordination between actors involved in these technologies/markets. Consequently, we may say that such channels are socio–techno–scientific in the sense that they fuse social, technical, and scientific issues of the technology in question. They are often seen as the most important policy mechanisms in terms of nurturing techno-economic innovation. Yet, as I will demonstrate, it remains controversial to what extent and how such nurturing of innovative activities actually results from RDD&I.

In all of the above-mentioned channels, wind energy was (and still is in the channels that remain active) an important, and in most channels *the* most important priority. An important lesson learned from comparing Denmark to other countries is that RDD&I spending translates only with great difficulty into techno-economic innovation and high levels of installed renewable energy capacity. For example, between 1973 and 1988, the United States (US) and Germany spent roughly US$380 million and US$79 million, respectively, on wind energy RDD&I with relatively little impact on national industry and market. Whereas Denmark, spending only US$15 million in the same period, came to dominate the world market for wind power in terms of wind turbine manufacturing and wind power penetration into the supply system (Righter, 1996; Heymann, 1995).

Many different reasons for these differences between RDD&I spending and industrial success have been put forward. Peter Karnøe, a Danish political scientist and organization scholar, has argued that the Danish manufacturers adhered to a bottom-up approach in contrast to the top-down approach found in the US and Germany. Thus, the Danish manufacturers all had previous experience in mechanical engineering and adhered to a step-by-step development in which they used their operational experience to obtain low-tech, reliable design solutions (Karnøe, 1990, 1991, 1993). Similarly, historian of technology Matthias Heymann has pointed to the Danish technological style that resulted from the values and beliefs of the non-academic engineers, technicians and artisans involved in Danish wind turbine design (Heymann, 1996, 1998).

In this ultra-brief – and obviously simplified – summary of their respective analyses, both Karnøe and Heymann seem to be suggesting that RDD&I played a minor role in the success of the Danish wind turbine manufacturers. This is only partly true, as both authors would readily acknowledge. For example, they both mention that the RDD&I work being done at the Test Station for (Smaller) Wind Turbines (the word smaller was removed from the name around 1985/6), established under the national energy research programme and the national subsidy scheme, was indeed vital in helping the wind industry to its current position of world market dominance. However, Per Dannemand Andersen, university-trained engineer and organization scholar, in his analysis of the relations between the Test Station and the manufacturers, found that most manufacturers thought that the research being done at the Test Station was of little immediate interest for the innovative activities of the manufacturers. Still, Andersen concluded that the personal contacts and communications, facilitated by the approval scheme for wind turbines, did play an important role in mediating design knowledge and operational experience between the manufacturers and the Test Station (Andersen, 1993).

I will leave aside for a moment the question of why the Danish wind turbine manufacturers have become so successful, and what role government RDD&I played in their success. Instead, I will look more closely at the two most important (in terms of funding) of the above-mentioned channels for wind energy RDD&I in Denmark, namely the national energy research programme (the projects on larger and smaller wind turbines, respectively) and the development programme for renewable energy (the Blacksmith Mill project). See Figure 5.1 for an overview of funding of these two channels.

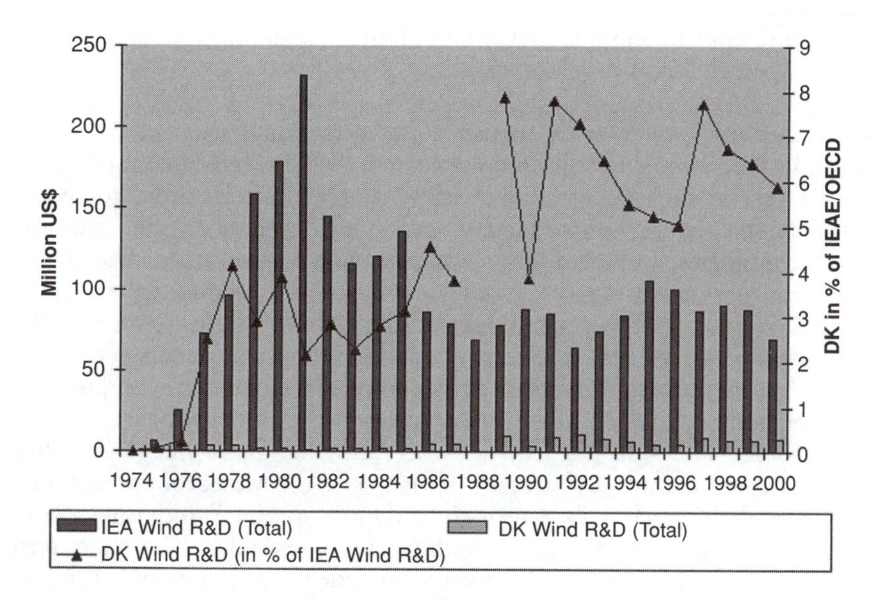

Note: All figures are in 2000 prices and exchange rates

Source: International Energy Agency, 2002

Figure 5.1 Wind energy R&D grants in IEA/OECD countries and in Denmark (including the national energy research programme and the development programme for renewable energy), 1974–2000

Larger wind turbines

The national research project on larger wind turbines formed part of the national energy research programme that began in 1976. The project on larger wind turbines formally ended in 1986 when it was joined with the project on smaller wind turbines. In practice, however, the separation between the projects on larger and smaller wind turbines still continued to exist into the late 1980s according to the 1991 evaluation of the national energy research programme (Energistyrelsen, 1993).

The national energy research programme followed as an immediate consequence of the first national energy policy, published in 1976, which introduced the idea of a highly diversified energy supply into Danish policy making (Handelsministeriet, 1976a). According to this policy, Danish energy supply suffered from its 90 per cent dependency on oil products and had to develop multiple sources of supply, such as oil, uranium, coal, natural gas and renewable energy sources (sun, wind, garbage, straw and so on). In order to obtain this diversified supply the policy presented an energy action programme that consisted of, among other things:

- a national heat plan for Denmark aiming at increased utilization of natural gas and cogeneration of heat and electricity;
- new energy savings by means of information, regulation, subsidies and control;
- increased coordination of energy and physical planning;
- energy research and development.

The 1976 energy policy envisioned a 3 per cent penetration of renewable energy into the national energy supply system by 1995, while also emphasizing the uncertainties such projections involved at the time. In order to reduce uncertainty, the policy promoted techno-economic developments in renewable energy technologies, and wind power was one of its top priorities. The Danish government predicted that large-scale wind turbine technology could be developed within 10–20 years, and wanted the Danish utilities to manage this development. Consequently, the research and development department of the Association of Danish Utilities, in collaboration with the engineering departments of the major utilities, was put in charge of the project on larger wind turbines. The project further involved the Department of Fluid Mechanics at the Technical University of Denmark, the Meteorology Section at Risø National Research Laboratory (including a special Wind Engineering Group) and a number of larger manufacturing companies. The short term objective was to initiate a realistic and practical development of larger, electricity-generating wind turbines. These larger wind turbines were to be produced by Danish manufacturing companies and operated by the utilities as a supplement to their existing power plants (Handelsministeriet, 1976b).

The project revolved around three, more or less separate, parts:

1 technology development;

2 wind turbine siting;
3 integration of wind power in the electricity supply system (Nielsen, 1984).

With respect to technology development, the first task was to refurbish and test the Gedser Wind Turbine. At the time the 200kW Gedser Wind Turbine, designed by Johannes Juul, was one of the few existing larger wind turbines with a considerable operation record of ten years, from 1957 to 1967. Comparing their test results with other wind turbine designs, the Risø researchers heading the refurbishment concluded that the design of the Gedser Turbine was one of the most promising solutions to the problem of designing a larger, reliable wind turbine rotor (Lundsager et al, 1979). On the basis of the Gedser Wind Turbine test programme and using new computer design codes based on aerodynamic blade-element theory, the utilities, the research institutions and the manufacturing companies together designed and constructed two experimental wind turbines at Nibe, known as the Nibe A and B turbines. The two wind turbines, both with 630kW generators and with a three-bladed, upwind rotor design similar to the Gedser Wind Turbine (Nibe A had a tip-stayed rotor whereas Nibe B had no stays), had different regulation systems (Nibe A had stall, and Nibe B pitch regulation) so that operational experience with both systems could be drawn from the project (Handelsministeriets og elværkernes vindkraftprogram, 1981).

In the early 1980s, a Danish wind turbine company, Danish Wind Technology, owned jointly by two power companies and the Ministry of Energy, was set up to develop the Nibe A+B design experiences into mature technology (Nielsen, 1984). However, although the company did deliver five 750kW wind turbines to the Masnedø wind energy park, owned and operated by the utility association of Eastern Denmark, it never succeeded in developing competitive large-scale wind turbine technology. Instead, the company marketed a number of small- and medium-sized wind turbine models before merging in 1989 with another wind turbine manufacturer, Vestas, becoming a sales department within the larger Vestas company organization (Madsen, 2000).

The project on larger wind turbines also included investigations into micro-siting methods for wind turbines. A group of Risø researchers developed the so-called wind atlas method to determine within a margin of ±5 per cent the expected wind electricity production for any given wind turbine model at any given site in Denmark. The researchers used the wind atlas method to identify and analyse the most profitable wind turbine sites in Denmark. From their analysis, the researchers concluded that up to 4TWh annually could be obtained if a larger number of larger wind turbines were to be located at these sites. At the time of the investigation, this amounted to 10–15 per cent of the expected electricity consumption in 1995 (Petersen et al, 1981).

This conclusion was followed up in an extensive investigation looking into the possibilities of siting a larger number of larger wind turbines (the equivalent of 4TWh annually). The Danish Planning Agency managed the investigation that employed the wind atlas method to generate geodetic wind energy maps of Denmark with the country divided into four energy classes. Around 400 potential sites were located and inspected, 209 of which were examined more closely as regarding wind regimes, landscape environment and planning conditions. These studies were made available to the local planning authorities and later formed the basis of the work of the Wind Turbine Siting Committee (see section on physical planning below) (Planstyrelsen, 1981–1986).

According to the evaluation of the national energy research programme carried out in 1986, notwithstanding the wind turbines of Danish Wind Technology, the project on larger wind turbines did not result in commercially viable wind turbine technology. From the point of view of the evaluators, this was due to a general lack of industrial involvement. The project had aimed at technology-push development, but failed because the companies involved did not have the internal resources necessary to pursue the development of larger wind turbines any further. Also, the managers of the individual investigations had been unable to make their results available to the companies in a comprehensible way. The evaluators therefore endorsed the fact that, from 1986 onwards, the project on larger wind turbines would become integrated with those on smaller wind turbines (Micheelsen et al, 1986).

Smaller wind turbines

Ironically, some of the major results from the project on larger wind turbines found application in the field of smaller wind turbines. The wind atlas method, for example, was used by individuals and cooperatives when considering which smaller wind turbine model to buy and where to site it. The manufacturers of smaller wind turbines used the results of the computer design codes developed by the Wind Engineering Group at Risø. And, finally, the operational guidelines on how to handle large wind power penetration into the power system are becoming more and more important as the Danish-built wind turbines have now become very large and their contribution to Danish power is almost 15 per cent.

The national research project on smaller wind turbines began in 1978 with the establishment of the Test Station for Smaller Wind Turbines at Risø. The primary objectives of the Test Station were to test the different wind turbine designs of the existing manufacturers of smaller heat- and electricity-generating wind turbines and also to assist these manufacturers in the design and development of new and improved wind turbine models. This also involved some research into wind turbine construction and wind turbine operation. In order to negotiate and to promote knowledge exchanges between the Test Station and the various actors in its immediate environment, including wind turbine manufacturers, wind turbine owners, the Danish Energy Agency and various energy organizations, a so-called contact group for interested parties was set up in relation to the Test Station. This group held regular meetings to assess and discuss the work of the Test Station (Handelsministeriet, 1978).

In 1979 the Danish Energy Agency authorized the Test Station to carry out wind turbine approvals in relation to the newly established subsidy scheme for renewable energy systems (see section on investment subsidies below). At the time nobody knew exactly how to perform such approvals, so the Test Station employees had to define wind turbine approval from rock bottom. From the beginning they were very conscious that the approval scheme could (and ought to) be used to communicate to the many manufacturers of smaller wind turbines the research and testing being done at the Test Station. The employees also decided to use the approval scheme to learn more about the immediate problems of the manufacturers and to frame their research around these problems. Also, some rules about safety and design had to be established in connection with the approval scheme. The Test Station employees wanted these rules to reflect their knowledge about wind turbine design and construction, but also to satisfy the needs of the manufacturers and wind turbine owners (Rasmussen and Jensen, 1999).

In the early years of the approval scheme, wind turbine approvals consisted of so-called system assessments in which the wind turbines were assessed with respect to overall functional and constructional safety. Lacking established codes of practice and standards of wind turbine construction, the assessments performed were relatively crude, and each wind turbine model had to be assessed on an individual basis. Still, in collaboration with the manufacturers and under supervision of the contact group, the Test Station employees gradually developed a practice of approval that, according to Per Lundsager, then manager of the Test Station, was the first step towards actual codes of practice for wind turbine construction. Moreover, this practice turned out to have significant influence on the safety and reliability of the first mass-produced smaller wind turbines in Denmark (Lundsager, 1981). After the end of the subsidies in 1989 the approval scheme developed into a formal type approval that aimed at raising the quality of Danish wind turbines (see section on government-assisted business development below).

Because of early criticism of its double research/approval role the Test Station was formally split up into a research and an approval section in 1981. The former was thereafter funded by research grants (most of which came from the national energy research programme), while the latter received funding partly from the Danish Energy Agency and partly from approval and testing fees paid by the manufacturers. In spite of the formal division the Test Station has maintained its physical unity (except now for a separate blade test centre at Sparkær and the new Test Station for offshore wind turbines being established at the west coast of Jutland). Moreover, at the time the division seemed to have little impact on the actual work being done. Thus, the employees of the Test Station continued to carry out research and development in the fields of aerodynamics and structural mechanics, applying their results to the development of the rules of approval. They also continued to support the manufacturers in developing more reliable wind turbines. According to the 1986 evaluation of the national research programme, the most important activity of the Test Station in the period from 1978 to 1986 was to promote the growing industry, primarily by means of wind turbine approval. The evaluators further concluded that both the Test Station and the manufacturers shared a trial-and-error approach to the development of the smaller wind turbines. As the manufacturers were starting up their own development departments and beginning to build larger and larger wind turbines in search for economies of scale, the evaluators advised the Test Station to embark on more fundamental research and internationally attractive projects such as rotor aerodynamics (Micheelsen et al, 1986).

Seemingly, this advice was taken seriously at the Test Station as the following evaluation concluded in 1991 that the funding from the national energy programme had helped create an internationally unique wind energy research environment at Risø with very strong international ties. The interviews performed as part of the 1991 evaluation also reported that the major manufacturers including Micon, Nordtank, Bonus and Vestas (the two former of which merged into NEG Micon, which then merged with Vestas) did carry out some development work themselves but very little actual research into general issues of wind turbine design. Danish manufacturers primarily kept informed on international research developments through their connections to the wind turbine approval and testing sections at the Test Station (Energistyrelsen, 1993).

During the 1990s and early 2000s, wind energy research was a major priority in the national energy research programme in Denmark, and still is. From 1991 onwards, Danish wind energy research has concentrated on (Energistyrelsen, 1997, 2002a):

- the Danish design concept, including rotor and blade design, control and regulation systems;
- new design concepts;
- integration of wind power into the power supply system;
- audiovisual effects of wind turbines on their physical environment;
- offshore wind turbines.

In total, from 1976 to 2001, the national energy research programme funded wind energy research with approximately DKK 300 million (running prices).

The Blacksmith Mill

In 1980 the Danish Board of Technology began looking into renewable energy and in 1982 formed the Steering Committee for Renewable Energy to be responsible for the development programme for renewable energy systems. In its nine years of existence the committee granted a total of DKK 224 million (running prices) to smaller and medium-sized firms and to various development and demonstration projects (Meyer, 2000).

The committee gave priority to wind energy research over other energy technologies, directing approximately one-tenth of its total funding to the Folkecenter for Renewable Energy (in the form of an annual basis grant) and to the so-called Blacksmith Mill design project carried out by the Folkecenter. The Folkecenter wanted the Blacksmith Mill design to become an alternative to the established wind industry in Denmark, including the research and development projects on larger and smaller wind turbines. The philosophy behind the Blacksmith Mill design concept was as follows. The Folkecenter would develop a prototype wind turbine, complete with design drawings and dimensioning, and this prototype was to be offered free of charge to anyone interested in manufacturing the Blacksmith Mill. The prototype was developed with small independent blacksmiths in mind, hence its name. Blacksmiths or others could also receive funding from the development programme of the Board of Technology for starting up their manufacture of Blacksmith Mills (Bjørgren et al, 1987).

However, according to three evaluation reports published in 1990, the Folkecenter had failed in establishing sufficient interest from potential manufacturers in the Blacksmith Mill concept. At the time of evaluation only two producers of the Blacksmith Mill existed, both of which had not even originally been blacksmiths. Moreover, these two wind turbine manufacturers were the two smallest in Denmark. In all, by 1990, only 185 Blacksmith Mills had been produced. None of the larger wind turbine manufacturers had shown any interest in the Blacksmith Mill concept. The evaluators concluded that one of the reasons why the Blacksmith Mill concept was a commercial failure was the lack of mutual involvement between the Folkecenter and the manufacturers. Even though the development efforts of the Folkecenter had been close to perfect, the Folkecenter had underestimated the fact that most manufacturers were unable to carry out the rest of the necessary development of the Blacksmith Mill concept. The manufacturers simply did not understand why their Blacksmith Mill designs failed, and communication between the Folkecenter and the manufacturers was lacking (Energisystemgruppen, 1990).

Besides the funding of the Folkecenter, the Board of Technology also gave grants to some of the first wind energy parks and to development projects within wind turbine blade technology with more impact for the larger wind turbine manufacturers. In 1988 the Board of Technology decided on a three-year action plan that included annual grants to the Folkecenter and one project on offshore wind turbines (Industri- og Handelsstyrelsen, 1988). Two years later, the Steering Committee for Renewable Energy was abolished, and the development programme of the Board of Technology transferred to the Board of Renewable Energy. In 1996 this board was closed down, and the development programme for renewable energy transferred to the Danish Energy Agency. From then on, the wind energy projects of the development programme for renewable energy were managed by the Special Committee on Wind Energy, also responsible for managing the national wind energy research programme (Meyer, 2000). In consequence of general cut-backs the government in office has now (2003–2004) decided to end the development programme for renewable energy. Among other things, this means that the Folkecenter now stands to lose its annual grant with retrospective application as from January 2002 (Folkecenter, 2002). Instead, the government is funding regional development activities (possibly including those of the Folkecenter, being situated in a region of low industrial activity) and promoting industrial R&D with tax deductions on development activities in smaller firms.

Investment subsidies

While RDD&I policy channels as those mentioned above may be said to aim more at technology push developments than at market pull, investments subsidies are usually oriented directly towards developing and strengthening of markets. Although the two terms, technology-push and market-pull, are a bit too rigid to capture the complexities of both RDD&I and market development efforts, they may serve as a first approximation. However, as will become clearer when looking more closely into the construction of a combined subsidy/approval scheme for wind turbine systems, Danish investment subsidies were intended to go closely hand in hand with RDD&I. Investment subsidies were seen as a means of making renewable energy technologies more financially attractive for buyers and investors, but also as a way in which to focus the research of Test Station into areas of immediate interest to the manufacturers, increasing the possibility of the latter profiting from the research results of the former.

From 1979 to 1989 the Danish government provided direct capital investment subsidies for wind turbine systems as a percentage of total investment costs (see Table 5.2).

Table 5.2 *The subsidy percentage for wind turbines. A total of 2567 wind turbines were subsidized with a total of DKK 276 million (running prices)*

	May 1979	Jan 1981	Jan 1982	Jan 1985	June 1985	Jan 1986	Jan 1989	Aug 1989
Subsidy	30%	20%	30%	25%	20%	15%	10%	0%

Source: Energimiljørådet, 1998

Originally, the subsidy scheme formed part of the national scheme of employment and was also supposed to supplement the ongoing energy research and development activities, especially those looking at smaller wind turbines. The subsidies were tax-free and only granted to systems that were officially approved. The stated purpose of the subsidy scheme was to further mass production of standardized renewable energy systems by means of a market-pull development. With respect to wind turbines, the ministry authorized the Test Station to carry out approvals and thus to manage the necessary standardization of wind turbines. The combination of subsidies and approvals would not only ensure a certain minimum constructional quality of the wind turbine systems subsidized, but also provide the Test Station and the Danish Energy Agency, managing the subsidies, with very important information on the market for wind turbines.

In 1981 the very rudimentary systems assessments (see section on smaller wind turbines above) were upgraded to system approvals. This meant that more formal rules of approval were being developed and that also the approval section of the Test Station would receive funding directly from the subsidy scheme. By the mid-1980s the approval rules had developed into more specific load and safety regulations, known as the load paradigm. Basically, the load paradigm defined the overall design characteristics of a typical Danish, three-bladed, upwind wind turbine with a rotor diameter of 5–25m, a rotor solidity of 7–15 per cent, and a medium blade-tip velocity of 35–50m/s. Based on the common experiences of the manufacturers and the combined approval and research experiences of the Test Station, the load paradigm also contained a semi-empirical calculation method for basic wind turbine design. The paradigm was deliberately conservative, that is it prescribed somewhat larger loadings than was thought necessary, in order to maintain a high degree of wind turbine reliability (Lundsager and Jensen, 1985). Thus, the load paradigm embodied not only a special Danish design, but also specific design approaches.

The combined subsidy/approval scheme helped translate the wind power policies of the government into an actual market for wind turbine systems, not only by raising market demand, but also by means of wind turbine standardization and increased market communication and transparency (Micheelsen et al, 1986). For example, the most important market actors, including the wind industry (organized in the Association of Danish Wind Turbine Manufacturers, now the Danish Wind Industry Association), wind turbine owners (the Association for Danish Wind Turbine Owners), the Energy Agency and the Test Station, met regularly in the contact group of the Test Station to discuss approvals, subsidies and more general issues of wind markets and wind technology.

Many more initiatives besides the subsidies, however, helped create, sustain and maintain the emerging wind turbine market. For example, in 1978 the owners' association initiated wind turbine statistics that included up to 85 per cent of all wind turbines in operation in Denmark. These statistics were published on a monthly basis in *Naturlig Energi*, the member magazine of the owners' association. This produced relative market transparency at a very early stage and, no doubt, gave everyone involved in the market important insights into the various wind turbine models on the market and into the differences between, and the importance of, different wind turbine sitings. Moreover, the service and maintenance packages that most manufacturers offered to wind owners when their standard two-year warranties were running out supplemented the market information supplied by the owners' statistics. The service and maintenance systems were vital for the manufacturers because they served to provide the manufacturers with detailed operation records from their own wind turbines. Moreover, they facilitated post-transactional communication between the manufacturers and their customers. Besides the existence of wind turbine statistics and of service and maintenance packages, the availability of finance and insurance schemes also played an important role in making wind turbines accessible for many prospective owners and in making actors on the emerging wind turbine market more knowledgeable about the financial state of the market (Madsen, 1986).

The scheme of subsidies was an important device for maintaining alignment between government energy policies and the growing wind turbine market. In late 1985, because of increasing private speculation in wind energy parks, the government added two criteria for obtaining wind turbine subsidies to the scheme of subsidies, the so-called residence and consumption criteria. Similar criteria had previously been part of the voluntary power purchase agreement between the wind turbine owners and the utilities, but had been removed from their second agreement of 1984 (see section on power purchase agreements below). The new residence criterion enforced by the government effectively limited the permitted geographical distance between the wind turbine and its owner(s). This criterion was believed to be important for maintaining the local connection between the wind turbine and its area of siting. The other criterion, the consumption criterion, was enforced in order to avoid wind turbine speculation, which in Denmark is generally thought to be a negative thing, often referred to as panty-turbines.[2] In order to reduce speculation in wind to a minimum the consumption criterion placed an upper limit on the amount of shares that owners of communal wind turbines could hold. According to the consumption criterion, each of the owners could only own a part of the wind turbine corresponding to a wind power production of approximately 8000kWh annually. For example, if the wind turbine produced, say, 81,000kWh annually, wind turbine owners were allowed to own a maximum of 10 per cent of the wind turbine in order to be eligible for investment subsidy. Also, the criterion limited the size of privately owned wind turbines to 150kW and further demanded that private wind turbines had to be installed on the owner's own property (Energistyrelsen, 1985, 1986).

Throughout the 1980s a few journalists tried to raise public concern about the investment subsidies (see, for example, Vestergaard, 1985). So, when in January 1989 the subsidy percentage came down to 10 per cent, even wind turbine manufacturers and wind turbine owners began to feel that it was better to end the subsidies altogether than to keep on hearing about the special treatment given to wind turbines and their owners. Consequently, subsidies were phased out the same year relatively quietly compared to previous reductions in the subsidy percentage (Madsen, 1999). Also, at the time, many of the above-mentioned actors were more deeply concerned about the establishment of the new type approval scheme for wind turbines in Denmark (see section on government-assisted business development below).

Renewables set-asides (quotas)

A renewables set-aside requires that a certain percentage of the total electricity generated, or a fixed number of MW-capacity installed, stems from renewable energy sources. This policy mechanism facilitates the growth and stabilization of markets for either renewable energy or renewable energy systems such as wind turbines. In Denmark the government and the utilities have made several agreements on renewables set-asides. Importantly, such quotas were used in relation to other mechanisms, primarily the direct investment subsidies mentioned above, but also government-assisted business development, physical planning and power purchase agreements. Generally, Danish renewables set-asides have been used as a means to further the high-end technological development of the Danish manufacturers by ensuring a market for their most recent and largest wind turbines.

The first of the many government-utility agreements on renewables set-asides was the so-called 100MW agreement of 1985. According to this agreement, the utilities committed themselves to install 100MW of wind power (more than double the amount of wind power already installed at the time) before 1990. The government and the utilities both viewed the 100MW agreement as complementary to the subsidy amendments (the residence and the consumption criteria mentioned above) that were enforced to stop private speculation in wind energy parks. They both feared that wind speculation would damage Danish wind power expansion in the long run. With the amendments to the subsidies and the first 100MW agreement, the government and the utilities effectively tried to split up Danish wind power expansion into two lines of development: On the one hand, smaller units, that is single wind turbines or wind turbine clusters, installed and managed by private and communal wind turbine owners, and, on the other hand, larger units, that is wind energy parks, installed and managed by the utilities (Anon, 1986a, b). However, in spite of such good intentions, communally owned wind energy parks accused of being wind speculations were being established after 1985, such as, for example, the Tændpibe Wind Energy Park at Ringkøbing on the west coast of Denmark, although in a slightly reduced form (Madsen, 2000).

Having decided to give up developing wind turbines on their own (as had been the original idea of the research project on larger wind turbines, see above), the utilities agreed to buy the wind turbines necessary in order to fulfil their part of the 100MW agreement from the Danish wind turbine manufacturers. Moreover, the utilities agreed to buy the largest wind turbines available at the time in order to obtain economies of scale, but also to support the technological development of the manufacturers. In this way the renewables set-aside was seen as a way of integrating market developments with technology developments, as was also the case with the combined subsidy/approval schemes (see section on investment subsidies above).

However, the utilities were accused by both wind turbine manufacturers and wind turbine owners of being unwilling to complete their part of the 100MW agreement with the momentum required (Madsen, 2000). The utilities, on their part, blamed the government and the municipalities for being unable to allocate appropriate sitings for the utility wind turbines. Therefore, with the second 100MW agreement, which came into being in 1991, long before the utilities had fulfilled their part of the first agreement, initiatives towards incorporating wind turbine siting into the physical planning of the municipalities were being taken (see section on physical planning below). In 1996 the Danish government imposed another 200MW renewables set-aside on the utilities, and in 1998 another 750MW offshore, the latter to be divided into five 150MW demonstration plants. The latter agreement has now been supplanted by an international bidding procedure. In late 2004, the Danish Energy Agency invited international tenders for two offshore demonstration plants at Horns Rev and Rødsand, for up to 200MW each, and five other offshore areas have been screened (Energistyrelsen, 2004b).

Government-assisted business development

When the subsidies were terminated in 1989 the Danish government was already working with the wind industry and the Test Station on a new type approval scheme, designed to assist the manufacturers in developing their products and their production systems towards higher standards of quality. Around the same time, the government was expressing increased interest in wind power – for example, in the new energy action plan of 1990, known as *Energy 2000*, which emphasized the role for renewable energy in the Danish power supply system (Energiministeriet, 1990). Among other things, *Energy 2000* focused on the reduction of CO_2 emissions and established a target of 1500MW of wind power to be installed before 2005 as one of the most important means of reducing Danish CO_2 emissions. By the end of 2001 almost 6500 wind turbines with around 2500MW of capacity had been installed in Denmark (Vindmølleindustrien, 2002).

The new type approval scheme was much in demand by almost everyone involved in wind power development in Denmark. The Test Station at Risø wanted to continue building up its wind turbine expertise by means of wind turbine approvals; the manufacturers were looking to implement new quality controls and wanted an official stamp of recognition with which to market their wind turbines at home and abroad; the wind turbine owners wanted the wind turbines on the market to be thoroughly tested and approved; the finance and insurance companies also wanted some kind of approval of the wind turbine systems financed/insured; and the Danish government wanted some means of market control. All agreed that a general boost of wind turbine quality was needed for three reasons at least.

Firstly, the technological development of larger and larger wind turbines, followed by an increasing optimization of components and materials, required more quality control in the construction, production, installation and operation phases of wind turbine manufacturing and wind power utilization. Secondly, the technical problems encountered by Danish wind turbines in California and on the home market – due to lack of quality control, lack of service and maintenance and inexpedient component choices, including components that did not even meet the specifications required – also added to the need for more quality control. Thirdly, since 1987, wind developers in the Californian market demanded product certification by Norske Veritas, the international classification, consulting and certification society, for wind turbine projects financed by Danish investors (Energistyrelsen, 1989).

All of the above-mentioned wind power actors, and many others, worked together designing the new type approval scheme. In order to classify as a type approval proper, the scheme had to include Danish engineering codes of practice for wind turbine design and construction, officially enacted by the Danish Engineering Society. Various ad hoc committees were established to build up the technical and organizational basis for the type approval scheme. These, and the Danish codes of practice for wind turbine design, were completed in 1991–1992 (Normstyrelsen, 1992).

The Danish type approval scheme now includes several sub-approvals, namely: design, production and installation approvals, each of which may be carried out by different bodies authorized to do so by the Danish Energy Agency. Also, an advisory committee – including representatives from the Danish Wind Industry Association, the utilities, insurance companies and the Danish Energy Agency – advises the agency in general issues of management. Moreover, a technical committee consisting of representatives of the authorized bodies takes care of the more specific technical and administrative problems (for more information about the Danish type approval system in both Danish and English, see the homepage of the approval scheme: www.vindmoellegodkendelse.dk).

The type approval scheme not only resulted in more detailed design codes and technical criteria for wind turbine construction, but also quality demands on management, production, service, maintenance, and so on. By promoting the establishment of the type approval scheme, the government sought to translate its policy of increased renewable energy production into techno-economic innovations, broadly defined. The type approval scheme assisted the development of the wind industry at a time when the Danish market was failing because of the end of the subsidies (see section on investment subsidies above), because of siting difficulties (see section on physical planning below) and because of uncertainties regarding payment for wind power electricity (see section on power purchase agreements below). Aiming at providing the Danish wind industry with a general boost in both quality and export, the type approval scheme supplemented another government-assisted business development initiative that was being established around the same time, namely the Guarantee Company of the Wind Turbine Manufacturers, a joint-stock company with a responsible capital of around DKK 50 million to serve as export finance guarantees for the wind turbine manufacturers (Madsen, 1999).

Physical planning

It is widely held that in a small and windy country like Denmark, physical planning is extremely important for wind power development. However, in the 1970s and 1980s wind power planning was more or less left in the hands of the 273 municipalities who all differed greatly with respect to wind power. In consequence the most attractive sites in the most obliging municipalities were beginning to fill up in the late 1980s, and wind turbine siting was becoming more and more difficult. In immediate consequence of Energy 2000 (see above) and in response to the siting difficulties of the utilities (see section on renewables set-asides above), the Danish government in 1991 established the Wind Turbine Siting Committee. The task of the committee was to follow up on the planning investigation carried out by the Danish Planning Agency under the auspices of the national wind energy research project by:

- locating and approving sitings for the second batch of 100MW utility wind power;

- considering how, by means of physical planning, the long-term energy policies of *Energy 2000* (1500MW of wind power installed before 2005) were to be realized (Vindmølleplaceringsudvalget, 1991).

The committee looked more closely into regional planning as a means with which to realize Danish wind energy policies. Based on its topological surveys the committee concluded that wind energy parks or wind turbine clusters with up to 2–5 wind turbines were more advantageous to Danish wind power expansion than were scattered wind turbines in terms of audio-visual impact on the landscape (Vindmølleplaceringsudvalget, 1992). In January 1994 the government distributed a circular on wind power planning to all municipality councils. The circular requested that all municipalities draw up municipal plan proposals with potential wind turbine sites (Miljø- og Energiministeriet, 1994). Three years later, 205 out of 273 municipalities had indeed made public such proposals including 5065 potential wind turbine sites totalling a capacity of 2318MW (Energi- & Miljødata, 1997).

Power purchase agreements

According to Redlinger et al (2002, p171), 'reliable power purchase contracts are perhaps the single most critical requirement of a successful renewable energy project'. Therefore, government policies promoting stable renewable energy prices for the future are very important to techno-economic developments in wind power. In Denmark, from 1979 to 1992, the wind electricity market was regulated by voluntary agreements between the utilities and the associations of wind turbine manufacturers and owners. These agreements obliged the utilities to purchase the wind electricity from wind turbine owners at a guaranteed minimum price while also distributing the cost of connecting wind turbines to the power grid (additional installations, grid strengthening, and so on) between the utilities and the wind turbine owners (Danske Elværkers Forening og Foreningen af Danske Vindmølleejere, 1981; Anon, 1984).

In the early 1990s the three parties (utilities, manufacturers and owners) failed to reach a third agreement because of disagreements regarding the per-kWh price for wind power and the distribution of grid strengthening costs. Consequently, in early 1992 the three parties were summoned at one day's notice to the Ministry of Energy and Commerce where the ministry principals simply informed them that the minister was going to solve their disagreements by law. This law maintained the previous payments for wind power (for which the manufacturers and owners were happy, but the utilities not) and stated that wind turbine owners would have to pay for connecting their wind turbines to the 10kV grid, whereas the utilities had to manage (and pay) for the necessary strengthening of the high-voltage grid including extra substations and service lines (Madsen, 1999; Anon, 1991–1992).

The 1992 law, which fixed the wind power tariff at 85 per cent of normal residential prices, was terminated in 2000 as a result of electricity liberalization.

Instead of the 85 per cent price, a fixed nominal payment for renewable electricity was enforced. This nominal tariff was to apply until a green market for renewable energy including a national Renewable Portfolio Standard would be able to facilitate green-power credit trading. The green market was seen as a way in which to promote wind power while also satisfying the European Commission that wanted to avoid the distortion of the free electricity market caused by different support schemes for renewable energy throughout the EU (Miljø- og Energiministeriet, 2001). However, because of widespread criticism of the perceived organization of this green market, a parliamentary majority decided on 19 June 2002 to postpone the green market, until further notice. Instead, the parties agreed to replace the nominal tariff by a combined subsidy/nominal tariff scheme for wind turbines erected after 1 January 2000 (Energistyrelsen, 2004a).

Environmental taxation

In order to provide a competitive advantage for renewable technologies, environmental taxation penalizes fossil fuel and/or adds to the payment received by producers of renewable energy. Environmental taxation in Denmark has been practiced in both ways. In 1984, with the intention of reducing electricity consumption, the Ministry of Taxation introduced a general per-kWh electricity tax on electricity consumption from which renewable energy was exempted. At the same time, for purposes of supporting renewable technologies, a per-kWh subsidy for producers of renewable energy was installed (Ministeriet for skatter og afgifter, 1984). In 1991, with the passing of the law on electricity production subsidies, this latter subsidy was changed into two separate subsidies, a CO_2 subsidy for all producers of renewable energy and an additional subsidy for producers of wind power (utility wind turbine excepted), waterpower, and biogas electricity (Skatteministeriet, 1991). These subsidies were again replaced by the fixed nominal tariff introduced in late 2000. With the 19 June 2002 parliamentary decision (see section on power purchase agreements above), however, the per-kWh CO_2 subsidy was reintroduced with the provision that the total tariff for wind power electricity must not exceed some upper limit (Energistyrelsen, 2002).

Conclusion

My final remarks concern the interrelations between policy making and techno-economic innovation. Clearly, most of the seven policy mechanisms discussed have interacted closely with one another, and, more importantly, this interaction was often the political reasoning behind their specific employment. Thus, policy making is not only about separately designing and implementing one or two of the mechanisms in question, but concerns the issue of how to combine several of the many mechanisms in order to make up a reasonable and sound whole. The best example from Danish wind power development is the arrangement of the approval and subsidy schemes. The subsidies supported the markets for renewable energy technologies, but also aligned policy makers, manufacturers, owners and researchers and thus helped spawn techno-economic organization and innovation. Also, combining subsidy amendments (the residence and consumption criteria) with renewable set-asides was a way of enrolling the utilities into Danish wind power techno-economic innovation as a developer of larger wind power units. In these ways and many others, relationships between different policy mechanisms were being established.

Meanwhile, policy making also interrelates closely with techno-economic innovation. One may even say that the seven policy mechanisms discussed all concern various ways in which to establish relations between policy and innovation. Policy making may promote certain innovative activities but can never guarantee that such activities are in fact realized or that they produce the effects desired. Thus, in a way, techno-economic innovation is policy making by other means and in another sphere. However, the converse could also be true, if only partly: that certain techno-economic innovations give rise to certain policy making possibilities is hardly controversial (just think of the wind turbine innovation that made possible the new type approval scheme in Denmark), but whether to conceptualize techno-economic innovation as a thoroughly political activity is debatable. Nevertheless, it seems not unreasonable to think of the Danish actors involved in techno-economic innovation as playing a major role in the political success that wind power has enjoyed in Denmark, and vice versa.

Notes

1 The author wishes to thank Professor Volkmar Lauber and Dr Dieter Pesendorfer, both of the University of Salzburg, for realizing and organizing a pleasant International Summer School on the Politics and Economics of Renewable Energy, 13–27 July 2002 in Salzburg. Also, I would like to thank Professor Lauber and all of the participants at the summer school for very constructive comments on a first draft of this paper.

2 According to the dictionary of present-day Danish, a panty-owner (trussereder) is a person who holds shares in a shipping company in order to evade taxes. The name originates from the fact that one of the first Danes to do so was also the owner of a lingerie company. The term panty-turbines is of course used in a derogatory sense, denoting all wind turbines owned by persons without genuine interest in renewable energy sources. The word was especially used in utility circles where the idea of privately and commercially owned wind energy parks was not accepted as the utilities wanted to operate wind energy parks, conceived of as smaller power plants themselves (see, for example, Anon, 1982; Röttger, 1984).

References

Andersen, P. D. (1993) *En analyse af den teknologiske innovation i dansk vindmølleinnovation – herunder Prøvestationen for Vindmøllers dobbeltrolle som forskningsinstitution og godkendende myndighed*, København, Copenhagen Business School

Anon (1982) 'Afregningsregler for vindmøller', *DEF. Medlemsorientering*
Anon (1984) 'Agreement concerning payment and connection of electricity-generating wind power systems', in folder: *DEF – Net og Afregning* in the personal archive of Birger T. Madsen
Anon (1986a) 'Vindmølleaftale undervejs', *El & Energi*, February, p5
Anon (1986b) 'Ønske til kraftværkerne: 100 MW på fem år', *Elsam-posten*, March/April, p4
Anon (1991–1992) 'Forslag til Lov om ændring af lov om udnyttelse af vedvarende energikilder m.v. og lov om elforsyning', *Folketingstidende. Tillæg B. Udvalgenes betænkninger*, vol 143(I), pp1239–1246
Bjørgren, J., Kruse, J., Maegaard, P., Skøtt, T. and Beuse, E. (eds) (1987) *Teknologiudvikling på en anden måde 1983–1987*, Hurup, Nordvestjysk Folkecenter for Vedvarende Energi
Callon, M. (1992) 'The dynamics of techno-economic networks', in Coombs, R., Saviotti, P. and Walsh, W. (eds) *Technological Change and Company Strategies*, London, Academic Press, pp72–102
Danske Elværkers Forening og Foreningen af Danske Vindmølleejere (1981) 'Aftale om afregning og tilslutning', *Naturlig Energi*, January, pp8–9
Energi- & Miljødata (1997) *Kommunernes vindmølleplanlægning – Status januar '97*, København, Planstyrelsen
Energimiljørådet (1998) *Undersøgelse af støtte til vedvarende energi*, Virum, Rambøll
Energiministeriet (1990) *Energi 2000. Handlingsplan for en bæredygtig udvikling*, København, Energiministeriet
Energistyrelsen (1985) 'Bekendtgørelse om statstilskud til udnyttelse af vedvarende energikilder m.v.', *Love og bekendtgørelser*, vol 34, pp803–804
Energistyrelsen (1986) 'Bekendtgørelse om statstilskud til udnyttelse af vedvarende energikilder m.v.', *Love og bekendtgørelser*, vol 35, pp319–321
Energistyrelsen (1989) 'Opbygning og organisation af en ny godkendelsesordning for vindmøller', in folder: *Ny Godkendelsesordning. Finansiering. Organisation. Teknisk Indhold* in the archive of the Wind Energy Department at Risø
Energistyrelsen (1993) *Evaluering af Energiministeriets Energiforskningsprogramr 1986–91*, 3 volumes, København, Energistyrelsen
Energistyrelsen (1997) *EFP: Det danske energiforskningsprogram*, København, Energistyrelsen
Energistyrelsen (2002) 'Aftale mellem Regeringen, Socialdemokratiet, Socialistisk Folkeparti, Det Radikale Venstre og Kristeligt Folkeparti', www.ens.dk/graphics/ENS_Forsyning/Politik/190602_energiaftale.pdf accessed in December 2004
Energistyrelsen (2004a) 'Wind Turbines', www.ens.dk/sw1105.asp accessed in December 2004
Energistyrelsen (2004b) 'Demonstration Program for Offshore Wind Turbines', www.ens.dk/sw1107.asp accessed December 2004
Energisystemgruppen (1990) *Evaluering af Teknologirådets støtteordning for vedvarende energi. Delprojekt 1 og 3*, Risø, National Research Centre
Folkecenter for Renewable Energy in North Jutland (2002) 'Folkecenter for Renewable Energy', www.folkecenter.dk/index_en.html accessed in December 2004
Handelsministeriet (1976a) *Dansk energipolitik 1976*, København, Handelsministeriet
Handelsministeriet (1976b) *Program for udbygning af dansk energiforskning og – udvikling. Første fase*, København, Handelsministeriet
Handelsministeriet (1978) *Program for udbygning af dansk energiforskning og – udvikling. Anden fase*, København, Handelsministeriet
Handelsministeriets og elværkernes vindkraftprogram (1981) *Større, elproducerende vindkraftanlæg*, København, Handelsministeriet
Hansen, H. C. (1985) *Poul la Cour – grundtvigianer, opfinder og folkeoplyser*, Askov, Askov Højskoles Forlag

Heymann, M. (1995) *Die Geschichte der Windenergienutzung 1890–1990*, Frankfurt a/M, Campus Verlag

Heymann, M. (1996) 'Technisches Wissen, Mentalitäten und Ideologien: Hintergründe zur Mißerfolgsgeschichte der Windenergietechnik im 20. Jahrhundert', *Technikgeschichte*, vol 63, pp237–254

Heymann, M. (1998) 'Signs of hubris. The shaping of wind technology styles in Germany, Denmark, and the United States, 1940–1990', *Technology and Culture*, vol 39, pp641–670

International Energy Agency (2002) 'Energy Research Statistics', www.data.iea.org accessed in October 2002

Industri- og Handelsstyrelsen (1988) *Vedvarende Energi. Teknologirådets handlingsplan for indsatsen indenfor området vedvarende energi i årene 1988, 1989 og 1990*, København, Ministeriet for Industri og Handel

Karnøe, P. (1990) 'Technological innovation and industrial organization in the Danish wind industry', *Entrepreneurship & Regional Development*, vol 2, pp105–123

Karnøe, P. (1991) *Dansk vindmølleindustri – en overraskende international success*, København, Samfundslitteratur

Karnøe, P. (1993) 'Approaches to innovation in modern wind energy technology: Technology policies, science, engineers, and craft traditions', *Policy Paper No 334*, Center for Economic Policy Research, Stanford University

Lundsager, P. (1981) 'Forslag til videreførelse og udbygning af Prøvestationen for mindre vindmøller, dækkende perioden 1981–1982', in folder *Strategi, Stabsmøde, Administration* in the Archive of the Wind Energy Department at Risø

Lundsager, P., Frandsen, S. and Jørgensen, C. J. (1980) 'Analysis of data from the Gedser wind turbine 1977–79', *Risø-M-2242*

Lundsager, P. and Jensen, P. H. (1985) 'Risø's design basis for small to medium size Danish windmills', *Risø-M-2531*

Madsen, B. T. (1986) 'Danish wind energy developments', in *danwea '86. International Wind Technology Exhibition, Herning, Denmark, June 11–14, 1986. Conference Proceedings*, Herning, The Association of the Danish Wind Turbine Manufacturers

Madsen, B. T. (1999) Interview conducted by the author on 9 December

Madsen, B. T. (2000) 'Den industrielle udvikling', in Beuse, E., Holdt, J., Maegaard, P., Meyer, N. I., Windeleff, J. and Østergaard, I. (eds) *Vedvarende energi i Danmark. En krønike om 25 opvækstår 1975–2000*, København, OVEs Forlag, pp149–165

Meyer, N. I. (2000) 'VE-udviklingen i Danmark – oversigt over et spændende og broget forløb', in Beuse, E., Holdt, J., Maegaard, P., Meyer, N. I., Windeleff, J. and Østergaard, I. (eds) *Vedvarende energi i Danmark. En krønike om 25 opvækstår 1975–2000*, København, OVEs Forlag, pp75–110

Micheelsen, B., Jensen, O., Johansson, M. Lauridsen, K., Nerenst, P. and Widell, K. E. (1986) *Evaluering af energiministeriets program for udbygning af dansk energiforskning og – udvikling*, København, Energiministeriet

Miljø- og Energiministeriet (1994) 'Cirkulære om planlægning for vindmøller', *Cirkulærer m.v.*, vol 50, p50

Miljø- og Energiministeriet 2001. 'Bekendtgørelse af lov om elforsyning. LBK nr 767 af 28/08/2001', www.retsinfo.dk/_GETDOCM_/ACCN/A20010076729-REGL accessed in December 2004

Ministeriet for skatter og afgifter (1984) 'Bekendtgørelse af Lov om afgift af elektricitet', *Love og bekendtgørelser m.v.*, vol 33, pp108–110

Nielsen, K. H. (2001) *Tilting at Windmills. On actor-worlds, socio-logics, and techno-economic networks of wind power in Denmark, 1974–1999*, unpublished PhD dissertation, History of Science Department, University of Aarhus

Nielsen, P. (1984) 'Store vindmøller i Danmark', in *Konference om energiministeriets forskningsprogramr den 21 november 1984 i Bella Centeret*, report no 10, København, Hans Møller Andersen ApS

Normstyrelsen (1992) *Norm for last og sikkerhed for vindmøllekonstruktioner, DS 472*, København, Teknisk Forlag
Petersen, E. L., Troen, I., Frandsen, S., Hedegaard, K. et al (1981) *Windatlas for Denmark*, Risø, National Research Centre
Planstyrelsen (1981–1986) *Om mulighederne for at placere mange store vindmøller i Danmark. I-VI*, 6 volumes, København, Planstyrelsen
Rasmussen, F. and Jensen, P. H. (1999) Interview conducted by the author on 18 November
Redlinger, R.Y., Andersen, P. D. and Morthorst, P. E. (2002) *Wind Energy in the 21st Century. Economics, Policy, Technology and the Changing Electricity Industry*, Basingstoke and New York, Palgrave
Righter, R. W. (1996) *Wind Energy in America. A History*, Norman and London, University of Oklahoma Press
Röttger, U. (1982) 'Enighed om vindmølleafregning', *El & Energi*, September, pp26–27
Skatteministeriet (1991) 'Lov om tilskud til elproduktion', *Love og bekendtgørelser m.v.*, vol 40, pp1637–1638
Thorndahl, J. (1996) 'Danske elproducerende vindmøller 1892–1962', *Polhem*, vol 14, pp324–391
Thorndahl, J. (2000) 'Johannes Juul – en rigtig vindelektriker', *Journalen*, vol 10, pp8–16
Vestergaard, F. (1985) 'Godt for private, men dyrt for samfundet', *Weekendavisen*, Copenhagen, 12 July
Vindmølleindustrien (2002) 'Capacity in Denmark', www.windpower.org/en/stats/capacityDK.htm accessed in December 2004
Vindmølleplaceringsudvalget (1991) *Betænkning I*, København, Planstyrelsen
Vindmølleplaceringsudvalget (1992) *Betænkning II*, København, Planstyrelsen

6

Germany: From a Modest Feed-in Law to a Framework for Transition

Staffan Jacobsson and Volkmar Lauber[1]

Introduction

The rate of diffusion for renewable power technologies is a central issue given the urgency of problems such as climate change and oil depletion. In Germany, renewable power technologies – especially wind turbines and solar cells – have spread particularly fast, so Germany has become a worldwide leader in this area. This was due to a regulatory framework put in place by the German parliament in several steps. The first feed-in law seemed destined just to induce the installation of a few hundred megawatts of small hydro plants and a modest number of wind turbines. In fact, it developed far beyond those expectations and set in motion a virtuous circle in which more and more elements of a true framework for transition fell into place, leading to visions of establishing an all-renewables power system some time during the second half of the 21st century. When total costs to society are considered, the costs of this transition are surprisingly modest. It is almost surprising how much political effort has to be invested to bring about such a change. But then this is not just a question of appropriate policy design but of overcoming vested interests and established modes of thinking about energy.

Analytical framework

This chapter draws on two strands of theory: the economics of innovation (as viewed by economists) and public policy analysis (as approached by political scientists).[2] Both are helpful in discussing the transformation that went on in the German renewable energy sector, a transformation that was the outcome of both a political process (that is the result of particular visions, of advocacy group coalitions and the like) and a particular design of public policy. The first section provides the analytical framework[3] for studying the various phases of renewable energy technology diffusion and system. The second section gives a summary overview of German politics and policies in this area. The third section contains a discussion of the financial and social costs of these policies. The fourth section finally draws the conclusions.

Now to the two technologies considered here. Fossil fuels in 1999 contributed about 9400TWh to electricity production worldwide. Large-scale hydropower and combustion of different types of biomass currently provide the bulk of the energy supplied by renewable energy sources. In 1999, these supplied roughly 2600TWh and 160TWh of electricity respectively, worldwide (UNDP, 2000;[4] IEA, 2001). In addition to these, the 'new' renewables – for example wind turbines and solar cells – are now diffusing at a quite rapid rate.[5]

Figures 6.1 and 6.2 (p127) show the global diffusion of wind turbines and solar cells. After an extensive period of experimentation, dating back decades[6] and lasting throughout the 1980s, the global stock of *wind turbines* grew very rapidly during the period 1990–2002 and reached a capacity of 32,037MW (about 47,000MW by the end of 2004). The stock of *solar cells* also grew at a high rate but the stock was more limited 2407MW in 2002 and approximately 4000MW by the end of 2004. For both technologies, the bulk of the stock was installed after 1995. In other words, we have been witnessing what may be the beginnings of a take-off period in the long-term diffusion of these technologies.

While the share of these technologies in the global energy supply is marginal at present – less than 0.5 per cent of the 15,000TWh of electricity generated in the world (Jacobsson and Bergek, 2003) – there are visions of wind power accounting for 50 per cent of Europe's electricity supply by 2030, and of solar cells supplying 20 per cent of it by 2050 (Garrad Hassan/Greenpeace, 2004; PV TRAC, 2004). The real issue is no longer the technical potential of these (and other) renewable energy technologies, but how this potential can be realized and substantially contribute to a transformation of the energy sector.

Yet a large-scale transformation process of this kind requires far-reaching changes, many of which date back several decades and involve political and policy support in various forms in pioneering countries. Drawing on a rich and very broad literature, we will outline elements of an analytical framework[7] that captures some key features of early phases of such transformation processes.

Some characteristics of such phases may be found in the literature on industry life cycles (for example Afuah and Utterback, 1997; Utterback and Abernathy, 1975; Van de Ven and Garud, 1989; Utterback, 1994; Klepper, 1997; Bonaccorsi and Giuri, 2000). It emphasizes the existence of a range of competing designs, small markets, many entrants and high uncertainty in terms of technologies, markets and regulation. We need, however, to understand the conditions under which this formative stage, with all its uncertainties, emerges in a specific country. We will outline four key conditions, or features, of early parts of such processes. These are institutional changes, market formation, the formation of technology-specific advocacy coalitions, and the entry of firms and other organizations.

First, as emphasized in the literature on 'economics of innovation', *institutional change* is at the heart of the process (Freeman and Louca, 2002). It includes alterations in science, technology and educational policies. For instance, in order to generate a range of competing designs, a prior investment in knowledge formation must take place and this usually involves a redirection of science and technology policy well in advance of the emergence of markets. Institutional alignment is also about the value base (as it influences demand patterns), market regulations and tax policies as well as much more detailed practices, which are of a more immediate concern to specific firms, as discussed for instance by Maskell (2001). The specific nature of the institutional framework influences access to resources and availability of markets, as well as the legitimacy of a new technology and its associated actors. As argued in the literature of both 'innovation systems' (for example Carlsson and Jacobsson, 1997) and 'transition management' (Rotmans et al, 2001), the nature of the institutional framework may therefore act as one of many mechanisms that obstruct the emergence of a formative stage and its evolution into a growth phase. Firms, therefore, compete not only in the market for goods and services, but also to gain influence over the institutional framework (Van de Ven and Garud, 1989; Davies, 1996).

Second, institutional change is often required to *generate markets* for the new technology. The change may, for instance, involve the formation of standards, such as the Nordic telecommunication operators' decision to share a common standard (NMT) for mobile telecommunications. In the formative phase, *market formation* normally involves exploring niche markets, markets where the new technology is superior in some dimension. These markets may be commercial and involve unusual selection criteria (Levinthal, 1998) and/or involve a government subsidy. A 'protected space' for the new technology may serve as a 'nursing market' (Ericsson and Maitland, 1989), where learning processes can take place and the price/performance of the technology improve (see also Porter, 1998). Nursing markets may, through a demonstration effect, also influence preferences among potential customers. Additionally, they may induce firms to enter, provide opportunities for the development of user-supplier relations and other networks, and, in general, generate a 'space' for a new industry to evolve in.

The importance of early markets for learning processes is not only emphasized in management literature, but also in the policy-oriented literature on 'Strategic Niche Management'. A particularly clear statement of this is found in Kemp et al (1998, p184):

> Without the presence of a niche, system builders would get nowhere... Apart from demonstrating the viability of a new technology and providing financial means for further development, niches help build a constituency behind a new technology, and set in motion interactive learning processes and institutional adaptation... that are all-important for the wider diffusion and development of the new technology.

Third, whereas individual firms and related industry associations may play a role in competition over institutions (Feldman and Schreuder, 1996; Porter, 1998), such actors may be but one part of a broader constituency behind a specific technology. The build-up of a constituency involves the 'entry' of other organizations as well as firms. It may involve universities but also non-commercial organizations (for example Greenpeace). Unruh (2000, p823) underlines the existence of a range of such organizations and the multitude of roles they play:

> ...users and professionals operating within a growing technological system can, over time, come to recognize collective interests and needs that can be fulfilled through establishment of technical...and professional organisations...These institutions create non-market forces...through coalition building, voluntary associations and the emergence of societal norms and customs. Beyond their influence on expectations and confidence, they can further create powerful political forces to lobby on behalf of a given technological system.

The centrality of the formation of constituencies is well recognized in the political science literature, in particular in the literature on networks (Marsh and Smith, 2000; Rhodes, 2001). Thus, Sabatier (1998) and Smith (2000) argue that *advocacy coalitions,* made up of a range of actors sharing a set of beliefs, compete in influencing policy. For a new technology to gain ground, *technology-specific coalitions* need to be formed and to engage in wider political debates in order to gain influence over institutions and secure institutional alignment. As part of this process, advocates of a specific technology need to build support among broader advocacy coalitions to advance the perception that a particular technology, for example solar cells or gas turbines, answers wider policy concerns. Development of joint visions of the role of that particular technology is therefore a key feature of that process. Hence, the formation of 'political networks' sharing a certain vision and the objective of shaping the institutional set-up is an inherent part of this formative stage.

Fourth, *entry of new firms* is central to the transformation process. Each new entrant brings knowledge, capital and other resources into the industry. New entrants experiment with new combinations, fill 'gaps' (for example become a specialist supplier) or meet novel demands (for example develop new applications). A division of labour is formed and further knowledge formation is stimulated by specialization and accumulated experience (for example Smith, 1776; Young, 1928; Stigler, 1951; Rosenberg, 1976). Finally, early entrants raise the returns for subsequent entrants in a number of ways, that is positive external economies emerge (Marshall, 1920; Scitovsky, 1954). In addition to the conventionally related sources of external economies (for example build-up of an experienced labour force and specialized suppliers of inputs) early entrants strengthen the 'political' *power* of a technology-specific advocacy coalition and provide an enlarged opportunity to influence the institutional set-up. Early entrants also drive the process of *legitimation* of a new field, improving access to markets, resources, and so forth for subsequent entrants (Carroll, 1997), and resolve underlying technical and market uncertainties (Lieberman and Montgomery, 1988).

The time-span involved in an early phase where these four features emerge may be very long. This is, for instance, underlined in a recent study of Israel's 'Silicon Wadis', which began a rapid period of growth in the 1990s after a history starting in the 1970s (de Fontenay and Carmel, 2001). Other examples are given in Geels (2002) and in Carlsson and Jacobsson (1997a).

A *take-off* into a rapid growth phase may occur when investments have generated a large enough, and complete enough, system for it to be able to 'change gear' and begin to develop in a self-sustaining way (Carlsson and Jacobsson, 1997; Porter, 1998). As it does so, a chain reaction of powerful *positive feedback loops* may materialize, setting in motion a process of cumulative causation. Indeed, as pointed out long ago by Myrdal (1957), these virtuous circles are central to a development process – as these circles are formed, the diffusion process becomes increasingly self-sustained and characterized by autonomous dynamics (Rotmans et al, 2001), often quite unpredictable in its outcome. All the four features of the formative phase are involved in such dynamics. For instance, the emergence of a new segment may induce entry by new firms, which in turn strengthen the political power of the advocacy coalition and enable further alignment of the institutional framework (which, in turn, may open up more markets and induce further entry etc).

Under what conditions a 'take-off' takes place seems to be extremely difficult to predict. A necessary condition is, however, that larger markets are formed – there must be an underlying wave of technological and market opportunities. Some ICT clusters have become successful by linking up to the US market (Breshanan et al, 2001) while the Nordic technological systems in mobile telephony grew into a second phase with the European GSM standard. As we shall see below, alterations in the *regulatory frameworks* have triggered a set of actions and reactions and propelled the diffusion process in the cases of wind power and solar cells in Germany. At the heart of the story that is to be told lies a 'battle over institutions'.

Wind energy and solar cells in Germany: politics, policies and their impact on diffusion

This section will deal with basic values and beliefs as well as processes leading up to policy making, the attendant policies, the impact of these policies on technology diffusion and subsequent feed-back loops to policy making.[8] Although we are analysing what with hindsight is an early phase in the diffusion process, we shall divide this into three sub-phases. The period covering 1974 to the late 1980s was a formative phase for both wind and solar cells. Important decisions in favour of market creation were taken beginning in 1988, and this policy was implemented during subsequent years. In 1990 a first take-off for wind arose while continuing the formative phase for solar cells. The year 1998 reinforced the take-off for wind and began a take-off period for solar cells. These three sub-phases are clearly seen in Figures 6.3 and 6.4, which portray the diffusion of these technologies in Germany. Whereas Germany accounted for a less than 1 per cent share of the global stock of these technologies in 1985 and

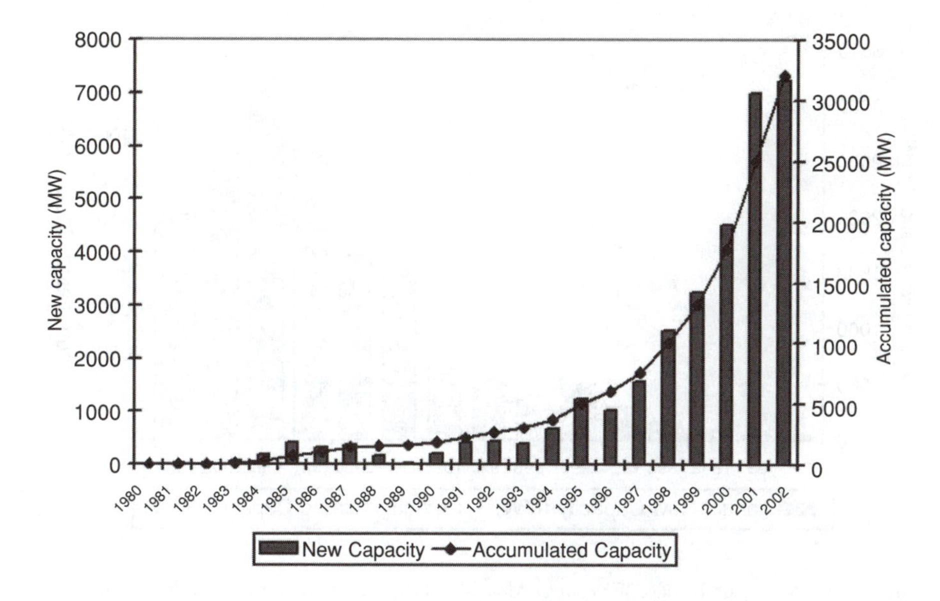

Source: Johnson and Jacobsson, 2001; Bundesverband Windenergie, 2003; *Windpower Monthly*, vol 19, no 1, January 2003, p58.

Figure 6.1 The global diffusion of wind turbines, 1980–2002

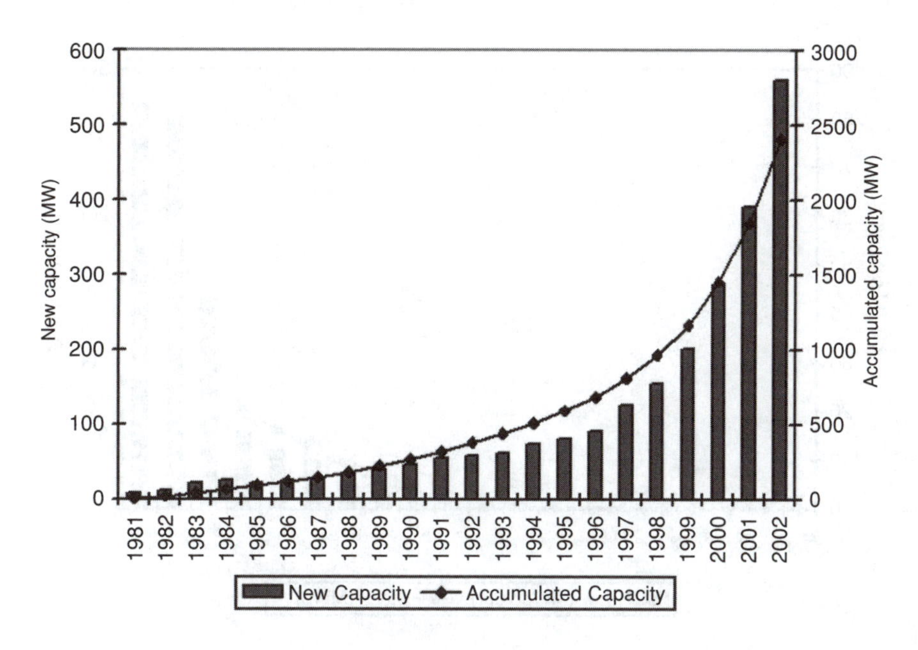

Source: Jacobsson et al, 2002; Schmela, 2003

Figure 6.2 The global production of solar cells, 1981–2002

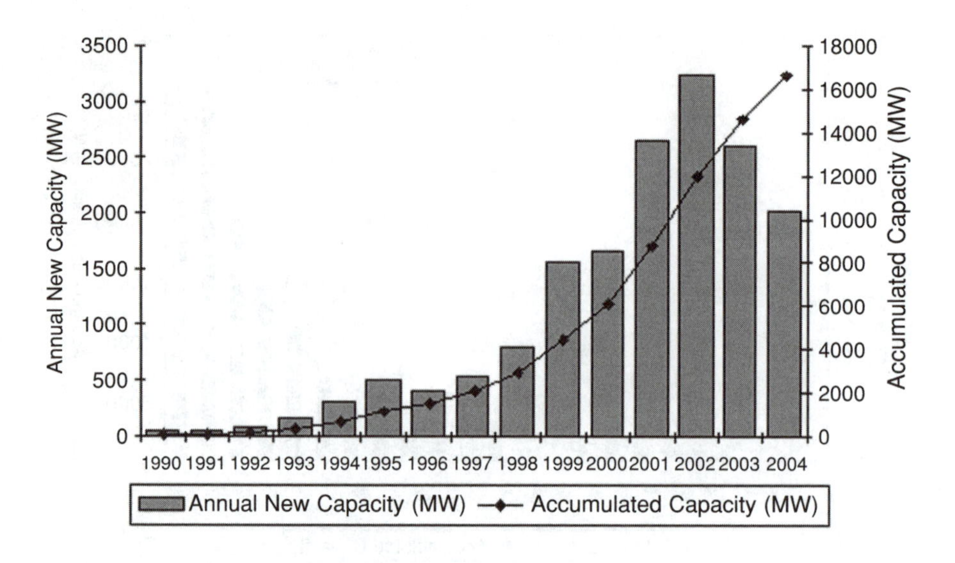

Source: Staiss, 2003, part II, p66; *Windpower Monthly*, 2003, vol 19, no 4, p78; *Windpower Monthly*, 2005.

Figure 6.3 Diffusion of wind power in Germany, 1990–2004

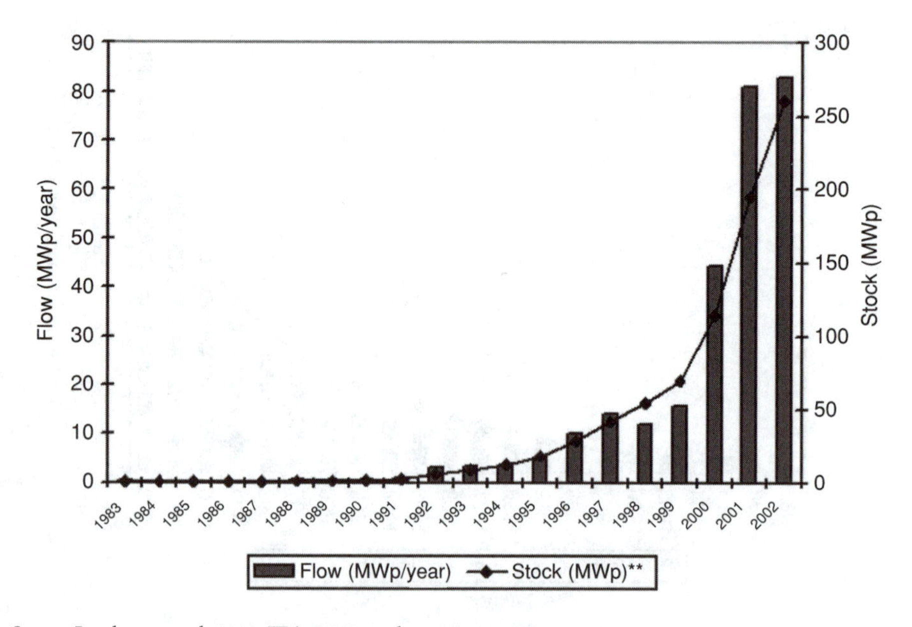

Source: Jacobsson et al, 2002; IEA, 2003; and www.iea-pvps.org.

Figure 6.4 The diffusion of solar cells in Germany, 1983–2002

1990 respectively, it came to play a prominent role in the global diffusion from the early 1990s. Indeed, at the end of 2004, Germany had about one third of the global stock of wind turbines or 16,628 out of 47,574MW (*Windpower Monthly*, 2005). As to PV power, at the end of 2003 Germany had more than one fifth of the global stock (430 out of 1809MW_p). By the end of 2004, this may actually have increased to as much as one third or more (IEA, 2004; EurObserv'ER, 2005).

1974–1988 – a formative phase of wind and solar power

The energy crises of the 1970s produced major rethinking in Germany as in many other countries. The main response there was to increase government support for hard coal and nuclear power use (Schmitt, 1983; Kitschelt, 1980). From the mid-1970s, however, nuclear power became increasingly controversial with the public; its rapid expansion led to many bitter confrontations and a policy of repression until the end of the decade. Many believed that the government should instead bank on energy efficiency and renewable energy. A first *Enquete Commission*[9] of the German parliament in 1980 recommended efficiency and renewables as first priority, but also the maintenance of the nuclear option (Meyer-Abich and Schefold, 1986). In 1981, the Federal Ministry of Research and Technology commissioned a five-year study, which drew a strong echo in the media when it was published around the time of the Chernobyl accident. It concluded that only reliance on renewables and efficiency would be compatible with the basic values of a free society, and that this would be less expensive than the development of a plutonium-based electricity supply as envisioned at that time (Meyer-Abich and Schefold, 1986). Against this background of strong pressure from public opinion, R&D for renewable energy sources was raised to a significant level – not as significant per capita as in other countries such as Sweden, Denmark and the Netherlands, but larger in total amount. In 1974, annual spending started with about DM 20 million. It reached a peak of DM 300 million in 1982 – the year when the government passed from the social democratic/liberal to a conservative/liberal coalition under Chancellor Kohl – and declined thereafter to a low point of 164 million in 1986 (the year of the Chernobyl accident). Further decline had been scheduled but was reversed at that point (Sandtner et al, 1997). Much publicly financed R&D was intended for developing off-grid renewable energy technologies for export to the developing world, not for the domestic market (Schulz, 2000).

Until the end of the 1980s and in fact beyond, renewable energy faced a political–economic electricity supply structure that was largely hostile. The electricity supply system was dominated by very large utilities relying on coal and nuclear generation. These utilities were opposed to all small and decentralized forms of generation, which they deemed uneconomic and foreign to the system. The two key ministries – Economic Affairs on one hand, Research and Technology on the other – offered only limited help. The Ministry of Economic Affairs was (and still is) in charge of utilities and, in fact, their chief ally. Both the Social Democratic-Liberal (before 1982) and the

Conservative-Liberal[10] governments (1982–1998) strongly supported nuclear and coal. This is clearly seen in the allocation of R&D funds, where R&D funding to nuclear power and fossil fuels dwarfed that of renewable energy technology (Figure 6.5).

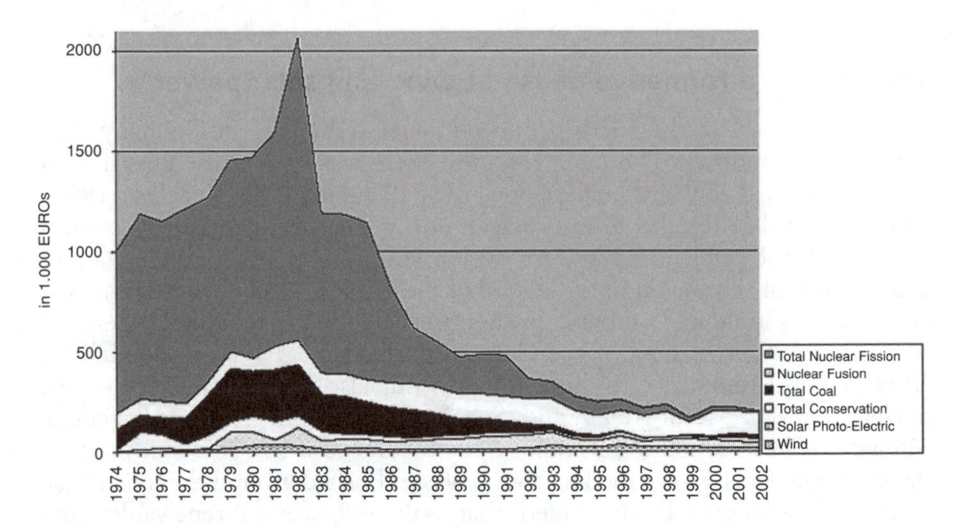

Source: IEA, 2003a.

Figure 6.5 Energy R&D in Germany, 1974–2002

Moreover, during the oil crisis, the government created powerful incentives for utilities to use otherwise non-competitive domestic hard coal. These incentives were paid out of a government fund financed by a surcharge or special tax on final customers' electricity prices. This surcharge varied between 3.24 per cent of that price in 1975–1976 and 8.5 per cent in 1989 (Bundesverfassungsgericht, 1994). At the same time, the Ministry of Economic Affairs – normally in charge of market creation programmes – did little for renewable energy sources. It only made use of the general competition law to oblige the utilities (then operating as territorial monopolies) to purchase electricity from renewable energy sources produced in their area of supply at avoided costs. However, the large utilities interpreted this so narrowly (as avoided fuel costs only) that the obligation had little effect.[11] The ministry resisted all demands for market formation with the slogan that energy technologies had to prove themselves in the market and that it was not prepared to subsidize technologies that were not mature.

At the same time, the Ministry of Research – the former Ministry of Nuclear Affairs renamed in 1962, whose tasks now came to include renewables – viewed its responsibility as one of only supporting research and development, and to a smaller extent demonstration. It was more generous in funding nuclear demonstration projects. By 1980, it had spent about DM 13 billion on nuclear RD&D (Kitschelt, 1980; Zängl, 1989). Under the prevailing distribution of responsibilities – which was jealously enforced by the much more powerful Ministry of Economic Affairs (Ristau, 1998) – it was allowed to support renewable energy technologies only in pre-market phases. There was little opportunity or willingness to bridge the gap between research prototypes and market-competitive products.

Yet, in this largely unfavourable political context, institutional changes occurred which began to open up a space for wind and solar power – a space that proved to be of critical importance for the future diffusion of these renewables. This institutional change largely related to the formation of government funded R&D programmes for these technologies.

These programmes provided opportunities for universities, institutes and firms to search in many directions, which was sensible given the underlying uncertainties with respect to technologies and markets. Some programmes may have pursued ambivalent goals; thus one of the purposes of the GROWIAN project of a large (several MW) wind turbine was allegedly to demonstrate that wind power was not viable (Heymann, 1999). However, the wind power R&D programme was large enough to finance most projects applied for and flexible enough to finance most types of projects (Windheim, 2000a). In the period 1977–1989, about 40 R&D projects were granted to a range of industrial firms and academic organizations for the development or testing of small (for example 10kW) to medium-sized (for example 200–400kW) turbines (elaboration in Windheim, 2000b).[12]

Much the same applied to R&D in solar cells. In the period 1977–1989, as many as 18 universities, 39 firms and 12 research institutes received federal funding (Jacobsson et al, 2002).[13] Although the major part of the research funding was directed towards cell and module development and the prime focus was on crystalline silicon cells, funds were also given to research on several thin-film technologies.[14] In addition, R&D funds were allocated to the exploration of a whole range of issues connected to the application of solar cells, such as the development of inverters. As a consequence, and in spite of the fringe status of that R&D, a broad academic cum industrial knowledge base began to be built up about 25 years ago for both wind turbines and solar cells.

In the 1980s, a set of demonstration programmes became part of the R&D policy. Investments in wind turbines were subsidized by several programmes (Hemmelskamp, 1998). At least 14 German suppliers of turbines received funding for 124 turbines in the period 1983–1991 (elaboration in Windheim, 2000b).[15] This programme constituted an important part of the very small national market in the 1980s – total installed power was just 20MW by the end of 1989 (Durstewitz, 2000). An early niche market was also found in 'green' demand from some utilities – reflecting the strength of the green movement (Reeker, 1999) – and from environmentally concerned farmers (Schult and Bargel, 2000; Tacke, 2000).

In solar cells, the first German demonstration project took place in 1983. This was wholly financed by the federal government and had an effect of

300kW$_p$, which was the largest in Europe at that time. In 1986, it was followed by a demonstration programme, which by the mid-1990s had contributed to building more than 70 larger installations for different applications. Yet, by 1990, the accumulated stock amounted to only 1.5MW$_p$ (see Figure 6.4, p128). Although the demonstration programme had only a minor effect in terms of creating a 'protected space', it was effective as a means of enhancing the knowledge base with respect to application knowledge. Hence, by that time, learning had taken place not only among four firms that actually had entered into solar cell production (for example AEG, MBB and Siemens), but also to some extent 'downstream' in the value chain.

In sum, this formative phase was dominated by institutional change in the form of an R&D policy that began to include, at the fringe, R&D in renewables. Although small in relation to R&D in nuclear and other energy technologies, it allowed for a small space to be opened for wind and solar power in which a range of firms and academic departments began a process of experimentation and learning. Small niche markets were formed and a set of firms were induced to enter. During this phase, utilities generally saw no promise in solar power. However, two major utilities were at that time close to supporting mass production of solar cells (30MW$_p$ annually) by Wacker Chemitronic. They withdrew because they were not sure whether the process developed by Wacker would stand up to competition from other firms (Mener, 2000, pp478–489).

In addition to these firms and universities, a range of other organizations was set up, organizations that were later to become key actors in advocacy coalitions for wind and solar power. These included conventional industry associations such as the German Solar Energy Industries Association, which was founded in 1978 (Bundesverband Solarindustrie, 2000). As importantly, environmental organizations that were independent of industry grew up to provide expertise and visions of the future. For instance, in 1977, at the height of the anti-nuclear power controversy, actors in the green movement set up the Institute of Ecology (*Öko-Institut*) in Freiburg to provide counter-expertise for their struggle with governments and utilities. This institute became very important for coming up with proposals for the development of renewable energy policies later on. In a similar vein, Förderverein Solarenergie (started in 1986), in 1989 developed the concept of 'cost covering payment' for electricity generated by renewable energy technology, a concept that was later applied in various feed-in laws at federal and local levels. A third type of association is Eurosolar, founded in 1988, which is an organization for campaigning *within* the political structure for support of renewables and which is independent of political parties, commercial enterprises and interest groups, yet counts several dozen members of the German parliament in its ranks (not only from red–green).

1988–1998 – take off for wind power but not for solar power

The accident in Chernobyl in 1986 had a deep impact in Germany. Public opinion had been divided about evenly on the question of nuclear power between 1976 and 1985. This changed dramatically in 1986. Within two years, opposition to nuclear power increased to over 70 per cent, while support barely exceeded 10 per cent (Jahn, 1992). The social democrats committed themselves to phasing out nuclear power; the Greens demanded an immediate shutdown of all plants.

Also in 1986, a report by the German Physical Society warning of an impending climate catastrophe received much attention, and in March 1987 Chancellor Kohl declared that the climate issue represented the most important environmental problem (Huber, 1997). A special parliamentary commission was set up to study this matter – the *Enquetekommission* on climate. The commission worked very effectively in a spirit of excellent co-operation between the parliamentary groups of both government and opposition parties. There was general agreement that energy use had to be profoundly changed. The matter was given increased urgency by the fact that the price of oil had declined again, so that further increases of fossil fuel consumption had to be expected unless serious measures were taken; at the same time, the price gap between renewable energy technologies and conventional generation grew larger (Kords, 1996; Ganseforth, 1996).

A series of proposals for institutional change was formulated, which included an electricity feed-in law for generation from renewables (Schafhausen, 1996). Pressure from parliament on the government to take substantial steps in favour of renewables increased, as evidenced by a variety of members' bills (Deutscher Bundestag, 1987, 1988a, 1988b, 1989, 1990a and 1990b). This was obviously reflecting a high level of public concern with this issue at that time. In 1987, two conservative deputies from northern Germany (where interest in wind power was already emerging), frustrated by the lack of activity by the government, took the highly unusual step of bypassing party leaders and presented, at a press conference, a draft bill in favour of renewable energy. The party's parliamentary leaders and energy experts strongly condemned this step, which however found unexpected support in the CDU/CSU party group and also attracted attention from the press. In the end, this served to strengthen the hand of the Minister of Research vis-à-vis the Minister of Economic Affairs, which the former used to announce a first 100MW wind programme for demonstration and market formation (whereas the Ministry of Economic Affairs had opposed such programmes out of principle). With the announcement of this programme, the dissenters declared to be satisfied and withdrew their bill. But in the meantime all parliamentary party groups developed an interest in the subject, closely followed by the media. At the end of 1989, another conservative MP, this time from southern Germany, at the end of his career and closely linked to hydropower interests, submitted another bill for which he sought cross-party support, even from the opposition. Support was surprisingly strong; from this emerged a more durable co-operation between conservative and Green MPs, who were particularly eager to achieve changes in this area; again this took the parliamentary leadership by surprise. In negotiations with the Ministry of Economic Affairs, their bill – designed to broadly support renewable energy generally – was limited to electricity, where it was expected to have only a very modest impact. Conservative parliamentary leaders who still thought that it would not gather a majority, ignored this bill. When they realized their mistake and formulated a counterproposal, it was already too late to stop the bill. Nor was the Ministry of Economic Affairs able to stop it by way of voluntary concessions on the part of the utilities; they simply refused outright. Neither of them made a major effort, thinking that the bill was of small significance and would impose only negligible costs on utilities (Kords, 1993).

Thus the government reluctantly – support only came from the Research Ministry, and the Environment Ministry under Töpfer – adopted several important measures. In 1988, the Ministry of Research launched two large demonstration cum market formation programmes. The first was directed at wind power and initiated in 1989. Initially, it aimed at installing 100MW of wind power – a huge figure compared to the stock of 20MW in 1989. Later, it was expanded to 250MW. The programme mainly involved a guaranteed payment, for electricity produced, of €0.04/kWh, later reduced to 0.03.[16] The second demonstration cum market formation measure was the 1000 roofs programme for solar cells. Furthermore, the legal framework for electricity tariffs was modified in such a way as to allow compensation to generators of renewable power above the level of avoided costs. Finally, the Electricity Feed-in Law was adopted, which was originally conceived mainly for a few hundred MW of small hydropower (Bechberger, 2000).

Remarkably, the Feed-in Law – the most important measure since it was conceived for a longer term – was adopted in an all-party consensus (though social democrats and greens wanted to go further in the support of renewable power).[17] As mentioned above, the basic concept of the Feed-in Law was put forward by several associations – Förderverein Solarenergie (SFV), Eurosolar and an association organizing some 3500 owners of small hydro power plants, many of whose members were politically conservative and able to effectively campaign for the new law in a larger association organizing small and medium-sized firms. It seems that actually passing the law did not require a large political effort, despite the opposition of the utilities, which were not entitled to receive any benefits under this law if they invested themselves in the new technologies (Ahmels, 1999; von Fabeck, 2001; Scheer, 2001). But then a few hundred MW of hydropower was hardly a serious matter, and in addition the big utilities were at that time absorbed in taking over the electricity sector of East Germany in the process of reunification (Richter, 1998).

The Feed-in Law required utilities to connect generators of electricity from renewable energy technology to the grid and to buy the electricity at a rate which for wind and solar cells amounted to 90 per cent of the average tariff for end customers, that is about DM 0.17.[18] Together with the 100/250MW programme and subsidies from various state programmes (DEWI, 1998), the Feed-in Law gave very considerable financial incentives to investors, although less for solar power since its costs were still very high compared to the feed-in rates. One of the declared purposes of the law was to 'level the playing field' for renewable power by setting feed-in rates at levels that took account of the external costs of conventional power generation. In this context, the chief member of parliament supporting the feed-in bill on behalf of the Christian Democrats in the Bundestag mentioned external costs of about 3–5 Euro cents per kWh for coal-based electricity (Deutscher Bundestag, 1990c).

These incentives – combined with additional incentives from the 100MW wind programme – stimulated the formation of markets and had three effects. First, it resulted in an 'unimaginable'[19] market expansion from about 20MW in 1989 to close on 490MW in 1995 (BWE, 2000).[20] Second, it led to the emergence of learning networks which developed primarily between wind turbine suppliers and local components suppliers due to the need of adapting the turbine components to the particular needs of each turbine producer. The benefits of learning also spilled over to new entrants (induced by market growth), since these could rely on a more complete infrastructure. Third, it resulted in a growth in the 'political' strength of the industry association organizing suppliers and owners of wind turbines who were now able to add economic arguments to environmental ones in favour of wind energy.

However, when the Feed-in Law began to have an impact on the diffusion of wind turbines, the big utilities started to attack it both politically and in the court system (basically on constitutional grounds) – unsuccessfully, as it turned out. This reflected more than just opposition to small and decentralized generation. First, no provision had been made to spread the burden of the law evenly in geographical terms; this came only in 2000. Second, the utilities were by this time marked by the experience of politically dictated subsidies for hard coal used in electricity generation. These subsidies had grown from €0.4 billion in 1975, the year the 'coal penny' was introduced, to more than €4 billion annually in the early 1990s (see above). Two-thirds of this was covered by a special levy on electricity; one-third had to be paid by the utilities directly but was also passed on to the consumers.[21]

Political efforts to change the law seemed at first more promising. In 1996, utilities association VDEW lodged a complaint with DG Competition (the subdivision of the European Commission that looks after fair competition) invoking violation of state-aid rules. The Ministry of Economic Affairs then proposed to reduce rates on the occasion of an upcoming amendment (the law had to be changed in any case in order to spread the burden of feed-in payments more evenly in geographical terms, and also because of liberalization), a measure supported by DG Competition. Even though the notification of the Feed-in Law to the European Commission had not drawn an adverse reaction right after its adoption, DG Competition now argued that feed-in rates should come down substantially along with costs, addressing particularly wind power (Salje, 1998; Hustedt, 1998; Advocate General Jacobs, 2000). The Ministry of Economic Affairs was happy enough with this support; its official line was that renewable energies were only 'complementary' and could not pretend to replace coal and nuclear generation.

All this led to insecurity for investors and stagnating markets for wind turbines from 1996 to 1998. Indeed, climate policy had suffered a general setback at the governmental level due to the financial and other problems resulting from German reunification (Huber, 1997). However, the issue was still strong with public opinion. Thus, a survey conducted in 1993 in 24 countries showed that concern over global warming was greatest in Germany (Brechin, 2003).

In any event, the big utilities' political challenge to the Feed-in Law failed in parliament (Ahmels, 1999; Molly, 1999; Scheer, 2001). In 1997, the government proposal to reduce feed-in rates mentioned above led to a massive demonstration bringing together metalworkers, farmer groups and church groups along with environmental, solar and wind associations; the Association of Investment Goods Industry VDMA gave a supportive press conference (Hustedt, 1997; Hustedt, 1998). The government failed to persuade even its own MPs. In a committee vote, the government proposal lost out by a narrow vote of eight to seven, and it seems that as many as 20 CDU/CSU members were determined to vote against the new rates in the plenary (Scheer, 2001). Clearly the new technology had by now acquired substantial legitimacy. As one CDU member and executive of the wind turbine industry put it: 'In this matter we collaborate with both the Greens and the Communists' (Tacke, 2000). The Feed-in Law was now incorporated in the Act on the Reform of the Energy Sector of 1997, which transposed the EU directive on the internal market for electricity.

When it became clear that the feed-in rates would remain unchanged, this removal of uncertainty resulted not only in a further expansion in the market for wind turbines (see Figure 6.3, p128), but also in the entry of larger firms into the wind turbine industry as well as into the business of financing, building and operating wind farms, strengthening the advocacy coalition yet again.

The second market introduction cum demonstration programme of the research ministry was focused on small solar cell installations, the 1000 roofs programme, for which it provided an investment aid of 60 to 70 per cent. Eventually, the programme led to the installation of more than 2200 grid-connected, roof-mounted installations with an effect of 5.3MW_p by 1993 (IEA, 1999; Staiss and Räuber, 2002). While the 1000 roofs programme was successful, the market formation that it encouraged was not large enough to justify investments in new production facilities for the solar cell industry, in particular as the industry was running with large losses (Hoffmann, 2001). The industry now expected that there would be a follow-up to the 1000 roofs programme, but no substantial programme emerged (Brauch, 1997). In 1993, Eurosolar proposed a 100,000 roof programme that in the subsequent year was taken up by the Social Democrats (Hermann Scheer, the first president of Eurosolar, is himself a Social Democratic MP). This proposal was, however, not supported by the party groups of the (Conservative/Liberal) government coalition (Scheer, 2001). If the industry was to survive, market creation had to come from other quarters. This led to intensified efforts to mobilize other resources, a process that demonstrated the high level of legitimacy that solar PV enjoyed in German society.

The most important help came from municipal utilities. In 1989 the federal framework regulation on electricity tariffs – the tariffs themselves are set at the *Länder* level – was modified in such a way as to permit utilities to conclude cost-covering contracts with suppliers of electricity using renewable energy technologies, even if these full cost rates exceeded the long-term avoided costs of the utilities concerned. On this basis, local activists petitioned local governments to enforce such contracts on the utilities. After much effort, most

Länder expressly allowed such contracts, and several dozen cities opted for this model, including Bonn and Nuremberg. As the process began in Aachen, this is known as the Aachen model (Solarförderverein, 2002; Staiss and Räuber, 2002).[22] It was carried by many activist groups and to some extent coordinated by some of the new associations such as Eurosolar and SFV (Solarenergie-Förderverein).

Additional help came from some of the *Länder* which had their own market introduction programmes, the most active being North Rhine-Westphalia. Some states acted through their utilities, which would subsidize solar cells for special purposes, for example schools (Bayernwerk in Bavaria, or BEWAG in Berlin). Some offered 'cost-oriented rates' which however remained below the level of full cost rates (thus HEW in Hamburg). Finally, in a major effort, Greenpeace gathered several thousand orders for solar cell rooftop 'Cyrus installations' (Ristau, 1998). Due to these initiatives, the market did not disappear at the end of the 1000 roofs programme but continued to grow (see Figure 6.4, p128).

Even though the size of the market was quite limited, these initiatives had two significant effects. First, they encouraged a number of new, often small firms to enter into and enlarge the industry. Among these were both module manufacturers and integrators of solar cells into facades and roofs, the latter moving the market for solar cells into new applications. Second, the large number of cities with local feed-in laws and a proliferation of green pricing schemes revealed a wide public interest in increasing the rate of diffusion – the legitimacy of solar power was apparent. Various organizations could point to this interest when they lobbied for a programme to develop yet larger markets for solar cells. As mentioned above, Eurosolar proposed a programme to cover 100,000 roofs in 1993 and, since 1996, the German Solar Energy Industries Association had worked towards the realization of such a programme (Bundesverband, 2000).[23]

Lobbying by the German solar cell industry also intensified. Siemens had at this time already started its production in the US and a second producer, ASE, had the opportunity of doing so with an acquisition of Mobil Solar. To continue production in Germany without any prospects of a large home market would clearly be questionable from a firm's point of view. ASE threatened at this time to move abroad if a market expansion did not take place (Hoffmann, 2001). A promise of a forthcoming programme was then given and ASE decided to invest in a new plant in Germany, manufacturing cells from wafers produced with a technology acquired from Mobil Solar. Production started in mid-1998 (ASE Press Release, 1998) in a plant with a capacity of 20MW (Hoffmann, 2001).

The decision to locate production in Germany implied a dramatic increase in the German industry's solar cell production. A second major investment was Shell's entry into the German solar cell industry through its investment in a new plant in Gelsenkirchen in 1998 (9.5MW, Stryi-Hipp 2001). Here too, a dialogue with policy makers preceded the investment (Zijlstra, 2001). Hence, in 1998, two major investments were made, which greatly expanded capacity in the German solar cell industry.

In sum, the initial 'space' given to wind and solar power in the 1970s and 1980s was now enlarged. In part, this was due to external changes (Chernobyl and the climate change debate) mediated by public awareness and the acceptance of the necessity to change the energy system. But it was also a result of the initial investments in the first formative period. Out of those investments came not only an initial knowledge base, but also an embryonic advocacy coalition consisting of industry associations, an infant industry and various interest organizations. A positive feedback from those early investments resulting in an ability of this coalition to shape further institutional change can be discerned (1990 Feed-in Law). Further feedback loops from market formation, through entry of various organizations, to an enhanced political power of the coalition and ability to defend favourable institutions (which then led to further market formation, entries, and so forth) was a key feature of the subsequent diffusion process for wind power in the 1990s. For solar power, the process of market formation was made more difficult by the high cost of photovoltaics, but through intensive work by the advocacy coalition, where the interest organizations Eurosolar and Förderverein Solarenergie plus Greenpeace played a key role, local market formation programmes were initiated and these were to become precursors to larger, federal programmes in the subsequent phase.

1998–2004 – take off for solar power, continued growth for wind power and new political challenges

In 1998, the Social Democratic/Green coalition, which replaced the Conservative-Liberal government, committed itself to a market formation programme for solar cells as called for by the PV industry and earlier on by Eurosolar and other organizations. The coalition agreement contained commitments to the introduction of an eco-tax on energy, to legislation improving the status of renewable energy, a 100,000 roof programme for solar cells and a negotiated phase-out of nuclear power; all these goals were realized by 2001 (Staiss, 2003). By January 1999, the 100,000 roofs programme (for about 350MW$_p$) was started, providing subsidies in the form of low interest loans to investors. For the sake of speed, the programme did not take the form of a law, but of a decree enacted by the Ministry of Economic Affairs. This ministry maximized bureaucratic obstacles at first, but relented after strong protests by parliamentary groups of the coalition (Witt, 1999b and 1999c). In 1999, 3500 such loans were granted for installations amounting to a mere 9MW$_p$. It was clear that everyone was waiting for a revision of the Feed-in Law, which however took some time to prepare.

Later in 1999, reform of the Feed-in Law began. After launching the trial balloon of a renewable energy levy that the utilities would be able to institute voluntarily (Witt, 1999a), the Minister of Economic Affairs – in charge of this matter – leaned in favour of a quota/certificate system. When it became clear that the minister was not prepared to respect agreements with the parliamentary party groups of the coalition, these groups seized the initiative and submitted a members' bill, which the ministry then tried to dilute and delay without much success, and which was finally adopted as the Renewable Energy Sources Act in March 2000 (Mez, 2003a).

The deputies, particularly the Greens, were inspired by the local feed-in laws for solar power and wanted to move this approach to the federal level. For that purpose they organized a process involving a very large, partly technology-specific advocacy coalition – various environmental groups, the two solar industry associations, the association of the machinery and equipment producers, VDMA, the metalworkers trade union IG Metall, three solar cell producers and politicians from some *Länder*, for example North Rhine-Westphalia (Pfeiffer, 2001). The unorthodox coalition even included a major utility (Preussen Elektra, which testified in favour of the new mechanism equalizing the burden of the law on the national level, although overall it would have preferred a quota system); as a result the big utilities were not united in their opposition. From these organizations and individuals, the Greens received help in terms of both information and support in influencing members of parliament.

The Social Democrats for their part had a strong industrial policy interest in rewriting the Feed-in Law (Eichert, 2001). They feared that the 1998 liberalization of the energy market would lead to a long-term decline in employment in the energy sector and in the associated capital goods industry, which had always been a point of strength of German industry. At this time, the German wind turbine industry had grown to be the second largest in the world and exhibited great dynamism (Bergek and Jacobsson, 2003). With liberalization, the price of electricity dropped, and with it, the remuneration for wind turbine owners. It was then feared that the incentive for further diffusion would be lost and that a less dynamic home market would hurt the German wind turbine industry. Strong renewables legislation, these deputies argued, would put German industrial structure and employment on a more sustainable basis, both environmentally and economically.

While the Federation of German Industries strongly opposed the law, key industrial association, VDMA (Equipment and Machinery Producers, counting about 3000 member firms with approximately one million employees), joined the ranks of its supporters – again demonstrating the increasingly broad legitimacy of renewables. The opposition parties (conservative CDU/CSU and the Liberals) were internally divided on many issues and unable to come up with a coherent alternative, though on the whole they argued for more competition and sometimes for state subsidies instead of passing on costs to final customers (Bechberger, 2000; Deutscher Bundestag, 2000a and 2000b). They also argued that the new law was bound to draw a state aid challenge from DG Competition, a point echoed by the Ministry of Economic Affairs. In fact, a special effort was made by the red–green members of parliament to ward off this possibility (rates declining over time; exclusion of state-owned utilities from the beneficiaries). After adoption of the law, DG Competition questioned its compatibility with EU rules; it withdrew its objection only in May 2002, even though the European Court in March 2001 had rejected a similar challenge in the case of *PreussenElektra* vs *Schleswag* (Lauber, 2001).

The Renewable Energy Sources Act made explicit the Feed-in Law's implicit commitment to take external costs into account. In fact, it provided three reasons for the special feed-in rates. First, it referred to the polluter pays principle with regard to external costs. The explanatory memorandum attached to the law explains that:

> most of the social and ecological follow-up costs associated with conventional electricity generation are currently not borne by the operators of such installations but by the general public, the taxpayers and future generations. The Renewable Energy Sources Act merely reduces this competitive advantage...

Second, the memorandum stresses that 'conventional energy sources still benefit from substantial government subsidies which keep their prices artificially low'. Third, the act purports to break the vicious circle of high unit costs and low production volumes typical of technologies for the generation of renewables sourced electricity (Federal Ministry of the Environment, 2000).

Under the new law, the rates of the tariff scheme were guaranteed to investors for 20 years (under the old Feed-in Law no such guarantee had existed). Also, utilities were no longer excluded from the benefits of the law; they were seen as important investors, especially for offshore wind farms, which require enormous investments. With regard to wind, rates varied with site quality. For at least five years from an installation date in 2000 or 2001 (nine years for offshore), the rate was to amount to €0.091/kWh and longer, depending on how far a turbine remained below the performance of a reference facility. For the first years of operation this meant an improvement of more than 10 per cent over the rate applicable under the previous system in 1998 and 1999 (Hirschl et al, 2002). This was compensated to various degrees by the later decline to €0.062/kWh. For turbines installed in 2002, these rates would be about 1.5 per cent lower, with the decline continuing at that annual rate (always for new installations only) for subsequent years, reinforced by inflation since rates are not adjusted to take it into account (Staiss, 2003). Overall this meant greater security for investors, particularly due to the 20-year guarantee mentioned above (Bönning, 2000). As a result, the diffusion of wind turbines was greatly stimulated (see Figure 6.3, p128).

With regard to solar, the improvement in incentives was much more dramatic. For 2000 and 2001, the new rates amounted to €0.506/kWh for solar cell facilities mounted on buildings, with a size of up to $5MW_p$ and for other facilities up to 100kWp. This rate was guaranteed until a cumulative capacity of $350MW_p$ was reached. All this would probably not have been obtained without the very considerable interest in paying for solar electricity as revealed by the numerous local feed-in laws (Scheer, 2001) as well as by survey data (Solarenergie-Förderverein, 1996). Here too the rate of compensation was set to decline every year for new installations, so that a solar cell unit installed in 2003 would receive €0.457/kWh for 20 years. The annual decline was to be about 5 per cent (Staiss, 2002).

In combination with the 100,000 roofs programme, the revised feed-in-law meant that solar cells became an interesting investment option for the first time. As is evident in Figure 6.4, p128), diffusion took off. A booming market attracted additional entrants that enlarged the industry further.[24] For instance, in 2000, there were ten firms showing roof integrated solar cells at an exhibition (Neuner, 2001), and Germany is seen as the world leader in roof integrated solar cells (Maycock, 2000). Also, the number of solar cell manufacturers rose from two in 1996 to six in 2000 and, as importantly, ASE announced that it would increase its capacity from 20 to 80MW (Schmela, 2001).[25] Eventually, it raised capacity to 50MW by the end of 2002 (under the name of RWE-Schott Solar).

Within less than three years – in mid-2003 – the 350MW_p ceiling was reached (150MW were allocated just in the first six months of 2003 under the 100,000 roofs programme; with this the programme ran out). Even though the ceiling for solar cell installations receiving the special Renewable Energy Sources Act rates was raised in 2002 to the figure of 1000MW_p, investment decisions slowed down greatly in the second half of 2003 as these rates proved insufficient without the low-cost loans of the 100,000 roofs programme. By that time, another amendment to the Renewable Energy Sources Act, to be adopted some time in 2004, was on its way. To secure the continuous growth of the photovoltaics industry, an advance law – a stopgap measure passed in anticipation of a more thorough reform – was adopted by Parliament just before 2003 ran out.

The Federation of German Industry (BDI) criticized the Renewable Energy Sources Act 2000 for creating exorbitant burdens, damaging German competitiveness and driving up electricity prices; the Utility Association (VDEW) pointed to extra costs resulting from the law to justify considerable price increases to final customers, increases which more likely resulted from a decline of competition. Nonetheless, pressure on renewables built up, amplified by the Ministry of Economic Affairs. Yet at the same time that ministry lost ground in terms of control over this policy area. In the parliamentary elections of 2002, the Greens had improved their support while the Social Democrats had declined; thus the Greens could claim a stronger position in government, and effectively secured the transfer of the competency for renewable energy from the Ministry of Economic Affairs (held by the social democrats) to the Environment Ministry (held by a Green). This also meant a shift in the parliamentary committee dealing with renewable energy, from the economic affairs committee to the environment committee.

Although no longer in charge of this policy matter, Economic Affairs Minister Wolfgang Clement from coal state North Rhine Westphalia joined the critics of the Renewable Energy Sources Act, and in summer 2003 a hardship clause was adopted supposedly to reduce the burden for those firms that could prove that their competitive standing was seriously affected. Only 40 firms were able to successfully invoke that clause by the end of 2003 (Witt, 2003; *Windpower Monthly* 2003; Deutscher Bundestag, 2004). Usually the utilities supplying industrial customers – for whom competition is intense – shift the burden to household and small business clients, whose burden is increased as a result (Bröer, 2003). The RESA amendment of 2004 however contained a redefinition of the hardship clause; subsequently, a much larger number of firms were able to claim the exemption (about 300 in early 2005; *Solarthemen*, 2005).

RESA was supposed to be revised following an evaluation after two years. The transfer of renewable energy competences from Economic Affairs to the Environment Ministry in 2002 created some delay, as expertise had to be built up anew. A first draft of the ministry amendment was submitted in the summer of 2003. But the hope for smooth passage of such an amendment (urgent particularly for photovoltaics, where in the meantime the end of the 100,000 roofs programme created the risk that PV markets might collapse) soon disappeared. As usual, the chief problem came from the Minister of Economic Affairs. By summer/fall 2003, Clement questioned the very principle of feed-in tariffs, apparently with the motive to secure a package deal for the protection of coal interests. Some Conservative and Liberal leaders – in particular Conservative leader Angela Merkel – also attacked the Renewable Energy Sources Act because its 'subsidies' supposedly represented a burden for the state budget (when in fact, since they are paid for by consumers, they do not even show up there). Coal and nuclear interests were fighting the law with new vigour, probably because their market shares were seriously threatened by now (with no growth in electricity demand expected). Undoubtedly they also seized on the ratification crisis of the Kyoto Protocol (after Bush's rejection and with Russia postponing ratification) as an opportunity to question the whole Kyoto philosophy. However, German public opinion was still strongly committed to climate policy and renewable energy (Brechin, 2003; Solarenergie-Förderverein, 2003). More importantly perhaps, these conflicts produced two new members of the renewables coalition: the German Confederation of Small and Medium-Sized Enterprises (BVMW) – representing about two-thirds of all employment – and service workers union ver.di.

In the end, Clement achieved some concessions for coal and some reductions in the rates for onshore wind, but the red–green majority proceeded to revise the government bill largely against his preferences. The leadership of the opposition parties also mobilized against the new bill (and in favour of nuclear power). In the upper chamber of parliament (*Bundesrat*), the Länder ruled by conservative-liberal governments, at first voted against the bill. Constitutionally, the *Bundestag* (lower chamber) could have simply overruled it. However, the red–green majority was eager to secure support for maintaining RESA beyond 2007 (the 2004 amendment calls for a new amendment by then). The conservatives did not form a common front on this issue. Some made their support conditional on the expiration of the act in 2007; some wanted a reversal of the nuclear phase-out; some criticized the newly set renewable power target of 20 per cent by 2020; and some actually supported the red–green amendment as it was. In Conciliation Committee, a compromise was achieved, which only provided for modest changes (exclusion of low-wind zones from the feed-in tariff); on this basis the bill was adopted in both houses. Most of the conservative *Länder* government voted in favour, despite the opposition to the act from their party leadership.

The main changes consist in a general strengthening of the legal position of generators vis-à-vis the utilities, inclusion of hydro plants up to 150MW, and significant new incentives for biomass (especially small plants with innovative technologies). On the other hand, there was the reduction of rates for onshore wind and the exclusion of low-wind zones from the benefit of the special feed-in rates (Bechberger and Reiche, 2004). Most important was probably an increase in photovoltaics rates, which for the first time made them commercially interesting without additional support. They had already been introduced by a special 'advance' law in late 2003 (as a stopgap measure to prevent a collapse of the market) and were incorporated in the 2004 Amendment. The result was a veritable solar boom in 2004, which seems set to continue, but which is likely to be constrained by a silicon shortage likely to last for two years or more. Within a time-span of a decade, annual installations in Germany evolved, from an average of about $10MW_p$ in the second half of the 1990s, to $40MW_p$ in 2000, $80MW_p$ in 2002, $150MW_p$ in 2003 and an estimated $300–600MW_p$ in 2004. Cumulative installations showed that Germany rapidly increased its share of worldwide installations from 13.3 per cent in 1997 to more than one third in 2004 (Hirshman, 2004; Eur Observ'ER, 2005). German production figures for solar cells showed a similar evolution: from $44MW_p$ in 2002, it went up to $209MW_p$ in 2004, with an estimated increase for another 50 per cent in 2005, though possibly constrained by the silicon shortage (Kreuzmann, 2005).

In sum, the red–green coalition that came to power in 1998, not only adopted the 'old' proposal of the 100,000 roofs programme early on but, drawing on a broad and increasingly strong advocacy coalition that now included VDMA, it also rewrote the Feed-in Law in a manner that was advantageous to wind and solar power. The diffusion of wind turbines took off again and that of solar cells soared. A clear feedback loop from early diffusion to subsequent ability to influence the political process shaping the regulatory framework can be discerned. Yet, the very success of that framework led to intensified efforts of coal and nuclear interest to change it – the 'battle' over the nature of institutions now moves into its third decade. A Conservative-led government may well make major changes in this area.

Financial flows and social costs: a modest price for transformation

The current renewable energy policy must be seen in a wider context. For the Conservative–Liberal government in office until 1998, renewable energy was at best 'complementary' energy rather than an alternative. For most of the red–green coalition, it is imperative that these energy sources replace other sources in the course of the 21st century. This is part of a climate strategy, which by 2020 should reduce CO_2 emissions by about 40 per cent, and by 80 per cent in 2050 (Jänicke, 2002; Bundesministerium für Umwelt, 2003). The red–green German government – though somewhat divided on the issue – and especially its parliamentary party groups want renewable power to grow from 6.25 per cent in 2000, to 12.5 per cent in 2010 and – a goal added in 2004 – to 20 per cent in 2020. By 2050, renewable energy (including imports) is envisioned to contribute over 60 per cent of total electricity demand (Bundesregierung 2002a; Bundesregierung 2002b). In this scenario, electricity from renewable energy sources is expected to require regulatory support until about 2020. After 2030 or 2035, it is expected to become cheaper than conventional generation, with a payback date some time before 2050 (Nitsch, 2002).

These visions, emanating mostly from the environment ministry, have led to important controversies. Not surprisingly, the Ministry of Economic Affairs – traditionally the advocate of conventional energy sources – arrived at cost estimates for an energy transition to renewables, which are up to ten times higher, though most of these costs are seen to occur in the transportation sector (Fischedick et al, 2002). Criticism also came from parts of the Conservative-Liberal opposition.[26] In June 2005, with the prospect of early parliamentary elections likely to bring about a changeover to a Conservative-Liberal (or just Conservative) government by autumn, Conservative leader Angela Merkel repeated her determination to redefine German energy policy for the benefit of coal, nuclear and the big utilities. The high cost of energy, she argued, had become a threat to German industry. It is interesting therefore to look at the financial flows as well as the social costs connected with the different forms of electricity generation. We will argue that the social (that is society's) price tag for conventional power generation is much higher than the private (that is as reflected in consumer electricity bills); that the support given to renewables is but a fraction of that given to 'conventional technologies'; and, finally, that the remuneration under current support policy is broadly equal to avoided social costs and, therefore, involves no or very small extra costs for society.

The social cost of power generation based on coal is much higher than the private. In calculating social costs, we need to consider both subsidies and external costs. In terms of 2003 Euros, *subsidies* to hard coal for electricity generation can be estimated very roughly at about €80–100 billion for the period 1975–2002,[27] and another 16 billion scheduled for the period 2005–2012 (Bundesverfassungsgericht 1994; Wachendorf, 1994; IEA, 2002; *Solarzeitalter* 4/2003, p57). Subsidies for soft coal amounted to about €1 billion annually (UBA/Wuppertal Institut, 2004). During the same time period, hard coal and lignite together caused *external costs* in the range of €400 billion or more, probably substantially more as external costs were considerably higher before the widespread use of flue gas cleaning (European Commission, 2003).[28] Total government funded R&D for coal amounted to €2.9 billion for 1974–2002 (IEA, 2003a).

Nuclear fission in Germany cost taxpayers some €14 billion in R&D funds since 1974 (IEA, 2003a; see also Figure 6.5, p130). This amount was spent 'to establish an internationally competitive industry', a goal which in the view of the government was not to be hindered by 'a premature and overstressed bias towards economic aspects' on the part of the utilities. It is true that most of these funds went to the development of 'advanced reactors' such as the high-temperature gas-cooled reactor or the fast breeder reactor (Keck, 1980, p316). However, at that time it was thought that advanced reactors relying on plutonium represented the future of nuclear power, since the uranium used in light water reactors would sooner or later become scarce (Meyer-Abich and

Schefold, 1986). For the purposes of the advanced reactor programme, the concept of 'R&D' was interpreted quite generously in order to facilitate financial support by the Federal Government, the programme was framed as an experimental development programme rather than a programme aimed at early commercialization (Keck, 1980, 323).[29] Finally, participation in the international nuclear fusion programme so far caused Germany R&D expenses of slightly more than €3 billion (IEA, 2003a), but this contribution will have to be multiplied many times over before fusion may actually generate electricity, estimated to occur not before 2050.

How do wind and photovoltaic power compare to all this? From 1975 through 2002, in terms of government R&D funds, wind received €0.47 billion, and solar cells €1.15 billion (IEA, 2003a; Sandtner et al, 1997; Räuber, 2002; Deutscher Bundestag, 2003; see also Figure 6.5). The red–green coalition so far has not modified energy research priorities substantially, even though Scheer and Fell – the parliamentary leaders of the coalition parties on renewable energy sources – are asking for an increase of R&D on those sources by a factor of ten (Eurosolar, 2003a; Frey, 2003; Siemer, 2003). There is also a cost resulting from market creation programmes. The 250 MW wind programme caused cumulative costs of €0.15 billion from 1989 through 2001 (Staiss, 2003, pII-27); to this the costs of the *Länder* programmes must be added, for example of Schleswig-Holstein and Lower Saxony (Paul, 2003). Most expensive so far is the 100,000 roofs programme; its cost was estimated at €0.1 billion for 2001 alone (Fischedick et al, 2002). Although this cost varies according to the prevailing interest rates (Genennig et al, 2002), it is safe to assume that annual cost in future is likely to be several times this amount, for a period of almost 20 years. Yet, we are speaking in terms of very small figures in the context of the energy sector. As to external costs, they were estimated in the ExternE project – an international project to assess the external costs of energy usage – to amount to 0.05 Euro cents/kWh for wind power and to 0.6 Euro cents for solar PV[30] (European Commission, 2003).

The largest flow of funds connected to renewables is in connection with the compensation under the Renewable Energy Sources Act. In 2002, this amounted to €2.2 billion (Deutscher Bundestag, 2003) for 24 TWh (*Umwelt*, 2003), which means an average feed-in rate of 9.1 Euro cents per kW_h. Compensation under this act will certainly grow for some time, and a 50 per cent increase of total compensation under the Renewable Energy Sources Act is expected between 2002 and 2005 (Deutscher Bundestag, 2003).

The difference between this compensation and that of the private cost of conventional power generation was about €1.45 billion in 2002. However, the relevant measure to consider is the social cost of that power. In other words, we need to relate the compensation under the Renewable Energy Sources Act to the social cost of generation power with conventional, coal-based technologies. For 2002, the cost of electricity generated from hard coal can be estimated at 9.9 to 12.5 Euro cents/kWh. This includes 3.4 to 3.8 cents direct generation costs (Staiss 2003, pI-248), 2 to 4.2 cents from coal subsidies (estimated on the basis of IEA, 2002; Statistik der Kohlenwirtschaft, 2003; for the higher figure see Janzing, 2004) and 4.5 cents[31] in external costs (European Commission, 2003).

For electricity from soft coal, the respective figure is 8.6 to 9.0 cents, allowing about 0.7 cents for subsidies.[32] For both sources, the cost of CO_2 permits will add another cent or more for each kWh (a trading price of €8 corresponds to about 1 cent) beginning in 2005. The 9.1 cents resulting from the Renewable Energy Sources Act mix of tariffs (see preceding paragraph), augmented by slightly more than 0.05 cents of external costs, are in fact below the cost of hard coal generated electricity, and almost equal to that generated from soft coal. As to wind power from turbines installed in 2004, the average rate over the 20-year period is somewhere near 7.0 cents including external costs (8.53 cents for the first five years or longer, coming down to 5.39 cents afterwards). There are two implications of this. First, if social costs are taken seriously – and this was one of the declared goals of both the Feed-in Law and of the Renewable Energy Sources Act – most renewable power would be in the competitive range right now.[33] Second, the remuneration under this act roughly compensates for the avoided social costs of coal-generated electricity, which means that in social terms, the extra cost to society appears to be negligible. The costs of emission trading reinforce the case for renewable power even more.

In short, taking into account all costs including subsidies and external costs, to increase the share of electricity covered by the Renewable Energy Sources Act appears as a well-founded choice for German society to take even in financial terms. And there are additional considerations in favour of such a choice. Security of supply is one of them. Being a technology leader also confers 'early mover' advantages, and the advocates of the German climate strategy view renewables sourced electricity as an area of strong export potential. Already renewable energy sources have created about 120,000 to 150,000 jobs; a further increase can be expected in the future. Also, the annual private cost per capita – about €18 in 2002 – seems far from exorbitant.[34] The calculation of the preceding paragraphs does not even include the benefits expected from renewable energy technology production in terms of industrial policy, of employment and exports, goals that are also laid down in the Renewable Energy Sources Act.

Conclusion

It might come as a surprise to see Germany among the leaders in the transformation of the energy system (here with regard to electricity). In the 20th century, Germany was one of the few large industrial states without oil resources and no large oil corporation of its own (Karlsch and Stokes, 2003). Partly for this reason, it came to rely with particular intensity on domestic coal, and later on nuclear energy. This was reinforced by the energy crises of the 1970s, where such a choice was imposed in a rather authoritarian fashion by Chancellor Helmut Schmidt, to be continued by his successor Helmut Kohl after 1982. But this choice led to intense controversies and the rise of a strong anti-nuclear movement in the 1970s, a strong environmental movement in the 1980s (especially over acid rain, largely from coal) and the first big Green party in Europe. Early on, renewable energy sources caught public attention as an alternative to the nuclear path towards a plutonium economy. Under pressure from a movement in favour of renewables, the above governments with some reluctance also supported the development of renewable energy sources, though not for domestic use at first.

Even this limited and ambivalent support fell on fertile ground, as there was a broad range of people just waiting to play an active role in developing the new technologies – as researchers, farmers, technicians, entrepreneurs, customers, investors and so on. For this reason even modest support was enough to create a space for wind and solar power to start out on a formative period. All four features of such periods were present: institutional change in the form of a changed energy R&D policy (although only on the margin), the formation of markets (although very small) in the form of protected niches, entry of firms and establishment of some of the elements of an advocacy coalition. Hence, all the four features were there, if only in an embryonic form while the existing structure remained intact. Yet, the value of this very first phase did not lie in the rate at which the new technology was diffused, or in the way existing structures such as the regulatory regime were altered (often they were not), but in the opportunities for experimentation, learning and the formation of visions of a future in which renewables would play a prominent role in electricity generation.

In the second half of the 1980s, Chernobyl, forest dieback (due to acid rain) and the emergence of climate change as a political issue led to strong demands for change from the public. These demands were mediated creatively – not by the government (which largely opposed them) nor by the parliamentary leadership of the governing coalition, but by backbenchers who were able to garner an impressive amount of support among their colleagues from all political parties, who on this issue were unusually cooperative. These MPs also learned to pressure and if necessary to bypass the government; in that sense Germany – like Denmark from the early 1980s to the early 1990s (Andersen, 1997) – also had its 'green majority' in parliament, at least in the energy field, that was prepared to bypass governments that were somewhat less 'green', except that in the German case this majority, although somewhat thinned by now, has held up for a decade and a half so far.

These demands led to the first important measures of market formation in the late 1980s. Large-scale demonstration programmes were initiated (250MW and 1000 roofs) which involved a very significant upscaling of the initial protected market space. The 1990 Feed-in Law gave additional and powerful financial incentives to investors in renewables. A first feedback loop from the investments in the formative phase to an emerging advocacy coalition capable of influencing the institutional framework can here be discerned. Indeed, with hindsight, the Feed-in-Law may well be seen as the first sign of a breach into an old structure.

With such a dramatic change in the institutional framework, wind power was able to move into a take-off phase characterized by very rapid diffusion.[35] Firms were induced to enter into the buoyant industry, learning networks evolved and the advocacy coalition was strengthened. Thus, virtuous circles, which involved all the four features, began to operate. The 'unimaginable' growth also led to an adjustment in beliefs. While Liberals and most Conservatives continued to see renewables as a 'complementary' source of energy, the social-democratic parliamentary group developed visions of a transition to renewables, which came close to those of the Greens. The legitimacy of renewable power gained additional strength in the political arena.

When the established actor network (utilities with the help of the Ministry of Economic Affairs and DG Competition) attempted a rollback of the Feed-in Law in the mid-1990s, they met with opposition from a coalition that had been strengthened by a rapid diffusion of wind turbines (something which led not only to a stronger renewable power industry but also strengthened the claim that renewable power was indeed alternative). This coalition was powerful enough to maintain regulatory continuity – one of the key criteria of success in this area (Haas et al, 2004). Thus, the advocacy coalition had gained enough strength to win battles over the shape of the regulatory framework – a second feedback loop from diffusion to the process of policy-making is here highly visible.[36]

Meanwhile, for solar power, a set of local initiatives provided enough protected market spaces for the industry to survive. Although small, these markets induced further entry of firms and revealed a strong legitimacy for solar power, which later helped the Greens and SPD to alter the regulatory framework in its favour.

When the red–green coalition took over in 1998, its parliamentary party groups – once more against the opposition of the Ministry of Economic Affairs – soon took measures to vastly increase the protected market space for solar power (100,000 roofs), to further improve the conditions for investors in wind power (in particular by reducing uncertainty) and to give investors in solar cells adequate financial incentives. In order to achieve this, the coalition drew new actors into this policy network, coming partly from the renewable energy sector (equipment producers, owners and operators of installations and their associations) and environmental NGOs, partly from 'conventional' associations such as investment goods industry association VDMA, or the metalworkers union, which had joined the coalition during the preceding years.

This institutional change accelerated wind power installation and brought an early take-off phase for solar cells as well. A virtuous circle was set in motion for solar power where the enlarged market encouraged yet more firms to enter and strengthen the coalition further, and others at least not to oppose it. This was already the case in 2000, when in a parliamentary hearing one of the largest utilities supported the revision of the Feed-in Law. In 2003/2004, the advocacy coalition was supplemented by new allies such as the union of service workers and the confederation of small and medium-sized enterprises (Eurosolar, 2003b), and prevailed against renewed opposition from nuclear and coal interests. In the passage of the 'advance' law on solar PV in late 2003, the CDU/CSU members of parliament gave their support. Despite the opposition of their leadership, several conservative *Länder* governments in the *Bundesrat* also supported the 2004 Act. It is likely that the photovoltaics boom of 2004 generated new support. The German solar power industry now stands a chance of catching up with that of Japan.

This suggests not only a growing acceptance of the regulatory regime but also an increasing acceptance of the visions of a different energy future that could come about within a few decades. Legitimacy of a new technology and visions of its role in future electricity generation are therefore not only a prerequisite for the initiation of a development and diffusion process, but also a result of that very same process. Legitimacy and visions are shaped in a process of cumulative causation where institutional change, market formation, entry of firms (and other organizations) and the formation and strengthening of advocacy coalitions are the constituent parts. At the heart of that process lies the battle over the regulatory framework.

To be successful, the diffusion must also be defensible on economic grounds. The comparison with other available sources shows that in terms of overall cost to society, renewable power is likely to be a perfectly reasonable choice, and one that will be amortized within a time-span that is not unusual for major infrastructure investments. It is clearly somewhat ironic that a major political struggle was required merely to 'get prices right' (and to get away from a choice of technology that was inferior from a social perspective), often against an opposition that claimed to be playing that very same tune. Even so, and despite the exceptionally high degree of legitimacy of renewable energy sources in German society, it may be difficult to maintain a supportive policy for the time period required, that is another two decades, against established actors that are still well-connected, particularly in a policy environment marked by liberalization and privileging considerations of short-term profitability over long-term strategies. From this perspective, it is encouraging to note the investment plans of E.on, one of the largest utilities in Germany (and worldwide). For 2005–2007, it was reported to plan €1.3 billion of investments in the conventional power sector – and €1.1 billion in renewable power facilities (not only in Germany; *Solarthemen*, 2004). It seems likely that successful exports of the wind and photovoltaics industries will contribute a momentum of their own. But as the Danish turnaround on renewable energy after the 2001 elections shows, such processes of diffusion remain fragile until renewable power has simply become 'normal'.

Notes

1 An earlier version of this chapter is being published in *Energy Policy* (Jacobsson and Lauber, 2006).
2 This chapter is a joint product of the two authors. The input by Jacobsson comes from a large project pursued together with Anna Bergek, Lennart Bångens and Björn Sandén (formerly Andersson). Jacobsson's input thus draws extensively from three of the papers written in that project. Key references are Bergek and Jacobsson (2003), Jacobsson and Bergek (2003) and Jacobsson et al (2002).
3 This section draws substantially on Jacobsson and Bergek (2003).
4 These data are for 1998.
5 Whereas we focus on these two technologies, we are aware of a larger range of

renewables that include for example wave power, new ways of using biomass (for example gasified biomass – see Bergek, 2002) and solar thermal.

6 Already in the 1930s, experiments with large (several hundred kW) wind turbines for electricity generation were undertaken in Germany, and the first solar cell was produced in 1954 by Bell Laboratories (Heymann, 1995; Wolf, 1974, cited in Jacobsson et al, 2002).

7 A longer discussion is found in Jacobsson and Bergek (2003) and in Carlsson and Jacobsson (2004).

8 A more detailed account of renewable power policy in Germany is given in Lauber and Pesendorfer (2004) and Lauber and Mez (2004).

9 Committee of the Bundestag (lower house) composed half of MPs, half of experts who also have the right to vote. Enquete commissions are set up irregularly to deal with major new policy issues turning very substantially on scientific expertise.

10 Conservative is used as synonymous with Christian Democratic (CDU/CSU).

11 Only some local utilities – Stadtwerke, that is municipal utilities – took a different course.

12 The numbers exclude funding given for the purpose of demonstration wind turbines. In addition, there was support for projects that could benefit all sizes of turbines.

13 These are estimates based on elaboration of data from Jahresbericht Energieforschung und Energietechnologien, various issues, Bundesministerium für Wirtschaft und Technologie.

14 These were: amorphous silicon (aSi), copper sulphide, cadmium selenide, cadmium telluride and copper indium diselenide (CIS).

15 According to Hemmelskamp (1998), 214 turbines were supported.

16 In addition, private operators, for example farmers, had the possibility to obtain an investment subsidy (Durstewitz, 2000a).

17 In the early 1990s, the Ministry of Economic Affairs actually demanded a very large support programme for renewable energies (about €0.75billion) but could not secure the necessary political support (Hemmelskamp 1999).

18 Generators were not required to negotiate contracts, participate in bidding procedures or obtain complicated permits; this simplicity was certainly essential for the success of this law (von Fabeck, 1998).

19 This was the word used by a central person in the evolution of the German wind turbine industry and market.

20 The bulk of the sales within the 100/250MW programme took place 1990–1995 and the programme accounted for most of the nearly 60MW that were installed in the years 1990–1992 (ISET, 1999).

21 In 1994, the Kohlenpfenning (coal penny) was held unconstitutional (Bundesverfassungsgericht 1994; Wachendorf, 1994).

22 In the same year, Bayernwerk introduced the first 'green pricing' scheme, which involved investment in a 50kW$_P$ plant. Shares were sold to about 100 people who paid a premium price for their electricity (Schiebelsberger, 2001). Many such schemes followed, for instance by RWE in 1996. About 15,000 subscribers eventually paid an eco-tariff (twice the normal tariff) for electricity generated by solar cells, hydropower and wind (Mades, 2001).

23 The late 1980s and the 1990s saw a veritable proliferation of renewable energy associations. For instance, an association for biogas (1992), one for biomass (1998) and yet another solar energy association (UVS, 1998). Most of these engage in lobbying and educational activities, sometimes also in exchange for information and experience.

24 Some firms also entered a few years earlier in response to the market formation following the local feed-in laws.

25 In 1998, domestic module production had covered less than one-quarter of a domestic demand of 12MW. Beginning in 1999 (demand 15MW, production 4.3MW), these figures increased steeply: 40 per cent of a demand of 66.5MW was covered in 2001. Estimates stand at around 70 per cent for 2002 and 2003. A survey of the industry carried out in 2003 listed four wafer manufacturers, eight cell producers and 21 manufacturers of modules, some of them highly specialized (Hirschl et al, 2002; *Solarthemen*, 2003).

26 In early 2004, CDU/CSU MPs were willing to support the government amendment to the Renewable Energy Sources Act on condition that a ceiling be introduced to limit feed-in payments in total volume, not in terms of extra cost; this ceiling was likely to be reached by 2010 or earlier (*Solarthemen*, 2004).

27 The actual figures may be higher as these figures do not seem to be adjusted for inflation.

28 A tax exemption for coal-generated electricity also needs to be mentioned here.

29 Tax breaks on undistributed profits for power plant decommissioning cost another €18 billion by 1998 (Mez, 2003b), and more since then. Extra costs to electricity consumers resulting from defective nuclear technology or simply expensive entrepreneurial decisions in this context were usually hidden in the electric rates allowed by sympathetic regulators in the days of territorial monopolies with privileged political connections (before 1998) and are therefore harder to identify (Mez and Piening, 1999). For the sake of perspective, it should also be added that total research spending on nuclear energy in OECD countries is estimated at about €150 billion, supplemented by about €300 billion in cross-subsidies from electricity tariffs, not counting damages or the cost of returning nuclear sites to their former state (Rechsteiner, 2003). There is also low insurance coverage for nuclear accidents.

30 The figures for solar PV in Germany are about ten years old and therefore problematic (Nickel, 2004).

31 This figure is in the middle of a range 3–6 cents indicated by ExternE.

32 These figures will go up as old coal plants need to be replaced, whereas the cost of generation per kWh of renewables sourced electricity will decline from now on if – as intended – solar cells will be introduced at a moderate rhythm.

33 Contrary to common assumptions, even current photovoltaics power may not be overpriced in market terms. See endnote 4 in the Introduction chapter to this publication.

34 As to the introduction of competitive mechanisms, their impact in Europe is quite limited so far and does not always point into the direction expected. Thus, prices for wind power seem to be considerably higher at present under Britain's Renewable Obligation system than in Germany, despite a more 'competitive' mechanism and much better wind conditions (Knight, 2003; Lauber on quota/tradable certificates systems in this volume; and Lauber and Toke, 2005).

35 Those measures were well designed in terms of regulatory design and impact, in particular the Feed-in Law. Bureaucratic entanglements and complex procedures were largely avoided.

36 Whereas Denmark in 1999 gave in under EU pressure and accepted liberalization of renewables sourced electricity as unavoidable, the German parliament stuck to its guns.

References

Advocate General Jacobs (2000) 'Opinion of Advocate General Jacobs delivered on 26 October 2000, Case C-379-98, PreussenElektra AG v Schleswag AG', European

Court of Justice
Afuah, A. and Utterback, J. (1997) 'Responding to structural industry changes: A technological evolution perspective', *Industrial and Corporate Change*, vol 6, no 1, pp183–202
Ahmels, H.-P. (1999) Interview with Dr Hans-Peter Ahmels, BWE, 8 October
Andersen, M. S. (1997) 'Denmark: The shadow of the green majority', in Andersen, M. S. and Liefferink, D. (eds) *European Environmental Policy: The Pioneers*, Manchester, Manchester University Press, pp251–286
ASE (1998) 'ASE initiates first industrial-scale manufacturing of innovative ribbon silicon solar cells in Alzenau, Germany', Press release, August 31
Bechberger, M. (2000) 'Das Erneuerbare-Energien-Gesetz: Eine Analyse des Politik-Formulierungsprozesses', *FFU-report,* 00-06, Berlin, Free University of Berlin
Bechberger, M. and Reiche, D. (2004) 'Renewable energy policy in Germany: Pioneering and exemplary regulations', *Energy for Sustainable Development*, vol 8, no 1, pp25–35
Bergek, A. (2002) 'Responses to a technological opportunity: The case of black liquor gasification in Sweden', mimeo, Göteborg, Department of Industrial Dynamics, Chalmers University of Technology
Bergek, A. and Jacobsson, S. (2003) 'The emergence of a growth industry: A comparative analysis of the German, Dutch and Swedish wind turbine industries', in Metcalfe, S. and Cantner, U. (eds) *Change, Transformation and Development*, Heidelberg, Physica-Verlag, pp197–228
Bonaccorsi, A. and Giuri, P. (2000) 'When shakeout doesn't occur. The evolution of the turboprop engine industry', *Research Policy*, vol 29, pp847–870
Bönning, C. (2000) 'Investitionssicherheit für Betreiber einer Anlage zur Erzeugung von Strom aus erneuerbaren Energien durch das Erneuerbare-Energien-Gesetz?', *Zeitschrift für neues Energierecht*, vol 4, no 4, pp268–271
Brauch, H. G. (1997) 'Markteinführung und Exportförderung Erneuerbarer Energien in der Triade: Politische Optionen und Hindernisse', in Brauch, H. G. (ed) *Energiepolitik*, Berlin, Springer, pp539–566
Brechin, S. R. (2003) 'Comparative public opinion and knowledge on global climatic change and the Kyoto Protocol: The US versus the world?', *International Journal of Sociology and Social Policy*, vol 23, no 10, pp106–134
Breshanan, T., Gambardella, A. and Saxenian, A.-L. (2001) '"Old economy" inputs for "new economy" outcomes: Cluster formation in the new Silicon Valleys', *Industrial and Corporate Change*, vol 10, no 4, pp835–860
Bröer, G. (2003) 'Härtefälle ohne harte Fakten', *Solarthemen*, vol 171, 2 of 6 November
Bundesministerium für Umwelt, Naturschutz und Reaktorsicherheit (2003) *Erneuerbare Energien in Zahlen – Stand: März 2003*, Berlin
Bundesregierung (2002a) *Umweltbericht*, Berlin
Bundesregierung (2002b) *Perspektiven für Deutschland: Unsere Strategie für eine nachhaltige Entwicklung*, Berlin
Bundesverband Solarindustrie (2000) *Solar Who's Who? The Compendium of the German Solar Industry*, Munich, Bundesverband Solarindustrie
Bundesverfassungsgericht (1994) BVerfGE 91, 186 – Kohlepfennig
BWE (2000) *Statistik Windenergie in Deutschland.* www.wind-energie.de/statistik/deutschland.html
Carlsson, B. and Jacobsson, S. (1997) 'In search of a useful technology policy – General lessons and key issues for policy makers', in Carlsson, B. (ed) *Technological Systems and Industrial Dynamics,* Boston, Kluwer Press, pp299–315
Carlsson, B. and Jacobsson, S. (1997a) 'Variety and technology policy – How do technological systems originate and what are the policy conclusions?', in Edquist, C. (ed) *Systems of Innovation: Technologies, Institutions and Organizations,* London, Pinter, pp266–297

Carlsson B. and Jacobsson, S. (2004) 'Policy making in a complex and non-deterministic world', *mimeo*, Industrial Dynamics, Chalmers University of Technology, Gothenburg and Weatherhead School of Management, Case Western University, Cleveland

Carrol, G. (1997) 'Long-term evolutionary changes in organisational populations: Theory, models and empirical findings in industrial demography', *Industrial and Corporate Change*, vol 6, no 1, pp119–143

Davies, A. (1996) 'Innovation in large technical systems: The case of telecommunications', *Industrial and Corporate Change*, vol 5, no 4, pp1143–1180

De Fontenay, C. and Carmell, E. (2001), 'Israel's Silicon Wadi: The forces behind cluster formation', *SIEPR Discussion Paper No 00-40*, Stanford Institute for Economic Policy Research

Deutscher Bundestag (1987) Drucksache 11/1175

Deutscher Bundestag (1988a) Drucksache 11/2029

Deutscher Bundestag (1988b) Drucksache 11/2684

Deutscher Bundestag (1989) Drucksache 11/4048

Deutscher Bundestag (1990a) Drucksache 11/6408

Deutscher Bundestag (1990b) Drucksache 11/7169

Deutscher Bundestag (1990c) Parliamentary record of plenary debate of 13 September 1990, pp17750–17756

Deutscher Bundestag (2000a) Wortprotokoll der öffentlichen Anhörung zum Entwurf eines Gesetzes zur Förderung der Stromerzeugung aus erneuerbaren Energien, Drucksache 14/2341

Deutscher Bundestag (2000b) Parliamentary record of plenary debate 14/91 of 25 February 2000, pp8427–8461

Deutscher Bundestag (2003) Drucksache 15/860

Deutscher Bundestag (2004) Drucksache 15/2370

DEWI (1998) ,'Wind energy information brochure', Wilhelmshaven, German Wind Energy Institute

Durstewitz, M. (2000) '250 MW Wind' programme, ISET, Kassel University

Durstewitz, M. (2000a) Private communication with Michael Durstewitz, ISET, Kassel University

Eichert, M. (2001) Interview with Marcus Eichert, Scientific Adviser to Volker Jung, member of the German Parliament (SPD), 4 October

Ericsson, W. and Maitland, I. (1989) 'Healthy industries and public policy', in Dutton, M. E. (ed) *Industry Vitalization*, New York, Pergamon Press

European Commission (2003) *External Costs. Research results on socio-environmental damages due to electricity and transport*, Brussels

EurObserv'ER (2005) 'Photovoltaic energy barometer, www.energies-renouvelables.org/observ-er/stat_baro/comm/baro166.pdf, accessed 12 June 2005

Eurosolar (2003a) 'EUROSOLAR-memorandum zur energieforschung', *Solarzeitalter*, vol 15, no 2, pp8–12

Eurosolar (2003b) 'Aufbruch in eine neue Zeit – Chancen der erneuerbaren Energien nutzen', Erklärung des "Aktionsbündnis erneuerbare Energien", *Solarzeitalter*, vol 15, no 3, pp12–13

Federal Ministry for the Environment, Nature Conservation and Nuclear Safety (2000), Act on Granting Priority to Renewable Energy Sources, Berlin; also published 2001 in *Solar Energy Policy*, vol 70, no 6, pp489–504

Feldman, M. and Schreuder, Y. (1996) 'Initial advantage: The origins of the geographic concentration of the pharmaceutical industry in the mid-Atlantic region', *Industrial and Corporate Change*, vol 5, no 3, pp839–862

Fischedick, M, Hanke, T., Kristof, K., Leuchtenböhmer, S. and Thomas, S. (2002) 'Da ist noch viel mehr drin, Herr Minister! Nachhaltige energiepolitik für Deutschland', *Wuppertal Spezial*, vol 22, Wuppertal

Freeman, C. and Louca, F. (2002) *As Time Goes By. From the Industrial Revolutions to the Information Revolution*, Oxford, Oxford University Press
Frey, B. (2003) 'Sparen bei der Forschung', *Solarthemen*, vol 164, 24 July, p1
Ganseforth, M. (1996) 'Politische Umsetzung der Empfehlungen der beiden Klima-Kommissionen (1987-1994) – eine Bewertung', in Brauch, H. G. (ed) *Klimapolitik*, Berlin, Springer, pp215–224
Garrad, H./Greenpeace (2004) Sea Wind Europe, www.greenpeace.org.uk/-seawindeurope, accessed 10 January 2005
Geels, F.W. (2002) 'Technological transitions as evolutionary reconfiguration processes: A multi-level perspective and a case study', *Research Policy*, vol 31, no 8–9, pp1257–1274
Genennig, B. et al (2002) 'Evaluierung des 100,000-Dächer-Solarstrom-Programms', report to Bundesministerium für Wirtschaft und Technologie, Forschungs- und Planungs GmbH Umweltinstitut Leipzig and Institut für Energetik und Umwelt GmbH, both Leipzig
Haas, R. et al (2004) 'How to promote renewable energy systems successfully and effectively', *Energy Policy*, vol 32, pp833–839
Hemmelskamp, J. (1998) *Wind Energy Policies and their Impact on Innovation – An International Comparison*, Institute for Prospective Technology Studies, Seville, Spain
Hemmelskamp, J. (1999) *Umweltpolitik und Technischer Fortschritt*, Heidelberg, Physica
Heymann, M. (1995) *Die Geschichte der Windenergienutzung 1890–1990*, Frankfurt, Campus
Heymann, M. (1999) 'Der Riese und der Wind. Zum schwierigen Verhältnis der RWE zur windenergie', in Maier, H. (ed) *Elektrizitätswirtschaft zwischen Umwelt, Technik und Politik. Aspekte aus 100 Jahren RWE-Geschichte*, Freiberg, Akademische Buchhandlung, pp217–236
Hirschl, B., Hoffmann, E., Zapfel, B., Hoppe-Kilpper, M., Durstewitz, M. and Bard, J. (2002) 'Markt- und Kostenentwicklung erneuerbarer Energien. 2 Jahre EEG – Bilanz und Ausblick', Berlin, Erich Schmidt Verlag
Hirshman, W. P. (2004) 'Evaluating the trend', *Photon International*, December, pp52–54
Hoffmann, W. (2001) Interview with Winfred Hoffmann, ASE, 5 March
Huber, M. (1997) 'Leadership and unification: Climate change policies in Germany', in Collier, U. and Löfstedt, R. (eds) *Cases in Climate Change Policy: Political Reality in the European Unity*, London, Earthscan, pp65–86
Hustedt, M. (1997) Parliamentary record of plenary debate 13/208 of 28 November 1997, p18975
Hustedt, M. (1998) 'Windkraft – Made in Germany', in Alt, F., Claus, J. and Scheer, H. (eds) *Windiger Protest. Konflikte um das Zukunftspotential der Windkraft*, Bochum, Ponte Press, pp163–168
IEA (1999) International Energy Agency. *Annual Report 1999, Implementing Agreement on Photovoltaic Power Systems*, Paris, IEA
IEA (2001) *Key World Energy Statistics from the IEA*. Paris, OECD/IEA
IEA (2002) *Energy Policies of IEA Countries. Germany*, 2002 Review, Paris, IEA
IEA (2003a) 'Energy Technology R&D Statistics', www.iea.org/stats/files/rd.htm, accessed 20 October 2003
IEA (2003b) 'IEA-PVPS website', www.iea-pvps.org, accessed 21 November, 2003
IEA (2004) IEA PVPS, www.oja-services.nl/iea-pvps/isr/22.htm, accessed 12 June 2005
ISET (1999) 'WMEP – Jahresauswertung 1998'. Institut für Solare Energieversorgungstechnik (ISET), Verein an der Universität Gesamthochschule Kassel, Kassel
Jacobsson S., Andersson, B. and Bangens, L. (2002). 'Transforming the energy system

– the evolution of the German technological system for solar cells', *SPRU Electronic Working Paper Series*, no 84

Jacobsson, S. and Bergek, A. (2003) 'Energy Systems Transformation: The Evolution of Technological Systems in Renewable Energy Technology', *mimeo*, Department of Industrial Dynamics, Chalmers University of Technology, Gothenburg, Sweden

Jacobsson, S. and Lauber, V. (2006) 'The politics and policy of energy system transformation – explaining the German diffusion of renewable energy technology', *Energy Policy*, forthcoming

Jahn, D. (1992) 'Nuclear Power, energy policy and new politics in Sweden and Germany', *Environmental Politics*, vol 1, no 3, pp383–417

Jänicke, M. (2002) 'Atomausstieg und Klimaschutzpolitik in Deutschland – 8 Thesen', in Bundesministerium für Umwelt, Naturschutz und Reaktorsicherheit (ed) *Energiewende: Atomausstieg und Klimaschutz.* Dokumentation der Fachtagung vom 15 und 16 Februar 2002, Berlin, pp5–9

Janzing, B. (2004) 'Subventions debatte – Wind billiger als Kohle', *Solarthemen*, vol 175, 15 January 2004, p6

Johnson, A. and Jacobsson, S. (2001) 'Inducement and blocking mechanisms in the development of a new industry: the case of renewable energy technology in Sweden', in Coombs, R., Green, K., Richards, A. and Walsh, W. (eds), *Technology and the Market, Demand, Users and Innovation*, Cheltenham, UK, Edward Elgar, pp89–111

Karlsch, R. and Stokes, R. G. (2003) *Faktor Öl: Die Mineralölwirtschaft in Deutschland*, Hamburg, Hoffmann & Campe

Keck, O. (1980) 'Government policy and technical choice in the West German reactor programme', *Research policy*, vol 9, pp302–356

Kemp. R., Schot, J. and Hoogma, R. (1998) 'Regime shifts to sustainability through processes of niche formation: The approach of strategic niche management', *Technology Analysis and Strategic Management*, vol 10, no 2, pp175–195

Kitschelt, H. (1980) *Kernenergiepolitik*, Frankfurt, Campus

Klepper, S. (1997) 'Industry life cycles', *Industrial and Corporate Change*, vol 6, no 1, pp145–181

Knight, S. (2003) 'Heat goes out of threat to cut support', *Windpower Monthly*, vol 19, no 10, pp38–39

Kords, U. (1993) *Die Entstehungsgeschichte des Stromeinspeisegesetzes vom 5. 10. 1990*, MA thesis in political science, Free University of Berlin

Kords, U. (1996) 'Tätigkeit und Handlungsempfehlungen der beiden Klima-Enquete-Kommissionen des Deutschen Bundestages (1987–1994)', in Brauch, H. G. (ed) *Klimapolitik*, Berlin, Springer, pp203–214

Kreutzmann, A. (2005) 'Complaints at the highest level. The growth of the German PV industry slows on account of lack of silicon', *Photon International*, January, pp12–17

Lauber, V. (2001) 'Regulierung von Preisen und Beihilfen für Elektrizität aus erneuerbaren Energieträgern durch die Europäische Union', *Zeitschrift für Neues Energierecht*, vol 5, no 1, pp35–43

Lauber, V. (2004) 'REFIT and RPS in a harmonised Community framework', *Energy Policy*, vol 32, no 12, pp1405–1414

Lauber, V. and Mez, L. (2004) 'Three decades of renewable electricity policy in Germany', *Energy & Environment*, vol 15, no 4, pp599–623

Lauber, V. and Pesendorfer, D. (2004) 'Success through continuity: Renewable electricity policies in Germany', in de Lovinfosse, I. and Varone, F. (eds) *Renewable Electricity Policies in Europe*, Louvain-la_Neuve, Presses Universitaires de Louvain, pp121–182

Lauber, V. and Toke, D. (2005) 'Einspeisetarife und quoten-/zertifikatssyteme: Erwartungen der Europäischen Kommission und Erfahrungen aus dem Vergleich zwischen Großbritannien und Deutschland', paper, Vierte Internationale Energiewirtschaftstagung, Vienna, 16–18 February

Levinthal, D. A. (1998) 'The slow pace of rapid technological change: Gradualism and punctuation in technological change', *Industrial and Corporate Change*, vol 7, no 2, pp217–247

Lieberman, M. and Montgomery, C. (1988), 'First mover advantages', *Strategic Management Journal*, vol 9, pp41–58

Mades, U. (2001) Interview with Uwe Mades, RWE Power, 7 March

Marsh, D. and Smith, M. A. (2000) 'Understanding policy networks: Towards a dialectical approach', *Political Studies*, vol 48, no 1, pp4–21

Marshall, A. (1920) *Principles of Economics*, 8th edition, London, Macmillan

Maskell, P. (2001) 'Towards a knowledge-based theory of the geographical cluster', *Industrial and Corporate Change*, vol 10, no 4, pp921–943

Maycock, P. D. (2000) *Renewable Energy World*, vol 3, no 4, pp59–74

Mener, G. (2000) *Zwischen Labor und Markt. Geschichte der Sonnenenergienutzung in Deutschland und den USA, 1860–1986*, Baldham, LK Verlag

Meyer-Abich, K. M. and Schefold, B. (1986) *Die Grenzen der Atomwirtschaft*, 3rd edn, Munich, Beck

Mez, L. (2003a) 'Ökologische Modernisierung und Vorreiterrolle in der Energie- und Umweltpolitik? Eine vorläufige bilanz', in Egle, C., Ostheim, T. and Zahrnhöfer, R. (eds) *Das rot-grüne Projekt. Eine Bilanz der Regierung Schröder 1998–2002*, Wiesbaden, Westdeutscher Verlag, pp329–350

Mez, L. (2003b) 'New corporate strategies in the German electricity supply sector', in Glachant, J. M. and Finon, D. (eds) *Competition in European Electricity Markets. A Cross-Country Comparison*, Cheltenham, UK, Edward Elgar, pp193–216

Mez, L., Piening, A. (1999) 'Ansatzpunkte für eine Kampagne zum Atomausstieg in Deutschland', *FFU-Report*, 99-6, Berlin, Forschungsstelle für Umweltpolitik

Molly, J. P. (1999) Interview with J. P. Molly, DEWI, 7 October

Myrdal, G. (1957) *Economic Theory and Underdeveloped Regions*, London, Ducksworth

Neuner, R. (2001) Interview with Mr Roland Neuner, Lafarge Brass Roofing Accessories, 5 March

Nickel, J. (2004) 'Neue EU-Studie mit alten Zahlen', *Photon*, January 2004, pp14–15

Nitsch, J. (2002) 'Die entwicklung der erneuerbaren Energien bis 2050', in Bundesministerium für Umwelt, Naturschutz und Reaktorsicherheit (ed) *Energiewende: Atomausstieg und Klimaschutz*, Dokumentation der Fachtagung vom 15 und 16 Februar 2002, Berlin, pp95–110

Paul, N. (2003) 'The Schleswig-Holstein story', *Renewable Energy World*, September–October 2003, pp34–45

Pfeiffer, C. (2001) Interview with Carsten Pfeiffer, Scientific Adviser to Hans-Josef Fell of the German Parliament (the Greens), 4 October

Porter, M. (1998) 'Clusters and competition. New agendas for companies, governments and institutions', in Porter, M. (ed) *On Competition*, Boston, A Harvard Business Review Book, pp197–287

PV TRAC (2004) 'A vision for PV technology in 2030 and beyond', http://europa.eu.int/comm/research/energy/photovoltiacs/vision_report_en.htm, accessed 15 January 2005

Räuber, A. (2002) 'Strategies in Photovoltaic Research and Development – R&D and History', in Bubenzer, A. and Luther, J. (eds) *Photovoltaic Guidebook for Decision Makers*, Heidelberg, Springer, pp215–241

Rechsteiner, R. (2003) *Grün gewinnt. Die letzte Ölkrise und danach*, Zurich, Orell-Füssli

Reeker, C. (1999) Interview with Carlo Reeker, BWE, 11 October

Rhodes, R. A. W. (2001) *Understanding Governance. Policy Networks, Governance, Reflexivity and Accountability*, Buckingham, Open University Press
Richter, M. (1998) 'Zwischen Konzernen und Kommunen: Die Strom- und Gaswirtschaft', in Czada, R. and Lehmbruch, G. (eds), *Transformationspfade in Deutschland*, Frankfurt, Campus, pp113–141
Ristau, O. (1998) *Die Solare Standortfrage*, Solarthemen, Bad Oeynhausen
Rosenberg, N. (1976) *Perspectives on Technology*, Cambridge, Cambridge University Press
Rotmans, J., Kemp, R. and van Asselt, M. (2001) 'More evolution than revolution. Transition management in public policy', *Foresight*, vol 3, no 1, February, pp15–31
Sabatier, P.A. (1998) 'The advocacy coalition framework: revisions and relevance for Europe', *Journal of European Public Policy*, vol 5, no 1, pp98–130
Salje, P. (1998) *Stromeinspeisungsgesetz*, Cologne, Heymann
Sandtner, W., Geipel, H. and Lawitzka, H. (1997) 'Forschungsschwerpunkte der Bundesregierung in den Bereichen erneuerbarer Energien und rationeller energienutzung', in Brauch. H. G. (ed) *Energiepolitik*, Berlin, Springer, pp255–271
Schafhausen, F. (1996) 'Klimavorsorgepolitik der Bundesregierung', in Brauch, H. G. (ed) *Klimapolitik*, Berlin, Springer, pp237–249
Scheer, H. (2001) Interview with Herman Scheer, Member of the German parliament and president of Eurosolar, 5 October
Schiebelsberger, B. (2001) Interview with Bruno Schiebelsberger, on, 9 May
Schmela, M. (2001) 'At Staffelstein, the glass is half empty', *Photon International*, April, p14
Schmela, M. (2003) 'A bullish PV year', *Photon International*, March, pp42–48
Schmitt, D. (1983) 'West German energy policy', in Kohl, W. (ed) *After the Second Oil Crisis*, Lexington, Lexington Books, pp137–157
Schult, C. and Bargel, A. (2000) Interview with Christian Schult and Angelo Bargel, Husumer Schiffswerft, 12 January
Schulz, W. (2000) 'Promotion of renewable energy in Germany', in Jasinski, P. and Pfaffenberger, W. (eds) *Energy and Environment: Multiregulation in Europe*, Aldershot, Ashgate, pp128–134
Scitovsky, T. (1954) 'Two concepts of external economies', *Journal of Political Economy*, vol 62, pp143–151
Siemer, J. (2003) 'Aktion umverteilen', *Photon*, June, p42
Smith, A. (1776) *The Wealth of Nations*, London
Smith, M. A. (2000) 'Policy networks and advocacy coalitions explaining policy change and stability in UK industrial pollution policy?', *Environment and Planning C: Government and Policy*, Vol 18, pp95–118
Solarenergie-Förderverein (1996) 'RWE befragt seine Kunden', www.sfv.de/lokal/mails/infos/soinf178.htm, accessed in October 2003
Solarenergie-Förderverein (2002) 'Die kostendeckende Vergütung', www.sfv.de/briefe/brief96_3/sob96302.htm and /www.sfv.de/lokal/mails/wvt/kvhistor.htm, both accessed in October 2003
Solarenergieförderverein (2003) *Solarbrief* 3/03, p45 (survey by TNS-EMNID).
Solarthemen (2003) vol 170, 23 October, p1
Solarthemen (2004) vol 176, 20 January, p2
Solarzeitalter (2003) no 4, p57
Staiss, F. (ed) (2002) *Jahrbuch Erneuerbare Energie 2001*, Radebeul, Bieberstein
Staiss, F. (ed) (2003) *Jahrbuch Erneuerbare Energien 02/03*, Radebeul, Bieberstein
Staiss, F. and Räuber, A. (2002) 'Strategies in Photovoltaic Research and Development – Market Introduction Programs', in Bubenzer, A. and Luther, J. (eds) *Photovoltaics Guidebook for Decision-Makers*, Heidelberg, Springer, pp243–256
Statistik der Kohlenwirtschaft (2003), www.kohlenstatistik.de, accessed 15 January 2003

Stigler, G. (1951) 'The division of labour is limited by the extent of the market', *The Journal of Political Economy*, vol 59, no 3, pp185–193

Stryi-Hipp, G. (2001) 'Shine on Deutschland', *New Energy*, April, pp40–41

Tacke, F. (2000) Interview with Franz Tacke, founder and former managing director Tacke Windenergie, 9 March

UBA (Umweltbundesamt)/Wuppertal Institut für Klima, Umwelt, Energie (2004) 'Braunkohle – ein subventionsfreier Energieträger?', www.umweltbundesamt.de, accessed December 2004

UNDP (2000) *World Energy Assessment. Energy and the challenge of sustainability*, Advance copy, United Nations Development Programme

Umwelt (2003) no 3, p589

Unruh, G. C. (2000) 'Understanding carbon lock-in', *Energy Policy*, vol 28, no 12, pp817–830

Utterback, J. M. (1994) *Mastering the Dynamics of Innovation: How Companies Can Seize Opportunities in the Face of Technological Change*, Boston, Massachusetts, Harvard Business School Press

Utterback, J. M. and Abernathy, W. J. (1975) 'A dynamic model of process and product innovation', *Omega*, vol 3, pp639–656

Van de Ven, A. and Garud, R. (1989) 'A Framework for Understanding the Emergence of New Industries', *Research on Technological Innovation, Management and Society*, vol 4, pp195–225

von Fabeck, W. (1998) 'Anmerkungen zu Gert Apfelstedt', *Zeitschrift für neues Energierecht*, vol 2, no 1, p42

von Fabeck, W. (2001) Interview with Wolf von Fabeck, Förderverein Solarenergie, 7 March

Wachendorf, H. G. (1994) 'Kohleverstromung in Deutschland', *Elektrizitätswirtschaft*, vol 93, no 12, pp672–680

Windheim, R. (2000a) Interview with Dr Rolf Windheim, Forschungszentrum Jülich, 7 March

Windheim, R. (2000b) Data supplied by Dr Rolf Windheim, Forschungszentrum Jülich, on projects funded within the federal R&D wind energy programme

Windpower Monthly (2003) vol 19, no 19, p26

Windpower Monthly (2005) vol 21, no 4, p66

Witt, A. (1999a) '"Zukunftspfennig" statt Gesetz?', *Solarthemen*, vol 55, 15 January, p1

Witt, A. (1999b) 'Ein Bonner Possenstück', *Solarthemen*, vol 56, 29 January, p1

Witt, A. (1999c) '100 000-Dächer Programm für alle?', *Solarthemen*, vol 57, 12 February, p1

Witt, A. (2003) 'Ausnahmen für Industrie', *Solarthemen*, vol 157, 10 April, p3

Wolf, M. (1974) 'Historic development of photovoltaic power generation', International Conference on Photovoltaic Power Generation, Hamburg, 25–27 September, pp49–65

Young, A. (1928) 'Increasing Returns and Economic Progress', *Economic Journal*, vol 38, pp527–542

Zängl, W. (1989) *Deutschlands Strom. Die Politik der Elektrifizierung von 1966 bis heute*, Frankfurt, Campus

Zijlstra, S. (2001) 'Interview with Mr. Sjouke Zijlstra', Shell Solar Deutschland, March 8

7

The UK Renewables Obligation

Peter M. Connor

An introduction to the nature of the renewables obligation

The Renewables Obligation (RO) is a form of the Renewable Portfolio Standard (RPS), a mechanism devised by the American Wind Energy Association and the Union of Concerned Scientists in 1996 (Rader and Norgaard, 1996). In simple terms, an RPS mechanism creates a demand for electricity from renewable energy sources by setting a minimum obligation to suppliers or consumers to obtain their electricity from these sources. In order to meet this obligation, consumers may acquire power from the cheapest available renewable energy generator from a specified list of technologies. The underlying theory is that the competitive process will act to minimize the costs of developing the new technology to those providing the subsidy.

The RO was brought into use in England and Wales on 1 April 2002 as a replacement mechanism for the Non-Fossil Fuel Obligation (NFFO). The NFFO had largely been discredited as a successful stimulus to increasing installed renewable energy capacity in the UK. Its adoption and implementation as the central mechanism for the support of renewable energy in the UK followed a long and convoluted process of consultancy with the industry and other relevant actors. The process, started by the Labour government in 1998, required both consultations regarding the desired nature of the mechanism to replace the NFFO, and multiple steps regarding proposals as to how the new mechanism would operate. This process effectively took four years, and implementation of the RO was further delayed by a year following a general election.

The UK's position

Only around 3 per cent of UK electricity is generated from renewable sources, with 46 per cent of this figure coming from large-scale hydropower projects, a resource that cannot be extended significantly further (Smith and Watson, 2002a). The UK has among the lowest level of renewable energy sources of the European Union (EU), ranking 14th out of the 15 EU member states in 2002 (ECOTEC, 2002). This low level of capacity exists despite the UK having some of the best renewable energy resources in Europe, including the best on- and offshore wind energy resources, and excellent wave and tidal resources. This low level of capacity can be linked to a number of factors, both directly related to UK renewable energy policy, and relating to the context in which such policy exists. Leading reasons for this deficiency include the large amount of indigenous fossil fuel resources in the UK, including North Sea oil and gas, and large coal deposits, rendering new sources of energy less necessary than in other nations with which the UK competes. The UK has been a net energy exporter for the last two decades, which is rare for both EU and historic G7 countries. The UK has also invested heavily in the development and installation of new nuclear energy facilities, a factor that has tended to draw funding away from investment in new renewable energy technologies. Somewhat perversely in light of this approach to nuclear, renewables in the UK have perhaps also suffered from the UK's commitment to the pioneering deregulation of its electricity supply industry, and to maximizing competition within the industry. Legislating to provide permanent and predictable subsidies from consumers on the scale of countries such as Denmark, Germany and Japan was thus rendered much less likely on ideological grounds. Instead the market-based RO was introduced with the idea that prices would be driven down over time. Despite this, a comparison of the average price paid for Renewables Obligation Certificates (ROCs) with the current level of German tariffs does not reveal any great disparity (Mitchell et al, 2006). This situation is rendered more striking by the fact that UK wind resources are far superior to those available in Germany.

Energy security as a whole has been argued to have become less of an issue in the UK since the privatization of the industry in 1990 up to 2000, as the 'dash for gas' led to overcapacity and an increasing number of nationally diverse suppliers have begun to provide fuel to the UK market. The introduction of the New Electricity Trading Arrangements (NETA) in 2001 acted to reduce operational capacity as a number of power stations were mothballed on economic grounds, and there is the potential for further impacts on the UK electricity supply industry. The main nuclear electricity generating company, British Energy, faced significant operational problems and continues to do so.[1] Issues surrounding the eventual replacement of nuclear power generation are likely to impact on UK energy policy whatever the outcome of British Energy's battle for survival as a company. The future role of nuclear power within the UK energy mix is one that is very much open to question. The UK's Energy White Paper (DTI, 2003b) leaves open the possibility of future nuclear build, though does not comment as to how this might be financed. It is possible that a nuclear future in the UK may depend on the success or failure of the RO.

As a response to climate change issues arising at the national, EU and G8 levels, the UK has committed to an indicative target of 10 per cent of electricity to be generated from renewable energy technologies (RETs) by 2010 in partnership with other EU countries (European Commission, 2001). Initially the upper commitment to be supported by the RO was 10.4 per cent, with this figure coming into use from 2011. This was to remain as the minimum figure up to at least 2027. However, by 2003 the government was already under increasing pressure to raise the upper target due to the absence of security perceived by potential investors. In late 2003 the government announced that the obligation would continue to increase in increments up to the figure of 15.4 per cent by 2015 linked to a national target of 15 per cent in that year (DTI, 2003a). The problem of security concerning long-term investment and a lack of long-term security within the RO is discussed later.

The annual increments for increase as additionally laid down are illustrated in Table 7.1. The estimated figures for capacity were not announced along with the percentage obligations for the years 2011–2016.

Table 7.1 *The level of the obligation*

Period	Estimated sales by licensed suppliers in Great Britain (TWh)	Total obligation (Great Britain) TWh	Total obligation as % sales (Great Britain) %
2001/2002	310.9		
2002/2003	313.6	9.4	3.0
2003/2004	316.2	13.5	4.3
2004/2005	318.7	15.6	4.9
2005/2006	320.6	17.7	5.5
2006/2007	321.4	21.5	6.7
2007/2008	322.2	25.4	7.9
2008/2009	323.0	29.4	9.1
2009/2010	323.8	31.5	9.7
2010/2011	324.2	33.6	10.4
2011/2012			11.4
2012/2013			12.4
2013/2014			13.4
2014/2015			14.4
2015/2016			15.4
2016/2017 to 2026/2027			15.4

Note: Table 7.1 details the annual increases in the obligation placed on suppliers in Great Britain, and the estimated volumes of electricity that will be required to meet these obligations.

Source: DTI, 2001.

The government is already under pressure to set a further target of 20 per cent by 2020. This figure was recommended in the Energy Review (PIU, 2002) carried out by the Performance and Innovation Unit (PIU), a body formed at the request of the Prime Minister to examine the 'long term challenges for energy policy in the UK'. The Government currently has in place an 'aspirational' target of 20 per cent of electricity to come from renewables, but this does not have any impact on the application of policy instruments. Increasing the RO target to 20 per cent by 2020 would increase security for developers, generators and renewable electricity consolidators, leading to increased prices for power purchase agreements and reducing capital costs by reducing investment risk attaching to renewable electricity projects.

The aims of UK renewables policy

There are currently five goals for UK renewable energy policy (DTI, 2001):

1 to assist the UK to meet national and international targets for the reduction of emissions, including greenhouse gases;
2 to help provide secure, diverse, sustainable and competitive energy supplies;
3 to stimulate the development of new technologies necessary to provide the basis for continuing growth of the contribution from renewables in the longer term;
4 to assist the UK renewables industry to become competitive in home and export markets and in doing so provide employment;
5 to make a contribution to rural development.

Historically, these issues have been addressed to highly varying degrees by the policies that are supposed to achieve them and it is clear that different priorities attach to each policy goal. It should also be noted that some of the policy goals are more recent than others; notably rural development was added only during the RO consultation process. This chapter will discuss the extent to which UK policies have addressed each of the policy goals listed, with particular reference to the new RO, and its likely relationship to them. To this end it will discuss the context in which the RO exists, other renewable energy policies that already exist or are likely to be adopted and the likely interplay of these with the RO. Finally, it will speculate about the potential for this collection of policies in achieving success with regard to the goals listed above.

While the specific goals of renewable energy policy are obviously fundamentally relevant to the determination of UK renewable energy policy, they are also subordinate to the overarching goals of UK energy policy as a whole. The Energy Review suggested that the fundamentals of UK energy policy can be summarized as being 'sustainable development, requiring achievement of economic, environmental and social objectives. It is also vital to maintain adequate levels of energy security at all points in time' (PIU, 2002, p32). These goals, and the priorities that attach to them, have significant implications for the approach applied for meeting renewable energy targets. The need for sustainable development clearly forms the basis for pursuing renewable energy targets, but this need is balanced by a wish to avoid excessive costs in achieving renewable energy policy goals, and also has an eye on the impacts that a switch to renewables will have on those actors in society who will be displaced by a shift to renewables, and the potential for social gain that might also result from the use of new technologies. Finally, security of supply issues are relevant both as a stimulus for renewables in that these act to reduce the level of fuel needed to be imported into the UK, and more negatively for renewables with regard to the support that other energy providers can command in order that security of supply is not undermined for the UK. The most obvious example of this is the £650 million loaned to the nuclear industry by the UK government in the autumn of 2002 in order to prevent the insolvency of British Energy, the nuclear power generator.[2]

It is important to note that the UK's commitment to ensuring competition and minimizing costs for electrical generation has meant that economic goals have taken precedence over social goals. This point that has been emphasized in the last two years with the introduction of NETA, which has further driven down the cost of electricity generation but has done so while undermining the viability of many renewable energy projects and with little consideration of how this disadvantage might be allayed (Ofgem 2000; Ofgem 2001a; Ofgem 2001b), and which has been of little benefit to the customer as most generators have acted to balance their reduced income through increases in costs via their supply companies.

UK renewable energy policy history

Prior to 1990, support for RE in the UK was largely confined to financial support for research and development concerning new RE technologies. Support for the creation of a market came with electricity privatization in 1990 (Connor, 2004).

UK renewable energy policy through the 1990s was dominated by the NFFO. This policy was originally adopted largely to meet the financing needs of the UK nuclear industry during the privatization of the electricity supply industry, effectively by creating a privileged market for nuclear energy. Subordinate to this, it also aimed to reduce the price of renewable energy technology in the UK through the use of a competitive market within specific renewable energy technology bands (Elliott, 1992; Mitchell, 2000). For example, wind energy projects competed only with other wind energy projects, and so on. A number of technologies were supported within the NFFO, though this was prone to some change between rounds.

Table 7.2 shows both the technology bands within each round of the NFFO and the prices paid for that technology. While achieving some reduction in the costs of some RETs, the NFFO was largely not that successful in meeting each of the supposed goals of UK policy and can even be regarded as subsidizing the nascent industries of international competitors (Mitchell, 1995). It failed to establish significant UK renewable energy technology manufacturing, and many of the projects that were successful in winning contracts did not come to fruition, due to overestimations in the price reductions that could be achieved, therefore undermining the results of the NFFO in achieving cost reductions. The absence of any form of penalty for failing to establish a contracted project encouraged developers to make low bids, which proved not to be realistic, and potentially this absence also meant that the mechanism could be used to submit bids knowingly below a realistic price in order to undercut competitors and deny them market share, without any cost to the successful bidder.

As a result of these problems it became apparent during the later tranches of the NFFO that the completion rate for projects was actually in decline, as demonstrated in Figure 7.1. Even technologies such as Landfill Gas (LFG), which had retained a high share in early rounds, saw a significant drop-off.

Table 7.2 *NFFO technologies and costs*

	NFFO-1	NFFO-2	NFFO-3	NFFO-4	NFFO-5
Technology Band	Cost Justification	Strike Price (p/kWh)	Average Price (p/kWh)	Average Price (p/kWh)	Average Price (p/kWh)
Wind	10.0	11.0	4.43	3.56	2.88
Wind sub-band	–	–	5.29	4.57	4.18
Hydro	7.5	6.0	4.46	4.25	4.08
Landfill Gas	6.4	5.7	3.76	3.01	2.73
M&IW[a]	6.0	6.55	3.89	–	–
M&IW[b]	–	–	–	2.75	2.43
Sewage Gas	6.0	5.9	–	–	–
EC&A&FW[c]	–	–	8.65	5.51	–
EC&A&FW[d]	–	5.9	5.07	–	–
EC&A&FW[e]	6.0	–	–	–	–
M&IW/CHP[f]	–	–	–	3.23	2.63
TOTAL	7.0	7.2	4.35	3.46	2.71

Notes:
a Municipal and industrial waste with mass burn technology;
b Municipal and industrial waste with fluidized bed technology;
c Energy crops and agricultural and forestry waste with gasification technology;
d Energy crops and agricultural and forestry waste with residual technologies;
e Energy crops and agricultural and forestry waste with anaerobic digestion;
f Municipal and industrial waste with combined heat and power.

Source: Mitchell, 2000.

Wind and Municipal and Industrial Waste (MIW) were initially the biggest potential sources of capacity, so reductions in the level of approval for planning saw both their respective completion rates fall and impacted heavily on overall rates.

Finally, the NFFO proved insufficient to ensuring the UK met its target of bringing 1500MW declared net capacity (DNC) of renewably generated electrical capacity online by 2000. By the time the RO came into practice in 2002, approximately 1000MW DNC was on line in the UK.

Perhaps the most significant barrier to significantly increased renewable energy capacity has been planning regulations. Problems relating to planning permission have particularly impacted on the expansion of wind energy, which, as the most mature of the new RETs could be expected to make up the bulk of new capacity. UK planning regulations – which in theory allow weight to be given to global benefits as well as any local benefits and disbenefits – have generally failed to take climate change issues into account. The system for granting planning permission has been particularly vulnerable to the actions of small anti-wind protest groups, most notably Country Guardians, which has led to a high failure rate for wind farm proposals. The fact that a decision regarding permission to build on a particular site was only made relatively late

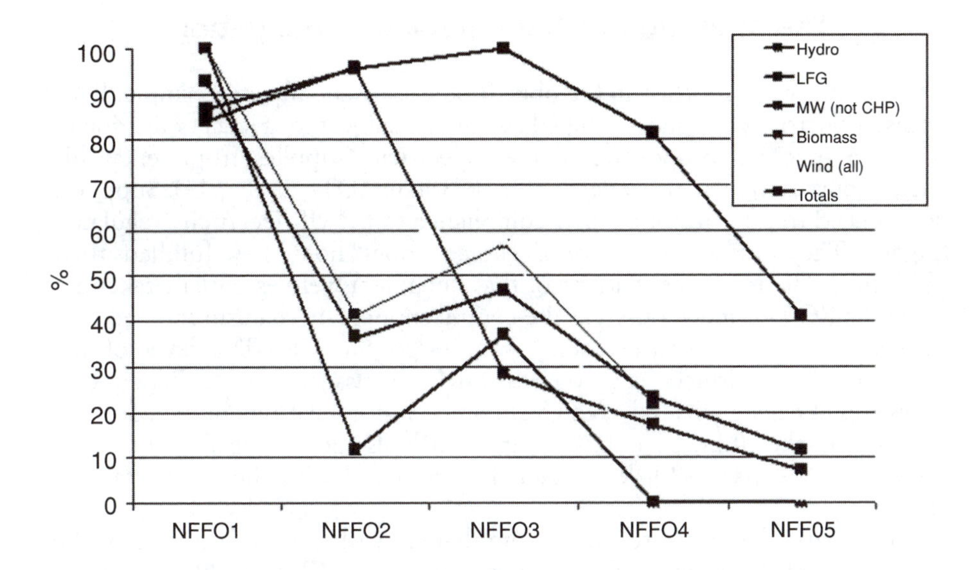

Source: Hartnell, 2003.

Figure 7.1 Overall completion rates for the NFFO

in the development process meant that it was possible for considerable investment to be lost on the part of developers. The government has attempted to address the problem with the introduction in late 2004 of new guidelines concerning the consideration of planning permission for renewable energy generation facilities (ODPM, 2004).

It should be noted that the NFFO operated only in England and Wales. The Scottish Renewables Obligation (SRO) and the NI-NFFO – separate, though similar mechanisms to the NFFO – operated in Scotland and Northern Ireland. The SRO was replaced with the Renewables Obligation (Scotland), which also came into operation on the same date as the RO began in England and Wales, 1 April 2002. The NI-NFFO has also terminated and has yet to be replaced. The provisions of the Utilities Act 2000 do not apply to Northern Ireland, though electricity generated from renewable sources there should be eligible to be sold in Great Britain providing it is produced in line with the regulations governing the RO. Following a consultation by the Department of Enterprise Trade and Investment Northern Ireland (DETI-NI) in 2003 (DETI-NI, 2003) the Government has announced that it intends to introduce the Northern Ireland Renewables Obligation (NIRO) from 1 April 2005. This would require suppliers to source 6.3 per cent of their electricity from renewables by 2012. This would bring Northern Ireland into line with the rest of the UK and potentially lead to the creation of a UK-wide trading market for ROCs (Action Renewables, 2004).

The operation of the renewables obligation

The RO as it was proposed in October 2000 was essentially quite simple. Each licensed electricity supplier in the UK would be subject to 'a legal obligation to supply a specified proportion of their electricity supplies from renewable energy sources to their customers in Great Britain' (DTI, 2000, p13). Suppliers would need to offer proof of their compliance to the UK electricity regulator, Ofgem. They would also have a choice about how they fulfilled their obligation. The recommendations of that document were essentially taken up when the RO became adopted as the central support mechanism for electrical generation from renewable energy sources in the UK. The level of the obligation is confirmed each year through the issuance of a Renewables Obligation Order, which takes the form of a statutory instrument issued under the terms of the Utilities Act (2000), and which also acts to legally enforce the obligation. A separate Order is issued in Scotland with the same practical effects.

Suppliers demonstrate their compliance with the RO through the production of Renewables Obligations Certificates (ROCs). ROCs can be earned through the supply of renewable energy purchased from generators, or alternatively can be purchased independently of their energy use. Suppliers are able to sell any excess ROCs to other suppliers.

The third option for suppliers to fulfil their obligation – and effectively the default option – is payment of the buy-out price. Companies unable, or perhaps unwilling, to find sufficient ROCs to fulfil their quotas are obligated to pay this price for every kWh of renewable energy they fall short of the RO specified fraction of the total energy they supply. The figure adopted for the buy-out price for the first buy-out period 2002–2003 was £30/MWh, and this figure will change annually in line with the retail price index.[3] Suppliers may pass on all costs of complying with the RO to their customers. The buy-out price acts as an upper limit on the overall amount that all consumers will have to pay out to finance the increased use of renewables, and limits the full burden that must be passed on to the consumer. Companies have six months after the end of each obligation period to make good on their obligation.

Those suppliers that fail to demonstrate that they have complied with the RO may be judged not to have met a relevant requirement of their licences. Ofgem may then impose a fine in line with the prevailing policy on such penalties.

As a further stimulus to meeting the obligation, the money collected as a result of suppliers meeting their obligation through payment of the buy-out price is distributed among all of those suppliers who submit ROCs. Suppliers receive a fraction of the total buy-out fund based on the number of ROCs they submit as a proportion of the total number of ROCs submitted. That is, if the total amount of eligible electricity supplied in a period was equivalent to 25TWh, a supplier presenting ROCs representing the supply of 2.5TWh would receive 10 per cent of the buy-out fund ((2.5/25) × 100) (DTI, 2001).

The mechanism allows limited banking of ROCs – that is, where ROCs are held over into the next accounting period – and a supplier may meet up to

25 per cent of their obligation in one period with ROCs issued in the immediately preceding period.

Borrowing against expectations of electrical generation in later obligation periods is not permitted.

The RO statutory consultation document estimates that the total cost to the consumer by 2010/2011 should be in the order of £779 million for that year for an estimated total of 33.6 TWh of renewable electricity in that year. The price rise is equivalent to a rise of 4.4 per cent on 1999 prices (DTI, 2001, p5).

The consultation document also confirmed that the obligation was intended to be in place from October 2001, until March 2026, thus making it one of the longest term commitments to the support of renewable energy devised in any country. The process was set back by one full year, with the RO finally coming into force on 1 April 2002 and is currently planned to run through to 31 March 2027, though it should be noted that this will be subject to its maintaining political support throughout that period.

The RO in practice

Smith and Watson point out that the market for ROCs created by the RO is, in theory, quite simple, as represented in Figure 7.2. The inclusion of a buy-out price acts to limit the operation of the market. If the market for RE is modest (for example Line T_{Mod}), then the price for ROCs will remain low (P_M). If the government acts to raise targets to more ambitious levels (for example Line T_{Amb}) and thus increases demand, the price will climb. As the price climbs above the buy-out price, then suppliers will decline to purchase ROCs.

The RO excludes already established large-scale hydro generation, and without hydro the supply of established UK renewable electricity generators is well below the 3 per cent figure set for the initial stages of the RO. Thus, in the initial years of the levy, demand will strongly exceed supply and the price will be forced upwards – this is within the intended model of the RO. The particular mechanism of the RO with its recycling of the funds paid into the buyout price pool to those who submit ROCs means that if supply can meet only half of demand – which at this stage is not an unreasonable approximation – then suppliers who submit ROCs will receive a return equal to the buyout price per kWh from the fund for each ROC. This figure will of course diminish as the number of ROCs submitted tends towards 100 per cent of those needed to fulfil the annual obligation. This may act to cause fluctuations in production to be magnified in the effect they have on ROC pricing, increasing uncertainty as to generator income and undermining the economics of project development.

Finally, a value attaches to the commodity value of the electricity produced by the generator, whether it is sold with the ROC or separately. The price it realizes remains subject to trading within the UK's New Electricity Trading Arrangements (NETA) as described later in this paper. The variability of the market price for electricity adds another source of uncertainty for trading in renewable electricity.

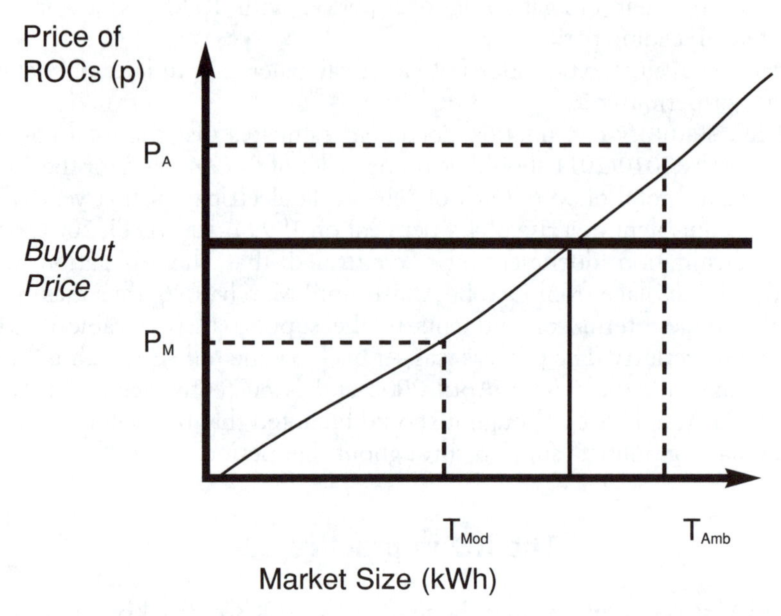

Source: Smith and Watson, 2002b

Figure 7.2 An ideal market in Renewable Obligation Certificates

Morthorst (2000) suggests that in order for a tradeable green certificate mechanism to operate optimally, the realized price must remain below the penalty price. As the realized price approaches the penalty price then the consumer is penalized through higher prices. In the UK system however, the use of recycling of the penalties means that the realized price can move beyond the penalty price, thus effectively placing more of a burden on those consumers who do not manage to obtain sufficient ROCs and forcing them to subsidize those consumers who do have achieve sufficient access to ROCs. The result should be that the overall burden on consumers as a whole is not increased, though there is the potential for it to fall more heavily on particular consumers. In a functioning marketplace, this should theoretically push consumers to those companies which can supply their needs to meet the quotas, and in theory, it would appear that this increased burden on those not meeting quotas should push suppliers to ensure increased access to renewably generated electricity, with the result that the system of recycling should prove to be more of a stimulus than a system where such recycling does not exist.

Demand is further pressured due to the exemption of renewables from the climate change levy of 0.43p/kWh, which applies to most other electrical generation in the UK.[4] The constituents of the total value of renewable generation eligible for RO in early 2003 are shown in Table 7.3.

Table 7.3 *Constituents of total value of renewable generation eligible for RO in early 2004*

	p/kWh	€¢/kWh
Energy	1.0–1.5	1.45–2.175
ROC	3.139	4.55
LEC	0.086	0.125
Recycled Green Premium	2.3	3.335

Notes: LEC = levy exemption certificate (climate change levy)
Conversion rate: 1p = 1.45€¢

While the values available seem reasonable, the generator does not necessarily capture them all. The recycled green premium for example increased in value by 50 per cent from the first to the second RO period. However, only a fraction of this was passed to the generator as typical power purchase agreements (PPAs) tend to leave most of the risk with the consolidator, and thus it is the consolidator who claims the bulk of any additional profit. Likewise, rising electricity prices – as have been seen recently in the UK – do not necessarily see the generator proportionately better off. Typically, generators within the RO receive around 6.0–7.0p/kWh from the combined revenue streams.

As a further price indicator for the total value of ROCs, one useful source is the value achieved by auctions of ROCs earned from contracts awarded under the later tranches of the NFFO, SRO and NI-NFFO. These remain eligible for payment for any electricity they produce. These obligations are fulfilled through a continuing fossil fuel levy on electricity bills, the effect of which is mitigated by the auctioning of the electricity by the Non-Fossil Purchasing Agency (NFPA) on a six-monthly basis; purchases at these auctions are eligible under the RO. The average prices that have been paid in auctions for electricity generated from NFFO contracts are displayed in Table 7.4, below.

Table 7.4 *NFPA auction prices*

Technology	Feb 01	Sep 01	Feb 02	Aug 02	Feb 03	Aug 03	Feb 04	Aug 04
Price	p/kWh	p/kWh	p/kWh	p/kWh	p/kWh	p/kWh	p/kWh	p/kWh
Small hydro	1.84	2.81	6.4	6.69	6.45	6.83	6.32	7.66
Landfill	1.92	2.84	6.74	6.76	6.56	7.13	6.23	7.86
MIW	1.59	2.21	2.27	1.93	1.64	2.42	1.79	3.13
Biomass	1.85	2.61	6.52	5.88	3.68	5.43	4.82	5.79
Wind	1.85	2.84	6.31	6.65	6.41	6.70	6.26	7.31
Average	1.89	2.81	6.44	6.5	6.26	6.81		

Note: Prices are for electrical output, and for appropriate technologies, inclusive of Climate Change Levy Exemption Certificates (LECs) and Renewable Obligation Certificates (ROCs).

Source: NFPA, 2003.

It can be noted from Table 7.4 that the value of the auctioned ROCs increased markedly from 2002, when the RO began. Prior to this date, purchases were generally made by actors keen to be seen to be using energy from renewable sources, and by suppliers using green schemes to meet these needs. It can also be noted that the auction value of ROCs since 2002 now exceeds the typical price agreed for NFFO contracts. The auctions by the NFPA are thus generating a surplus. In early 2003, this surplus totalled around £60 million and this figure has continued to grow. These funds have been allocated to application as capital grants to support increased uptake of offshore wind and biomass technologies.

Whether this value of ROCs rises or falls will be dependent on the rate with which new renewable generation is brought online. A host of complicating economic and social factors will dictate this. One key factor will be how fast the UK system can respond to increased demand for renewably generated electricity. Table 7.1 shows how the obligation will increase annually; if increases in the obligation outstrip the speed with which new plants are brought online it is entirely possible that the price will initially rise. In theory, this should provide further stimulus for new development, though in practice there may be a range of complicating factors. The time taken to advance new projects from conception to generation is variable. New wind turbine capacity, for example, might typically require a minimum of 18 months, though this can be longer depending on problems relating to planning permission, the securing of financing and the availability of wind resource information, among other issues. Furthermore, a high failure rate of projects at the planning stage in the last few years has meant that increases in capacity cannot be predicted with any certainty. The last year has seen a significant increase in approval for wind turbine siting, though there are strong regional fluctuations. Scotland, for example, has seen high rates of approval while Wales has continued to have very low rates. While the application of planning regulation to renewables is a matter of some concern in the UK, no firm conclusion has been reached by the UK government as to how it might be addressed. The issue is currently being addressed with the review of Planning Policy Guidance Note PPG22, which provides guidelines as to how planning authorities should assess the benefits of new renewable energy generation capacity. A consultation document for the revised guidance, PPS 22, was published in November 2003 and the document itself issued in August 2004. It should be noted that Scotland is subject to different planning guidelines; there guidance is provided through National Planning Policy Guideline, NPPG-6 (Scottish Executive, 2000), produced in 2000, alongside Planning Advice Note 45. Both these sets of documents provide a clearer view of the full range of potential benefits, both local and global, of increased renewable capacity that must be taken into account in assessing the worthiness of an RE planning application.

A further issue regarding contracts under the RO concerns how much of the value of an ROC will be passed on to the generator, which seems likely to be the primary factor determining whether new renewable energy projects will be economically viable. The potential for uncertainty in the long-term market seems likely to induce developers to accept lower prices in order to secure longer contracts. If this should occur it seems likely that this will place further pressure on investors, and further reduce the potential for smaller investors to get involved with the development of new capacity.

The completion of the first annual operational period of the RO, 1 April 2002 – 31 March 2003, plus the six-month period which follows to allow companies to demonstrate their compliance with their obligation, 1 April 2003 – 30 September 2003, saw the coming to light of operational problems within the RO.

The first problem related to the operational failure of two supply companies during the course of the first operating period. Both TXU and Maverick moved into administration, leaving both with obligations that they were unable, or more particularly that their administrators were unwilling, to meet. The implication of companies unable to meet their obligations is that an amount lower than expected is entered into the buy-out fund and thus less money passes to those companies that have met their obligations through the submission of ROCs. Effectively this reduces ROC value; Ofgem estimates that the failure of TXU and Maverick led to estimated reductions in payments through the recycled buy-out fund of 0.4p/kWh and 0.1p/kWh, respectively (Santokie, 2004). This has short-term financial implications for these companies and long-term implications for the perceived value of ROCs, which is potentially damaging to the future marketplace.

The end of the six-month compliance period brought to light another problem. Five supply companies failed to comply with their obligations within the period, both reducing the funds in the buy-out fund and breaking the terms of their licences. Despite their willingness to pay, it became apparent that it would not be legal to place their money in the fund – and thus also to distribute it – once the compliance period was over. Ofgem managed to come to an agreement with the companies to pay the funds over and has made amendments to the regulations such that such late payments can be placed into a particular fund and distributed equitably. The Government is currently considering the introduction of a mutualization scheme, which will force all suppliers to pay additional monies into the buy-out fund should individual suppliers default on their payments in any given annual period (DTI, 2004).

The problem of security within the RO

Fundamentally the RO is a market-based instrument; that is, it aims to create a demand in order to stimulate new energy providers with a product specific to meeting that demand. Competition to meet this demand means – at least in theory – that the most efficient projects are successful in seizing market share. In order to maximize the benefits of competition to them, buyers wish to be able to move from one supplier to another at will. If a generator wishes to attract a buyer into a longer contract they must offer a lower price in order to offset any future benefits the buyer may be able to capture by moving to a cheaper supplier. The problem with this is it applies to the creation of a market for renewably generated electricity is that RE generators are essentially long-term investments – capital costs are high and profits tend to accrue only in the long term. RE generators – and by default their investors – are left with two choices. Accept either a low paying contract in the longer term, which may offer insufficient return over too long a period or a higher paying deal in the shorter term, which leaves them without any long-term guaranteed income. The presence of the buy-out price, and of the specific level at which it is set, means that a limit is placed on the point at which companies with obligations to meet decide it is in their interests to opt not to buy ROCs into the future, but instead risk paying the buy-out price should they not be able to meet their future obligations. Because RE is still perceived as a relatively risky investment, it is difficult to secure investment under these conditions. Essentially, the unpredictability of future prices means that it is difficult to secure Power Purchase Agreements, as it is not possible to hedge against them.

Specifically, the situation in the UK is that larger companies in the industry can underwrite their own projects against their own value. However, even for the larger companies, there is a limit to how far this kind of financing can stretch. Smaller companies on the other hand are generally unable to secure sufficient funds to underwrite any significant capacity. The danger is that this may mean there are no investors willing to invest significant amounts in RE in future years. A lack of new investment could effectively bring RE expansion to a halt in the UK. It is likely that new financing paradigms will be required if RE is to flourish, with a particular need apparent at the level of larger projects such as offshore wind.

Naturally this has been the subject of some discussion in the UK. Insecurity relating to investment in the UK market was regarded as being further exacerbated by a number of other factors. The early lack of expansion of the RO target beyond 2010 represented one source of insecurity, and the extension of the RO to 15 per cent by 2015 can be regarded as a response to this. Nevertheless, as time moves on there is increasing pressure to adopt a further target for 2020 in order to add additional investment security. Difficulty with securing finance is regarded by many industry commentators as a key factor in determining the likely success of the RO. Issues which have the potential to undermine prices relating to the RO threaten the availability and cost of capital and threaten to undermine the RO as a whole.

One major source of insecurity with regard to the RO has been the level of tinkering that the RO has been subject to by the UK government. A technical review of the obligation was carried out in 2003, with changes to the regulations concerning co-firing of biomass in fossil fuel stations undergoing alteration from April 2004. These effectively widened the scope for using biofuels in this way, thus increasing the number of ROCs that could be earned and reducing the overall value of ROCs. A further consultation document was published in September 2004 with the idea of introducing some changes in the operation of the RO from April 2005 (DTI, 2004).

The RO is to be subject to a full review in 2005–2006; while this is seen as necessary to ensuring that the RO is operating towards achieving its intended goals, it also implies that it will be subject to change should any problems come to light. Significant changes are expected, which may impact on the price of

ROCs. This potential for change increases regulatory risk, effectively making investment more expensive. It should also be noted that the RO has already been subject to some change.

What the RO excludes

The introduction of the RO, and the particular style that has been adopted for its operation, both act to discount the use of other available options for stimulating the growth of RE capacity in the UK. These options include both the use of entirely alternative mechanisms to the RO, and the use of particular instruments within the adopted RO.

Alternative mechanisms: Renewable Energy Feed-In Tariff

Historically, the most significant mechanism for stimulation of RE has been the Renewable Energy Feed-In Tariff (REFIT). A REFIT mechanism acts to encourage increased capacity through the payment of a fixed sum for any electricity generated from renewable sources. It has historically been the favoured central support mechanism for RE in Denmark and Germany, and its use in practice would appear to indicate that it may be better suited to achieve some policy objectives than other mechanisms, while being weaker at other policy concerns. Notably, the REFIT mechanism does not have a reduction in the costs to the consumer as a priority, an aim that can be regarded as the central aspiration of an RPS mechanism. This seems to be central to its rejection as an option in the UK, where this aim is regarded as fundamental to any energy policy. Indeed, it would appear that a REFIT-type mechanism was not even considered as a potential option. Certainly, no discussion of its use is evident in the many consultation documents that prefaced the final uptake of the RO as the UK mechanism of choice. More of the comparative advantages and disadvantages of the RO and a general REFIT mechanism will be discussed later in this paper.

Alternative options for the operation of the RO

The choices made regarding the operation of the RO are also of some interest in terms of determining the priorities of UK RE policy.

One of the most significant of the choices regarding the operation of the RO is the decision not to use technology banding. Banding would mean the adoption of separate obligations relating to particular technologies. The decision not to employ a banding system is perhaps predictable in terms of overall UK RE policy, given that its exclusion acts to keep the mechanism as 'near-market' as possible, and thus in theory anyway, acts to minimize costs to the consumer. The use of banding as part of the NFFO meant that it could not be ruled out from the RO consultation process though, and a number of organizations that submitted opinions to the RO consultation process favoured the use of banding within the RO.

There are a number of drawbacks to the decision not to use technology bands. Primary among these is that, as the RO currently operates, it effectively cuts off those technologies that are currently outside the mechanism, that is, those that are currently less economic. It is possible that some of these currently less competitive technologies, most notably those on the margins of being economic, may be brought back within the mechanism, dependent on how the rate of expansion of the obligation compares with the rate with which opportunities become available. However, the majority of those technologies initially outside the obligation are unlikely to move within the obligation unless there are considerable reductions in the costs of a technology through the stimulation of its use outside the UK, or if prices are brought down due to the use of another policy instrument. The likeliest course is that as the RO progresses, those technologies initially outside its scope are likely to be pushed further out. This will impact not just on the simple costs of the technology, but also on factors such as the development of support industries and the growth of the national knowledge base in the new technologies. Each of these will impact on the UK's long term abilities to develop both the ability to utilize new RETs, but also to exploit new RETs as internationally competitive industries, supposedly a key aim of UK policy. Technologies unlikely to be stimulated by the RO include photovoltaics (PV), tidal stream and a number of other less mature RETs. The alternative is to employ additional mechanisms that address some of the failings of the RO.

Another instrument that might have been included within the RO is the use of a minimum price, as was to occur in the Danish variation of the RPS mechanism (Morthorst, 2000). This is again to be expected of UK policy, but the use of such a lower limit might have acted to change the way the mechanism is likely to operate in practice. Within the Danish system there was a need to bridge the policy change from a REFIT to an RPS style mechanism and the minimum price might be justified as an attempt to provide increased security for continued expansion of wind generating capacity during this period. This does not rule out that increased security would also be welcome in the UK situation. As with banding though, such a minimum price acts to raise costs and interferes with the operation of the market. The Danish system differs from the UK system in that considerable capacity already existed and that some of this would be more likely to be capable of operating at low cost after years of public subsidy; thus the minimum would have acted to ensure existing capacity enjoys what is regarded as a reasonable return, and that new capacity would not be completely priced out of the market by established capacity. The low level of currently existing capacity in the UK, particularly in comparison with the demand to be created by the RO, means that new capacity is unlikely to be ruled out in this way, and, it should be noted, any minimum price would be likely to remain irrelevant in the UK for some time to come, even were it to be politically acceptable.

This leads to what may prove to be a particular problem with the operation of the RO, most notably in comparison with mechanisms such as the REFIT. The RO effectively compels suppliers to furnish their customers with a minimum fraction of renewables in their electrical supply. The onus is to do this at the minimum cost to these consumers, with simple economic theory dictating that the higher costs rise, the more likely the supplier is to lose consumers to other, more efficient competitors. As already noted, the use of a competitive mechanism reinforces a desire on the part of suppliers to attempt to secure the greatest possible degree of flexibility in securing renewably supplied electricity, on the grounds of being able to move to cheaper opportunities as they arise. The result is reductions in the length of contracts available to RET developers.

Effectively, the RO acts to limit security for those considering investment in new renewable energy projects. This reduction is further compounded by the possibility that the existence of the mechanism itself may not be politically secure, as mentioned above. The likely consequence of this reduction in security of investment is the inhibition of investment by smaller investors, and the limitation of investment to larger investors who are more likely to be capable of utilizing their own capital to finance projects. This has the potential to reduce the total level of investment going into RETs in the UK, an area in which the sector has historically experienced significant problems (Mitchell, 1994). It seems likely that this will have an impact on involvement in small-scale projects – an area that still has the potential to deliver significant capacity increases – in terms of access to financing.

The perceived security of the RO is also undermined by the political nature of the RO mechanism. While any support mechanism has the potential to be removed, it is possible to place obligations within legislation such that payments are effectively guaranteed over an extended period once a project is generating. Withdrawal of the RO at any point however, would completely and immediately remove the market for RE it had established, undermining the sector totally. The only guarantee attached to the RO is that the UK government has stated that it currently has no plans to curtail the mechanism before 2027. Clearly this is subject both to changes in political priority, and most notably, to changes in the administration. There is some potential for repeal of the law, though the effects of this are less than the other regulatory risks associated with the RO so far. This acts to further reduce security in a market where the return on investment (ROI) is concentrated in the long term due to high initial capital costs, and thus where long-term support is necessary to stimulate investment.

Long-term security is further undermined by the government's refusal to commit to a 20 per cent target for renewables by 2020. By adopting the 20 per cent target only as an aspiration as the government did in 2003, the willingness of companies to make long-term commitments is further undermined. The adoption of a 15 per cent by 2015 target in late 2003 goes some way to addressing the problem. However, the government has also committed itself to a review of the RO in 2005–2006 in order to address any problems inherent in the mechanism. However, this in itself would appear to be causing a problem in that it suggests the potential for future change in the mechanism, thus again undermining stability.

Historically, it has been clearly demonstrated that problems with the security of support mechanisms have impacted very negatively on development of increased capacity. Massive drop-offs in domestic sales blighted the eventually aborted change from the REFIT mechanism to an RPS mechanism in Denmark in early 2000, and significant reductions in the German installation rate followed threats to the REFIT system from the German utilities and the European Commission in the mid-1990s, as demonstrated in Figure 7.3. This is further emphasized in the work of Jacobsson and Bergek, where considerable evidence is presented that the key characteristics for any RE policy should be that support is 'powerful, persistent and predictable' (Jacobsson and Bergek, 2002). The RO fails to meet each of their criteria; it is unpredictable due to the risks associated with both the price that generators will be able to get for their production and volume as generators have no guarantee as to how much of their production will actually be purchased. It is less powerful than tariff mechanisms in the sense that it is much easier for generation projects at the margins not to come to fruition than under the certainty of the guaranteed tariff price. Finally, the RO is not persistent due to the variability inherent in much of its income streams.

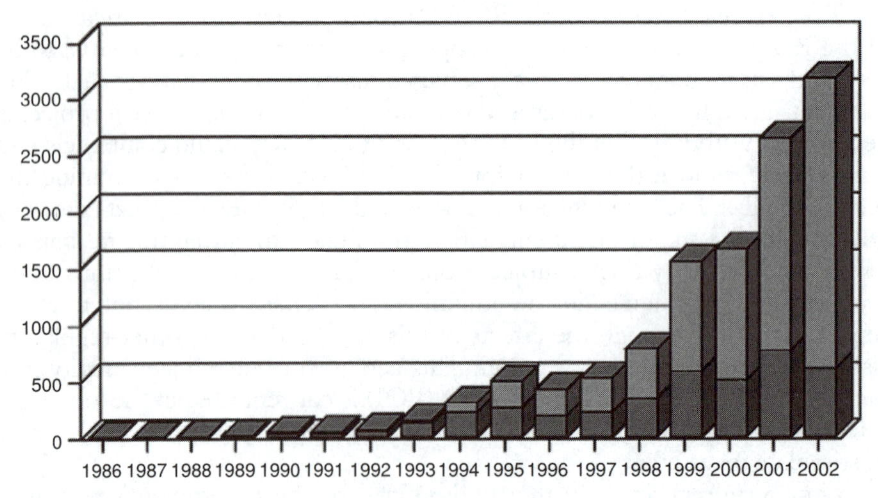

Note: Dark grey represents coastal turbines, grey inland turbines.

Source: IWR, 2003.

Figure 7.3 Annual wind turbine installation in Germany

Further effects of the RO

While some of the advantages and disadvantages of the RO have been raised above, it is worthwhile to look at some other aspects of the policy in order to gain a fuller picture of what it will mean in practice.

One further issue relating to the use of the RO is its impact on green electricity markets. As with a number of European nations and US states, the UK has a burgeoning market for green electricity, that is, a market where consumers volunteer to pay a premium for electricity from certified renewable energy sources. By setting a quota for renewables, the RO mechanism will act to drive up demand for renewably sourced electricity. This demand is not likely to be met very easily by new capacity, particularly in the early years of the mechanism. As described above, the result of this will be that the purchase of electricity from all renewable sources will effectively be increased to account for the 3.0p/kWh buy-out price, plus whatever figure accrues from the return of ROCs, the avoidance of the climate change levy and benefits associated with avoiding buying units from other sources to meet the full demand of their customers. Any voluntary purchasers of green power will have to compete to make their purchases in this market, with the result that the premium charged for such programmes will be considerably increased. There is clear evidence that the number of consumers willing to join green power schemes directly correlates with the size of premiums charged (Holt, 1996; Fouquet, 1998; Madlener and Fouquet, 1999). It seems likely that such voluntary purchasing will be threatened by the RO. However, while it is perhaps worth noting that the end of a voluntary green power market might mean a slight reduction in funds available to finance new RET from that particular source, the application of the obligation to all consumers more than balances this. It is possible to argue that this will deny those that might wish to fulfil more of their electricity needs from renewable sources, but the balance is more RE on stream overall, and a reduction in the amount of free-rider behaviour apparent in the system. Thus the loss is perhaps best viewed as being outweighed by the overall potential benefits for the finance and growth of renewable energy capacity, which is largely the aim of green power schemes. Both of these observations touch on complex ethical arguments however and are somewhat outside the scope of this paper. It is also likely that there will be some exceptions, and that some green power schemes which do not charge a premium to consumers may continue to prosper due to the reduced costs of acquiring consumers of this nature and their increased 'stickiness' rendering them more attractive to supply companies.

Additional UK RE policy

The RO, it is important to note, will not exist in a vacuum, nor will some of the problems it creates, or at least fails to contend with, go unaddressed. During 2001, the UK government committed considerable sums to a number of additional funding mechanisms relating to RE.

Grants

A particular departure for UK policy is the establishment of a grant system to fund development of specific RETs. Areas eligible for funding include offshore wind, energy crops and the use of RE in households and communities. The DTI claims that offshore wind will be supported with £74 million and bioenergy will receive £66 million. The government initially made 18 licences available for offshore wind farms, and grants aimed to support a number of these in becoming established. A second round of offshore licences, aimed at promoting larger sites and establishing up to 7.2 GW of offshore generating capacity were made available in 2004. This round is not supported by grants, and it is possible that this will be a significant stumbling block to development of projects. The availability of a further £40 million in grants for round three of the offshore wind licensing scheme was announced in 2004.[5] The capital grants available for bioenergy are aimed at promoting 'the efficient use of biomass for energy, and in particular the use of energy crops by stimulating the early deployment of biomass fuelled heat and electricity generation projects. It will do this by awarding capital grants towards the cost of equipment in complete working installations' (DTI, 2002). A DTI document published in July 2002 listed the following availability of funding:

- £18 million per annum for the government's enhanced renewable energy research and development programme, available through the DTI;
- £74 million support for offshore wind, with £64 million of this available through the DTI and the rest from the National Lottery's New Opportunities Fund (NOF);
- £30 million from the DTI and £36 million from the NOF for bioenergy capital grants, plus a further £29 million from the Department of Environment, Food and Rural Affairs (DEFRA) in the period up to 2002;
- a further £10 million for innovative PV schemes;
- £10 million for renewable use in the community from the DTI, plus up to £50 million from DEFRA aimed largely at increasing CHP usage;
- a £5 million fund for wave and tidal project demonstration;
- £4 million for utilities to investigate new metering and control technologies (DTI, 2002).

The introduction of a grants system on such a scale is an interesting departure for UK RE policy in that such a policy has historically been rejected in the UK on the grounds that it acted to distort the market to too great an extent. That grants have now been made available would thus seem to indicate both a policy volte-face, and the acknowledgement that perhaps a greater degree of interference with the market might be necessary if the full range of renewable energy policy goals are to be achieved, both with regard to the stimulation of capacity and with regard to the stimulation of the volumes of funding needed to be invested in the industry if it is to become competitive. The use of grants recognizes that some technologies are left stranded by the RO and their use is effectively an attempt to kick-start the use of such technologies, with an aim of bringing down costs, creating institutional pathways and allowing new entrants to the sector to gain experience in the market.

Carbon Trust

A further policy mechanism that may have important implications for renewables is the formation of the Carbon Trust. The Carbon Trust is a not-for-profit company limited by guarantee[6] and established by the UK Government in 2001. It was established, according to Prime Minister Tony Blair, to 'take the lead on low carbon technology and innovation in this country, and put Britain in the lead internationally' (Blair, 2000). The Trust strategy document expands these to three main aims:

1 to ensure that UK business and the public sector contribute fully to meeting ongoing targets for greenhouse gas emissions;
2 to improve the competitiveness of UK business through resource efficiency;
3 to support the development of a UK industry sector that capitalizes on the innovation and commercial value of low-carbon technologies nationally and internationally.

The Trust has three main programmes; the most interesting of these with regard to renewables is the Low Carbon Innovation Programme (LCIP). The LCIP currently has an annual budget of around £20–30 million. Initial proposals suggested the Trust should have a considerably more substantial budget. The budget may yet increase, dependent on the perceived level of success of the Trust. While it is possible that the role to be adopted by the Carbon Trust may be a useful one, the scale of the funding available through the LCIP remains relatively small when compared with the scale of support that has been provided for renewables in other countries. Renewables will not receive the entire LCIP budget, but stand to receive about half of it, in the approximation of the Director of the LCIP programme, suggesting a figure of £10–15 million annually. While this might be a useful stimulus for new technologies in the UK it does not compare favourably with the scale of programmes employed elsewhere to stimulate industrial development of renewable energy technology. The Environment and Energy Saving Programme of the Deutsche AusgleichsBank's European Recovery Programme in Germany, for example, has made available billions of euros in 'soft' loans to support the growth of both capacity and a German manufacturing sector, and is only one ingredient in the German approach to stimulating growth in both renewable energy installation and industrial development.

However, it must be noted that in the first year of the LCIP's operation, 2001–2002, the Trust spent only £6 million of the £30 million it received from the CCL. This can be related to both the length of time it took to establish the programme for applications, and perhaps more worryingly, that it was unable to attract enough bids that it regarded as of sufficient quality to receive assistance through the 'Foundation Programme', which is the initial operating title of the LCIP. The Trust thus returned around £24 million of its CCL funding of £30 million for that year. A reticence to give out funding where a project is insufficiently able to provide assurance of a reasonable return, either fiscally or in the form of a carbon return clearly makes sense in preserving funding and ensuring that funds are not wasted inappropriately. However, continued difficulties in finding appropriate investment opportunities and in failing to dispose of all funds would seem likely to lead to cuts in the available funding, though the timescale in which problems might occur is impossible to estimate.

It is likely that RETs will be a major focus of the LCIP, and it is notable in terms of the impacts of the RO discussed above that the funding will only be made available to those aspects of technologies with no other sources of funding. It should also be noted that it will not provide funding for those technologies which are already at the competitive stage. The Trust will instead make distinctions between the states of advance of technologies, and the likely 'carbon return' that they provide in comparison with the required level of funding. The term 'carbon return' refers to an estimate for the potential of the technology applying for support to act to reduce carbon emissions. The likely reduction is pegged against the required investment in order to achieve these reductions. Technologies will effectively be placed in a banding grid. Technologies will compete for funds against other technologies at the same level of development, that is, technologies at the research level will compete for funds only against other technologies also at the research stage, with the same applying for technologies at the development and demonstration phases. Supporting technologies at the disseminations phase is more controversial, and is inherently more difficult due to the high costs that can be associated with enabling a technology to make the jump to being economically competitive that this phase requires.

Perhaps the most notable feature of the LCIP, and the methodology for the application of its funding, is the use of 'carbon returns'. Since such estimations will be subject to considerable room for subjective judgements, there is considerable scope for funding projects in such a way that UK projects are favoured. This can be regarded as a particular shift in the UK government's approach to the support of new renewable energy technology.

As with the use of grants, the funding of the LCIP demonstrates a break with traditional UK policy relating to the development of new RETs. While still attempting to remain as close to the use of market instruments as possible, the use of the LCIP to stimulate the research, development and demonstration of RETs and other environmental technology perhaps indicates some acceptance of the need to reform the way that these sectors are aided in the UK. Failure to attract private funding to support RD&D has been a long-term problem for industrial development of UK renewable energy technology manufacturing. While the RO will bring in funding to the sector, as with the NFFO, it seems likely that this will be spent on establishing projects for direct returns to the developer, rather than on stimulation of the UK industrial sector. The additional programmes running alongside the RO can thus be regarded as essential to UK RE industrial policy.

The UK market context in which renewables exist

As well as the specific mechanisms that impact on the development of renewables in the UK, consideration must be given to the overall regulatory framework in which these mechanisms exist. These can exert significant influence on the economics of renewables. These issues can effectively be split into two areas with respect to their effects. These are, first, the structure of the marketplace for the trading of electricity, and second, the economic regulatory framework which governs the operation and incentivization of distribution network operators.

The introduction of the New Electricity Trading Arrangements (NETA) in the Utilities Act (2000) brought with it a number of problems pertinent to the economics of renewable electricity generation in the UK. The idea underlying NETA as a trading mechanism is that it should be cost-reflective and promote reliability in power supply. However, its particular set-up acts to place a number of barriers to the greater use of smaller generators; in total these barriers sum to form a significant block to desired RE growth.

NETA is a bilateral trading system, with trades carried out from the long-term – often years ahead – through to on-the-day transactions and as close to real-time as possible through the Balancing Market (BM). National Grid Transco (NGT) is the system operator and is responsible for balancing electricity supply; NETA is administered by NGT's wholly owned subsidiary, Elexon. Governance of Elexon is dictated by the Balancing and Settlement Code (BSC). Both generator and supplier engaging in bilateral trade must be signatories to the BSC.

The BM is informed of any trading one hour before 'real-time'. If the buyer or seller fails to meet their obligations an imbalance occurs. The system operator then has to trade to mitigate for the imbalance. The costs of doing so – the imbalance settlement price – must be paid for by the trader who defaulted on their commitment, for each half-hour.

Small generators are vulnerable to three sets of problem within this system:

1. Disproportionate costs stemming from scale – most notably the costs of BSC membership necessary to trade in the market. The membership fee does not vary with generation capacity and thus tends to impact on smaller companies more significantly than on larger competitors. As RES/DG projects tend to be smaller this has effectively meant that in general they have to sell their electricity to larger traders who can afford membership, thus placing DG/RES operators at a disadvantage
2. Costs specific to smaller generators now – such as consolidation and changes in the gate closure period – that are expected to reduce as NETA settles down following its introduction. A change in the gate closure period – that is the period between notification and despatch – from 3.5 hours to 1 hour has seen a reduction in the costs associated with unpredictable delivery. This change stemmed from an approved modification to NETA. Increased consolidation may allow RE generators to secure better prices in trading

3 Costs relating to lack of proper cost-reflectivity and which are unlikely to change within the current context of NETA.

Effectively, NETA is systematically favourable to larger generators, which tends to mean it is less favourable to RE systems that are typically smaller. NETA is supposed to be based on cost-reflectivity, with any intermittency of generation effectively punished by the generator having to meet the additional costs. The nature of renewables means they sometimes have to meet intermittency costs. There is contention however as to whether NETA is truly cost-reflective for renewables, with a number of reports conflicting with the Ofgem's assertion that it is (ILEX and Strbac, 2002; PIU, 2002). It has been suggested that the cost imposed on smaller generators by NETA is actually greater than the costs of their intermittency to the system.

There is a system for modifying NETA through the agreement of a number of specific stakeholders in order that NETA be allowed to evolve to suit the needs of the market. Proposed modifications are presented to a modifications panel made up of representatives of the regulator, of customers and of industry. After consideration by industry experts in a modification group, the Panel rules on a modification's adoption. The relevant Secretary of State has 14 days to strike down any approved modification. Overseen by Elexon, considerable numbers of modifications have so far been proposed and accepted.[7] This does not mean there are likely to be significant changes to NETA.

Key to understanding the use of NETA is to comprehend that it is purely an instrument aimed at maximizing the efficiency of the market, that is, to drive down wholesale electricity prices as much as possible. It gives no consideration to other UK energy policy goals outside that of ensuring the supply of electricity as cheaply as possible.

The framework of economic regulation in which renewable energy exists

There are a range of problems arising from the way in which regulation of distribution networks impacts on the economic performance of renewable energy, and more generally, of distributed generation (DG). Simply, generators, which are directly connected to the distribution networks, are subject to a set of regulations governing the operation of that network. These are different from those regulations that govern the operation of transmission networks and the generators that connect to them.

Currently, the way that the distribution network operator (DNO) is incentivized in terms of its revenue and thus what directs its behaviour, its investment patterns and the way in which the network develops, tends to favour centralized over distributed generation. Altering the system such that it allows DNOs more flexibly to develop their networks can be seen not as an environmental argument but a purely economic one; the costs and benefits of DG are not properly monetized within the current system employed in the UK. Clearly however, there are environmental implications alongside the economic benefits. The current system has significant implications for the connection and operation of DG/RES on those networks.

There are a number of problems in the UK relating to DNO incentivization. First, it should be noted that future DNO revenue stems from DNO spending on their Regulatory Asset Base (RAB), that is, from the infrastructure spending that they have made and which has been approved by the Regulator. This forms the basis for their allowed rate of return. DNO income is then largely based on the number of customers attached to their network and to the volume of electricity traded over their network. That is, there is a focus on capital rather than operational expenditure. The benchmarking system governing operational expenditure only incentivizes its reduction, not its use for more innovative uses of the network. The result is a lack of innovation in the way that networks are managed; effectively DNOs carry out the same tasks each year and try to improve the efficiency with which they carry them out in order to wring a margin from the gains. The system effectively prevents them from investing in other potentially beneficial avenues by failing to provide any incentives for them to profit from the improved efficiency that these changes might bring. Mitchell and Connor (2002) have made a more detailed analysis of these issues.

One fundamental barrier that negatively affects the economics of DG in the UK is the use of deep connection charges. This means that a new generator connecting to the distribution network has to meet the full costs of extending the network and of strengthening the existing network in order to be connected. This is out of step with the situation in most EU countries, and most notably of those that have enjoyed the most success in promoting renewable energy. The problems such charging leads to are multiple. First, it means that the generator has to meet very large capital costs in what is already a very capital-intensive industry. Second, deep connection also leads to the first-comer problem, wherein essentially, the first developer to build their generator in a location has to meet the full network upgrading costs, but then has no way of recouping the costs from later developers who build in the same area and take advantage of the upgraded network. A third problem with deep connection arises from the fact that all the network investment lies with the generator. As no investment is made by the DNO, nothing is accrued to the DNO's Regulatory Asset Base (RAB), with the effect that the DNO has no interest in the upgrading of the network, as it is not incentivized by any revenue relating to it.

The system means there are no price signals available to indicate better sites, that competition is reduced and that the costs of investment are not transparent to generators. Each of these factors, as well as others, acts as a barrier to DG/RES development. It should also be noted that generators connecting to the transmission grid pay only shallow connection costs, that is, the immediate costs of linking to the network, with the network operator having to meet the majority of the costs of extending the network to meet the generator. The network operator may then add its costs to its resource asset base and makes a profit by charging for the use of the network.

Conclusion

The basis for UK renewable energy policy can perhaps be best summed up with a simple quote made in 2000 by the then Secretary of State for Trade and Industry, Stephen Byers, that the policy of the UK's Department of Trade and Industry with regard to renewable energy should be 'one of action but not of direct intervention'. That is, any interference with the market will occur through as near-market instruments as possible. The adoption of the RO mechanism is a striking example of the application of this general policy to actuality. The doggedness inherent in adhering to this basis for policy would seem to tie the UK into a limited range of choices with regard to its policy, with the almost automatic rejection of choices that do not meet the near-market instrument criteria inherent to UK industrial philosophy. Particularly, this ideology would seem to rule out the application of significant public funds to the particular stimulation of new national industry, as opposed to the stimulation of new markets, as has proved to be fruitful with regard to the state backing of the wind turbine manufacturing industry in Denmark. Thus the UK gives no consideration to the potential return at the national level of an investment in the development of new renewable energy technology, meaning less is given to their establishment than is the case where policy choice is not constricted by such an ideology. Consequently, it is harder for a UK sector to compete against the industries of competitors that have garnered greater support in their own domestic marketplace.

While it is possible that the RO will succeed in achieving some of the goals of UK policy, most notably in increasing renewable energy capacity and reducing greenhouse gas emissions, it also seems likely that it will fail to address others with an equal level of application. However, as distinct from that period in which the NFFO was the central mechanism for the support of RE, the wider range of additional policy instruments being employed alongside the RO may go some way to addressing a fuller range of policy goals.

In the future, those looking back on the performance of the RO may do well first to examine the full range of goals that the RO is meant to achieve. As a choice, the RO is clearly aimed at stimulating RE capacity while minimizing the cost to the consumer – it is entirely possible that it may enjoy some success in achieving this goal. It is worth noting however that present prices in the UK for RES electricity seem to be comparable with those in Germany, but that Germany also benefits from the creation of at least 35,000 jobs in the wind energy sector alone. The instruments existing alongside the RO have more direct effects on stimulating industry, and it is difficult to judge both the effects of these, and thus also the relative efficiency of their use. It will be interesting to note which nations have both the highest costs – and the greatest returns – once currently nascent industries have achieved maturity.

Notes

1 British Energy hit a financial crisis in September 2002 and was bailed out by a loan from government on the grounds of national energy security. The loan appears to have been at rates not especially favourable to the company. Further problems saw an October 2003 rescue package that meant the shareholders exchanged £700 million of new bonds and 97.5 per cent of their equity in exchange for passing on their liabilities of £1.4 billion. Its problems continued through 2004 as it failed to take advantage of rising electricity prices.
2 It should be noted that the terms of the loan could be regarded as being favourable to the government and perhaps less so for the company itself.
3 The figure for the period 2003–04 is £30.51, for 2004–05 it is £31.39.
4 While the exemption is set at this level, the reality is that the advantage is really only worth 20 per cent of this figure (0.086p/kWh), due to industrial exemptions from the full rate.
5 See www.dti.gov.uk/energy/renewables/support/guidance_notes.pdf.
6 A company limited by guarantee is an alternative type of incorporation used primarily for non-profit organizations that require corporate status. A guarantee company does not have a share capital, but has members who are guarantors instead of shareholders. The guarantors give an undertaking to contribute a nominal amount towards the winding up of the company in the event of a shortfall upon cessation of business.
7 A thorough explanation of the modification process can be found at www.elexon.co.uk/ta/modifications/modifications.html.

References

Action Renewables (2004) *A Study into the Renewable Energy Resource in the Six Counties of Northern Ireland*, Belfast, Action Renewables

Blair, T. (2000) *Richer and Greener*, CBI/Green Alliance Conference on the Environment, London, UK Government

Connor, P. (2004) 'Renewable Electricity in the United Kingdom: Developing Policy in an Evolving Electricity Market', in De Lovinfosse, I. and Varone, F. (eds) *Renewable Electricity Policies in Europe: Tradeable Green Certificates in Competitive Markets*, Louvain, Presses Universitaires de Louvain, pp243–300

DETI-NI (2003) *Towards a New Energy Strategy for Northern Ireland*, Belfast, Department of Enterprise, Trade and Investment

DTI (2000) *The Renewables Obligation Preliminary Consultation*, London, HMSO

DTI (2001) *Renewables Obligation Statutory Consultation*, London, DTI

DTI (2002) *Renewables Funding*, Aberdeen, Renewables UK

DTI (2003a) *Government Gives Green Light to Renewable Future*, London, DTI

DTI (2003b) *Our Energy Future – Creating a Low Carbon Economy*, London, TS

DTI (2004) *The Renewables Obligation Order 2005 Statutory Consultation*, London, Department of Trade and Industry

ECOTEC (2002) *Renewable Energy Sector in the EU: its Employment and Export Potential*, Birmingham, UK, ECOTEC/DG Environment

Elliott, D. (1992) 'Renewables and the privatization of the UK ESI: A case study', *Energy Policy*, vol 20, no 3, pp257–268

European Commission (2001) 'EU Renewables Directive', 2001/77/EC

Fouquet, R. (1998) 'The United Kingdom demand for renewable electricity in a liberalised market', *Energy Policy,* vol 26, no 4, pp281–293

Hartnell, G. (2003) *Renewable Energy Development 1990–2003*, London, RPA

Holt, E. A. (1996) *Green Pricing Experience and Lessons Learned*, US DOE Energy Efficiency and Renewable Energy (EERE)

ILEX and Strbac, G. (2002) *Quantifying the System Costs of Additional Renewables in 2020*, Manchester, UMIST/ILEX

IWR (2003) 'Windenergiemarkt 2002 Deutschland', Münster, Germany, International Economic Forum for Renewable Energies, www.iwr.de/wind/markt/iwrwind markt2002.pdf, accessed in January 2005

Jacobsson, S. and Bergek, A. (2002) 'Transforming the energy sector: The evolution of technological systems in renewable energy technology', Conference on the Human Dimensions of Global Environmental Change, Berlin

Madlener, R. and Fouquet, R. (1999) 'Markets for Tradeable Renewable Electricity Certificates: Dutch Experience and British Prospects', 3rd Energy Economics Conference, St John's College, Oxford, UK, BIEE

Mitchell, C. (1994) *The Renewable Energy NFFO – A Case Study of the Barriers to Energy Technology Development*, Science Policy Research Unit, Brighton, Sussex University

Mitchell, C. (1995) 'The renewables NFFO: A review', *Energy Policy*, vol 23, no 12, 1077–1091

Mitchell, C. (2000) 'The England and Wales non-fossil fuel obligation: History and lessons', *Annual Review of Energy and the Environment*, vol 25, pp285–312

Mitchell, C. and Connor, P. (2002) *Review of Current Electricity Policy and Regulation – UK Case Study*, Coventry, CMuR

Mitchell, C., Bauknecht, D. and Connor, P. M. (2004) 'Quotas versus subsidies – Risk reduction, efficiency and effectiveness – A comparison of the renewable obligation and the German feed-in law', *Energy Policy*, in press

Morthorst, P. E. (2000) 'The development of a green certificate market', *Energy Policy* vol 28, no 15, pp1085–1094

NFPA (2003) Press Release, NFPA

ODPM (2004) *Planning Policy Statement 22: Renewable Energy*, London, Office of the Deputy Prime Minister

Ofgem (2000) 'An overview of the new electricity trading arrangements', 31 May, OFGEM/DTI, www.ofgem.gov.uk/elarch/retadocs/Overview_NETA_V1_0.pdf, accessed in January 2005

Ofgem (2001a) 'The new electricity trading arrangements: A review of the first three months', London, Ofgem, www.ofgem.gov.uk/temp/ofgem/cache/cmsattach/377_31aug01.pdf, accessed in January 2005

Ofgem (2001b) 'Report to the DTI on the review of the initial impact of NETA on smaller generators', www.ofgem.gov.uk/temp/ofgem/cache/cmsattach/103_31aug01.pdf, accessed in January 2005

PIU (2002) *The Energy Review*, London, Performance Innovation Unit

Rader, N. A. and Norgaard, R. B. (1996) 'Efficiency and sustainability in restructured electricity markets: The renewables portfolio standard', *The Electricity Journal*, vol 9, no 6, pp37–49

Santokie, F. (2004) 'ROCs under review', *Environmental Finance*, September 2004, pp20–21

Scottish Executive (2000) 'NPPG6: renewable energy developments', National Planning Policy Guideline-6, www.scotland.gov.uk/library3/planning/nppg/nppg6.pdf, accessed in January 2005

Smith, A. and Watson, J. (2002a) *The Renewables Obligation: Can it Deliver?* Brighton, SPRU, University of Sussex

Smith, A. and Watson, J. (2002b) *The Challenge for Tradeable Green Certificates in the UK*, Successfully Promoting Renewable Energy Sources in Europe, 6–7 June 2002, Budapest, Hungary

8

The Design and Impacts of the Texas Renewables Portfolio Standard

Ole Langniss and Ryan Wiser

Introduction

The renewables portfolio standard (RPS) – a policy instrument that ensures that a minimum amount of renewable energy is included in the portfolio of electricity resources – has become increasingly popular in energy policy and research circles worldwide. The concept of an RPS is deceptively simple: it is a requirement for retail electricity suppliers (or, alternatively, electricity generators or consumers) to source a minimum percentage of their electricity needs from eligible renewable resources. To add flexibility and reduce the cost of meeting the requirement, tradeable renewable energy certificates (REC) can be used to track and verify compliance.

The RPS has been recognized by some as perhaps the ideal way to encourage renewable energy development in competitive markets: the RPS aims to ensure that renewable energy targets are met at least cost and with a minimum of ongoing administrative involvement by the government (Rader and Norgaard, 1996; Haddad and Jefferiss, 1999; Berry and Jaccard, 2001; Morthorst, 2000). Detailed recommendations for the proper design of an RPS have been provided (Rader and Hempling, 2001; Timpe et al, 2001; Mitchell and Anderson, 2000; Schaeffer et al, 2000; Wiser and Hamrin, 2000; Schaeffer and Sonnemans, 2000; Espey, 2001; Lorenzoni, 2003; Langniss, 2003; Wiser et al, 2004). Others have sought to project the costs and impacts of RPS requirements.

Most of these recommendations and cost estimates have had to rely on theoretical principles, however, as practical experience in the application of the RPS has been limited. RPS policies have been established by legislation or regulation in the countries of Australia, Austria, Belgium, Italy, Sweden and the United Kingdom (UK), but experience with the actual operation of the policy has only just begun.[1] RPS policies, and related mandates, have recently become the most popular form of support for the commercial application of renewable energy technologies in the United States (US). As of December 2004, 18 states had developed renewable energy portfolio standards or mandates, covering over 40 per cent of total US customer load.[2] Figure 8.1 identifies the states in which RPS policies have been established, as well as their terminal renewable energy purchase requirements. The establishment of a national RPS has also been discussed in the US, but has so far failed to gain the critical support needed in the US Congress.

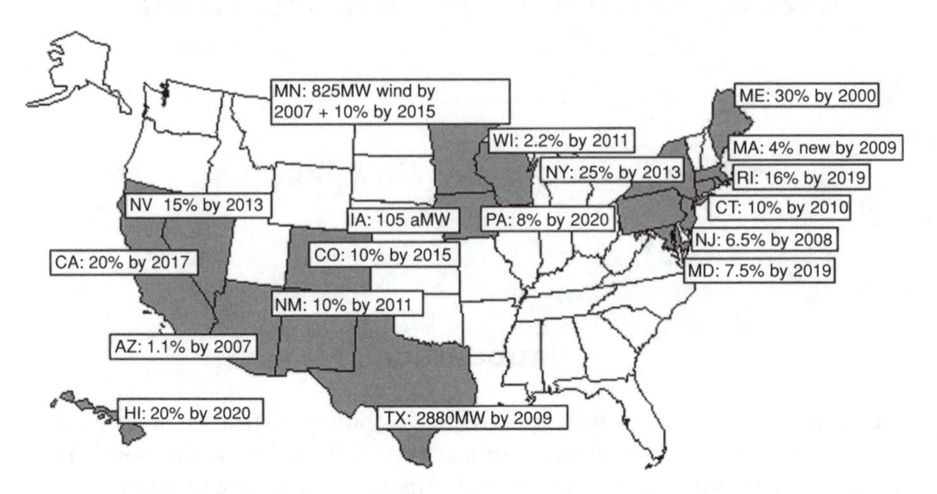

Figure 8.1 Existing state RPS policies (status December 2004)

Many of these state RPS policies were developed as a component of electricity reform. In more recent years, however, RPS policies have increasingly been developed on a stand-alone fashion. Figure 8.2 identifies when state RPS policies have been established (or substantially revised), and illustrates the fact that these policies are a relatively new occurrence in the US.

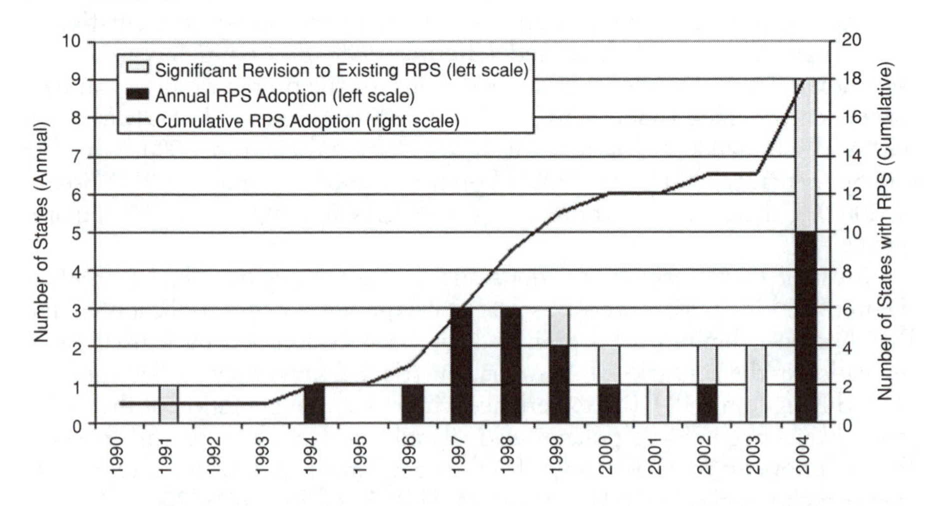

Figure 8.2 The date of adoption of, or major revision to, state RPS policies

It is frequently said that states serve as 'laboratories' for policy experimentation, and this is certainly the case with RPS programmes, where no two states have designed their policies in the same way. Nonetheless, though experience remains somewhat limited and the design of these policies is diverse, the effect of state RPS policies is beginning to be felt. Approximately 2300MW of new renewable energy capacity had been developed through 2003 as a result of renewable portfolio standards or related mandates. The Union of Concerned Scientists projects that these existing state policies could help support approximately 25,000MW of new renewable generation by 2017. Though this represents just 2.5 per cent of total US electricity demand, it is significant growth by recent standards in the US.

RPS policies in the US have begun to have positive effects, but replacing existing renewable energy policies with an as yet somewhat untested approach in the RPS is a risky business. Some countries – including Germany, Spain and Denmark – have had particular success in driving clean energy development with attractive 'feed-in' tariffs. And experience in several US states shows that a poorly designed RPS does little to increase renewable generation (Rader, 2000).

Despite this risk, several US states – most notably Texas – have begun to demonstrate that a well-crafted and implemented RPS can deliver on its promise of strong and cost-effective support for renewable energy with a minimum of ongoing administrative intervention by the government. This chapter describes, in some detail, the successful efforts of Texas in designing and implementing an RPS policy. Though experience even in Texas is still limited, and some challenges have been faced, the Texas RPS has already fostered substantial renewable energy development, surpassing the achievements of any other RPS developed to date.

The anatomy of the Texas RPS

In 1999, the Texas government – under then governor George W. Bush – established an RPS within the restructuring of the state's electricity market (Table 8.1).[3] Detailed RPS regulations were subsequently established by the Texas Public Utilities Commission.[4] The RPS is intended to encourage the development of new, environmentally beneficial resources and thereby reduce the environmental impacts of power production, and contribute to the development of rural areas by creating new renewable energy business opportunities. Resistance towards the RPS was significant among some sectors – large industrial customers especially. Helping to overcome this resistance was the fact that the RPS was only a small part of the overall restructuring legislation in which it was embedded, that the renewable and environmental advocacy communities argued forcefully and collaboratively for the RPS, and that public surveys showed overwhelming support for renewable energy.

The Texas RPS requires the installation of 2000MW of new renewable capacity by the year 2009, in addition to preserving the 880MW of renewable energy already on line.[5] This translates to approximately 3 per cent of present

Table 8.1 *Texas details and RPS timeline*

STATE POPULATION	
20 million (1999)	
ANNUAL RETAIL ELECTRICITY SALES	
318 million MWh (2000)	
FUEL MIX	
39% coal, 49% natural gas, 11% nuclear, 1% renewable	
TIMELINE	
RPS legislation	May 1999
RPS rulemaking begins	June 1999
RPS rulemaking ends	December 1999
REC system established	July 2001
RPS begins	January 2002
RPS ends	December 2019

electricity consumption.[6] This goal is modest relative to the enormous potential for renewable energy development in Texas and the requirements of a truly 'sustainable' electricity supply.[7] Nonetheless, it represents a marked increase in renewable energy capacity in the state. It also represented one of the most ambitious state renewable energy policies in the US at the time of establishment, at least in terms of capacity additions.

Intermediate new renewable capacity goals in Texas are 400MW by 2003, 850MW by 2005, 1400MW by 2007 and finally 2000MW by 2009. These capacity goals are translated into megawatt-hour-based energy requirements by using an average capacity factor of all eligible renewable plants; its value was initially set at 35 per cent and is adjusted over time based on actual plant performance.[8]

Electricity retailers that serve markets open to competition are obliged to fulfil their portion (based on yearly retail electricity sales) of the renewable energy requirement by presenting RECs to the regulating authority on an annual basis. The obligation began in 2002 and ends in 2019. The tradeable RECs are issued for each MWh of eligible renewable generation located within or delivered to the Texas grid. With the exception of renewable power plants with a capacity smaller than 2MW, which are eligible irrespective of their vintage, the REC trading programme is restricted to facilities erected after 1 September 1999. A wide variety of renewable technologies are eligible. Table 8.2 summarizes the design features of the policy.

Early achievements: The Texas wind rush

Renewable energy development

Though RPS obligations did not begin until 2002, the announcement of the RPS in 1999 and the subsequent completion of implementing regulations propelled Texas to become one of the largest renewable energy markets in the

Table 8.2 *The Texas RPS: Design details*

Design element	Design details
Renewable energy purchase obligations	capacity targets of 400MW of eligible new renewables by 2003, 850MW by 2005, 1400MW by 2007 and 2000MW by 2009 and through 2019
	annual energy-based purchase obligations beginning in 2002 and ending in 2019 derived based on capacity targets and average capacity factor of renewable generation (initially set at 35%)
Obliged parties	all electricity retailers in competitive markets (80% of total Texas load) share the obligation based on their proportionate yearly electricity sales; publicly-owned utilities must only meet the RPS if they opt-in to competition
Eligible renewable energy sources	new renewable power plants commissioned after 1 September 1999 and all renewable plants less than 2MW capacity, regardless of date of installation
	power production from solar, wind, geothermal, hydro, wave, tidal, biomass, biomass-based waste products and landfill gas are eligible
	purchases of renewable energy from plants larger than 2MW and built before September 1999 may count towards a supplier's REC obligation, but are not tradeable
	power must be located within or delivered to the Texas grid
	renewable energy sources that offset (but do not produce) electricity (for example solar hot water, geothermal heat pumps), and off-grid and customer sited-projects (for example solar) are also eligible
Tracking and accounting method	tradeable renewable energy certificates with yearly compliance period
	3-month grace period after compliance period allowed for fulfilment
Certificates	issued on production, unit 1MWh, 2 years of banking allowed after year of issuance, borrowing of up to 5% of the obligation in first two compliance periods allowed, development of web-based certificates tracking system[a]
Regulatory bodies	Texas Public Utilities Commission establishes RPS rules and enforces compliance; ERCOT Independent System Operator serves as REC trading administrator
Enforcement penalties	the lesser of 5(US)¢ or 200% of mean REC trade value in compliance period for each missing KWh

Note: a Some countries, notably Denmark, have considered establishing a price floor for RECs. No US RPS has included this design feature.

US in 2001, with approximately 900MW of wind power plants erected. Table 8.3 depicts the development of new renewable energy capacity registered under the RPS scheme. An additional driver for the fast development in 2001 came from the federal production tax credit (PTC), which was then slated to expire at the end of the year. The voluntary green power market has also been significant in Texas, and some of the new capacity that has come online has been used to serve that demand (as a result, it would be a mistake to attribute all of the growth in capacity solely to the RPS). As shown in Table 8.1, the initial rapid pace of development has slowed considerably since 2001, in part because the RPS targets for 2003 (400MW) and 2005 (850MW) have already been met.[9]

Table 8.3 *Status of new renewable energy capacity under the Texas RPS*

	2001 (MW)	2002 (MW)	2003 (MW)
Wind generation capacity	862	942	1140
Hydro generation capacity	10	10	10
Biomass generation capacity	0	5	5
Landfill gas generation capacity	0	31	31
Solar generation capacity	0	0.17	0.17
Total REC generation capacity	**872**	**989**	**1186**

Source: ERCOT, 2004

Technology selection and cost reductions

Wind power projects are the most competitive of all RPS-eligible renewable energy technologies in Texas at the moment, as untapped landfill gas resource opportunities are limited and hydro resources are nearly fully exploited. Solar generation as well as traditional forms of biomass energy are too costly in Texas to compete with wind power at this time. Most of the new and planned wind power plants are located in West Texas, where average annual wind speeds of 8 m/s are common and capacity factors can exceed 35 per cent. The sizeable purchase obligation under the RPS also allows wind projects to gain the economies of scale necessary for deep cost reductions. Combine this factor with the outstanding wind power resource and with the federal ten-year 1.8(US)cent/kWh PTC, and wind power projects in Texas are able to deliver power to the grid for less than 3(US)¢/kWh.[10]

That the initial RPS targets have been exceeded may therefore come as little surprise: wind power in Texas, with the PTC, is able to compete on purely economic grounds against new natural gas facilities, even with relatively low natural gas prices. This is even truer when the volatility and recent escalation of natural gas prices are considered (Bolinger et al, 2004). With early over-compliance with the purchase standard and compliance costs that are at low levels given the competitive pricing offered by renewable generators, there are beginning to be calls for increasing the policy's renewable electric capacity goals.

Long-term contracting

An equally important achievement under the Texas RPS is that the larger obligated electricity suppliers have been willing to sign long-term (10–25 year) contracts for RECs and the associated electricity, with smaller and less creditworthy electricity suppliers purchasing RECs on a short-term basis from those larger suppliers. Without long-term contracts, renewable energy developers are faced with the unenviable position of developing merchant renewable energy projects with highly uncertain returns (Wiser and Pickle, 1997; Helby, 1997; Langniss, 1999 and 2003). Similarly, electricity retailers risk not being able to procure the requisite number of RECs by year's end or only being able to procure credits at astronomical prices due to supply constraints or market manipulation.

Long-term contracts, on the other hand, ensure developers a stable revenue stream and access to low-cost financing, while delivering to electricity retailers a reliable stream of renewable electricity at stable prices. In fact, though renewable developers have been able to choose between REC-only sales and sales that combine the RECs and electricity, virtually all of the long-term contracts to date have covered both the certificates and the electricity. This clearly demonstrates the importance of reducing revenue risk on the part of developers. Large retail electricity suppliers have also had a strong incentive to bring renewable energy projects on line quickly under long-term contracts and with locked-in prices: with the PTC for wind power initially slated to expire at the end of 2001, it was feared that REC prices would rise in the future.

A final component of the long-term contracting process in Texas deserves mention. To shield risk on the retail suppliers' end that REC costs will increase and to reduce the risk that the supplier will fail to comply with the RPS, contract terms strongly penalize project construction lags and operational problems. This can be clearly seen in Table 8.4, where we list the standard contract provisions for two utilities as expressed through RFP documents.[11] Unlike competitive bidding situations in the UK under the Non-Fossil Fuel Obligation (NFFO) and in California under its system-benefits charge policy (Mitchell, 2000; Bolinger et al, 2001), there is little incentive in Texas for developers to propose projects that do not have a high probability of completion.[12] In fact, such bidders will either be unsuccessful in garnering a contract or could face severe penalties if they were able to secure a contract but were subsequently unable to deliver under that contract. This may be an important advantage to the RPS approach.

With renewable electricity prices hovering around or below 3(US)¢/kWh and numerous closely matched projects vying under each competitive solicitation, competition for cost-competitive renewable energy supply in Texas is working.[13]

Certificates tracking system

A final milestone of achievement in Texas has been the development of a web-based platform for the administration of the REC programme. This platform – which allows for the issuance, registration, trade and retirement of RECs – was

Table 8.4 *Elements of typical renewable energy contracts*

Proposed Provisions	TXU	SPS
Requested product	RECs or RECs and associated energy	RECs or RECs and associated energy
Quantity	Approx. 500,000MWh/year total; 1000MWh/year minimum quantity of individual proposals to minimize administrative burdens	Approx. 123,560MWh/year total; no minimum quantity of individual proposals
Term	10 years;[a] start date must be before 2002	15 years; start date must be before 2002
Options for term extension	Buyer may opt twice for four additional years	None[b]
After termination	Option to purchase facility at fair market value	No provisions
Annual amount	Fixed over the contract term; must sell all electric production including excess amount to buyer (if bid for RECs and associated energy)	Fixed over the contract term; must sell all electric production including excess amount to buyer (if bid for RECs and associated energy)
Contract purchase price	One price for the entire term; price may vary for each option period	Fixed by contract for every year
Definition of excess amount	> 105% of contracted amount	> 110% of contracted amount
Purchase price for excess amount	50% of the usual contract price	50% of the usual contract price
Penalty for under-performance	5(US)¢/kWh payment for consistent production less than annual amount	5(US)¢/kWh payment for consistent production less than annual amount
Security required once a project is shortlisted for contract consideration	Irrevocable letter of credit or comparable for 2 years, 0.5(US)¢/kWh based on yearly production	Irrevocable letter of credit or comparable for 2 years, 0.5(US)¢/kWh based on yearly production
Security required once a purchase contract has been finalized	0.5(US)¢/kWh based on yearly production to cover under-performance penalties, and so forth	5(US)¢/kWh based on yearly production to cover under-performance penalties, and so forth
Construction requirements	Projects only selected if have demonstrated business and technical expertise to deliver on time and within contract requirements	Projects only selected if high probability of timely construction; monthly progress reports; penalties for not meeting construction milestones
Operation requirements	Adequate staff for operation required; transmission and ancillary services handled by buyer if RECs and associated energy; timely maintenance and status updates	Joint development of operating procedures; timely maintenance and status updates; minimum performance requirement (> 90% availability)

Note: a Terms as short as five years appeared to be allowed in initial documentation, later to be replaced with a ten-year term. b The possibility of a three-year extension was included in the RFP, but later abandoned in the model contract.

Source: Public requests for proposal documents from two Texas utilities, TXU and SPS. We note that these are proposed contract requirements. Actual contracts may differ somewhat.

established in May 2001. The platform facilitates the tracking of RPS compliance, but does not provide the 'market making' function of a certificate exchange, as this function is left to the private marketplace, as are REC brokering and financial markets.

As noted earlier, a substantial amount of the RECs in Texas are initially bundled in long-term 'electricity plus certificates' forward contracts through bilateral trades. These RECs, however, are then sometimes sold through brokers to small electricity suppliers on an unbundled basis. In fact, a large number of REC brokers are active in the Texas market. Prices for brokered REC-only transactions started at well under 1(US)¢/kWh in the initial phase of the RPS, but have since risen and stabilized at levels between 1.2 and 1.5 (US)¢/KWh (Evolutions Market Data, Figure 8.3). These prices are supported not only by the RPS, but also by active voluntary green power markets in Texas.

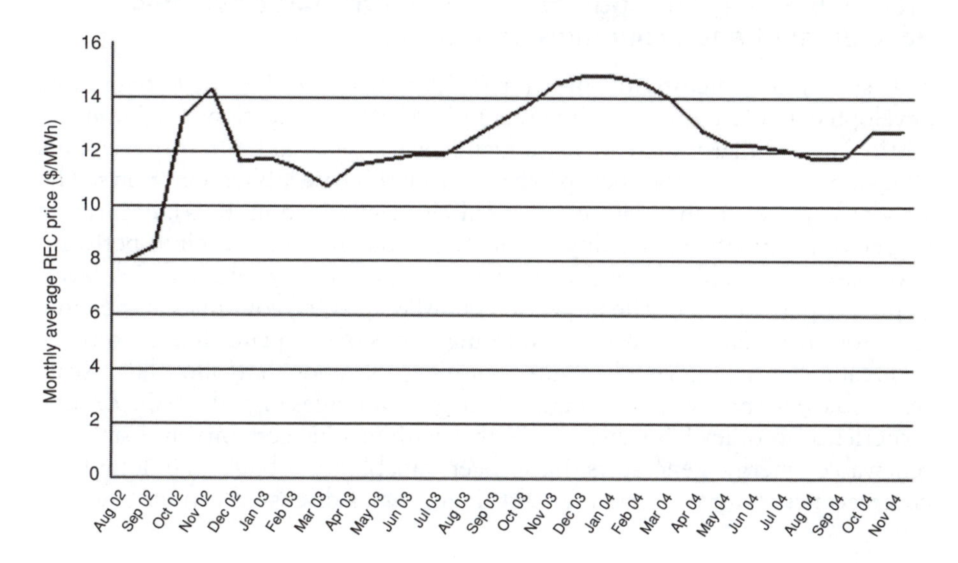

Source: Evolution markets.

Figure 8.3 Monthly average REC prices for brokered REC-only trades

Success factors: The devil is in the details!

Though there are numerous ways of effectively structuring an RPS, certain fundamental policy design principles must be followed if an RPS is to function at low cost and with maximum impact. Of particular importance is that the RPS must provide sufficient confidence to renewable energy developers and retail electricity suppliers to ensure long-term, least cost investment in renewable energy facilities. A number of other state RPSs have failed or appear likely to fail in this respect (Wiser et al, 2004). The early successes of the Texas RPS, on the other hand, can be largely attributed to several positive design and implementation features of the policy.

Strong political support and regulatory commitment

Strong legislative support for the Texas RPS and a committed Public Utilities Commission charged with implementing the RPS ensured that the policy's design details were carefully crafted.[14] Such strong support and commitment have not been evident in several other US states' RPS policies, where implementation details are often poorly designed and languish in uncertainty.

Predictable long-term purchase obligations that drive new development and economies of scale

The size and structure of the Texas RPS ensured that new renewable development was required to meet suppliers' REC obligations beginning in 2002. The standard increases gradually over time, and offers developers adequate time to develop their projects before the REC obligation grows. The standard applies to the majority of retail electricity load in Texas, ensuring a degree of competitive neutrality. Capacity targets are translated into performance-based renewable electricity purchase obligations to encourage high levels of project performance. The target, at 2000MW in 2009, continues at the same level for an additional ten years, offering projects adequate time to recover their capital costs. Intermediate targets are sizeable enough to allow large-scale renewable energy development and, through economies of scale, reduce costs. Experience in other US states shows that without this certainty and stability, renewable energy generators have been unable to obtain the long-term contracts necessary to access low-cost financing, as discussed further below.

Credible and automatic enforcement

Without adequate enforcement, electricity suppliers will surely fail to comply with the RPS. In this environment, renewable energy developers will have little incentive to build renewable energy plants. The enforcement rules of a number of US RPS policies are vague in their application, and in some cases adequate enforcement is clearly lacking. In Texas, however, retail electricity suppliers that fail to meet their RPS obligations are faced with sure and strong penalties: the penalty for non-compliance is set to the lesser of 5¢(US) per missing kWh or 200 per cent of the mean trade value of certificates in the compliance period. It does not pay to delay compliance, and retail suppliers have sought to ensure their ability to comply by inserting penalty provision in their renewable energy contracts so projects come online on schedule and operate within specifications. There are only several instances of supplier non-compliance, most of which are rather minor. The strong political commitment to the policy and an effective enforcement mechanism also provide the support necessary to support low-cost, long-term contracting. While the 5(US)¢/kWh penalty also acts as a cost cap to the policy, there is no evidence that this cap will be reached, with RECs currently trading far below this value.

Flexibility mechanisms

Though enforcement of non-compliance will be swift and sure, adequate flexibility is built into the policy to ensure that suppliers have every opportunity to meet their obligations in a cost-effective fashion. A yearly compliance period, a three-month 'true up' period, REC banking for two years after the year of issuance, a six-month early compliance period in 2001,[15] and allowance for limited REC borrowing in the initial years of the policy all offer the necessary flexibility. Given the degree of over-compliance that has occurred in the initial years of the Texas RPS, it appears as if REC banking in particular has been commonplace.

Certificate trading

Though certificate trading may not be absolutely essential for the effective design of a state RPS, a REC system eases compliance demonstration and tracking, improves liquidity in the market, provides additional flexibility to suppliers, and lowers the overall cost of policy compliance. The smaller and less creditworthy retail suppliers are absolutely dependent on being able to purchase unbundled RECs in the Texas market. The Texas RPS, meanwhile, features the first such REC tracking system in operation in the US (several other tracking systems have since come into operation, or are in development).

Favourable transmission rules and siting processes

Though the RPS is the principal driver in the growth of the Texas renewable energy market, other features of the Texas market facilitate RPS compliance at low cost and with limited hurdles. First, with a world-class wind resource and limited wind power siting constraints, wind projects can be built in large increments, capturing cost reductions due to economies of scale. Second, the state has established favourable transmission planning and costing approaches that benefit renewable generation. As in many European countries (but unlike other US states), grid expansion costs are paid by Texas electricity customers rather than by the power plant operator. Moreover, fees to recover the embedded costs of existing and new transmission infrastructure are placed on electricity consumers based on a flat fee, or postage stamp approach independent of the location of production or consumption (congestion costs will also be charged in addition to the flat fee). Scheduling rules and requirements for intermittent generation are also relatively favourable.

In part because of these policies, however, severe transmission capacity limits have initially constrained wind power generation in West Texas (ERCOT, 2003a). With the present regulatory situation, few incentives are provided to power plant operators to consider grid limits in their siting decisions. As a result, wind power generators were compensated with some $9 million for curtailments in 2002 funded through general transmission charges, and much effort and expense is being put into the strengthening of the transmission grid (ERCOT, 2003b). These added costs of the RPS policy have not been insignificant.

Production tax credit

Finally, the federal PTC for wind projects also significantly reduces RPS compliance costs. Moreover, the fact that the PTC was originally only available for plants erected before the end of 2001,[16] and that it was feared REC prices would increase in the future if the PTC was not extended, provided every incentive for early RPS compliance and long-term contracting between retail electricity suppliers and renewable energy projects.

The need for long-term contracts: Results from other US states

One of the most enduring lessons from US RPS experience is simply that renewable energy projects generally require long-term sales contracts to obtain financing and to deliver energy at reasonable cost. The provision of such long-term contracts in Texas and several other states is a key reason for their successful experience with RPS policies.

Other states have not been so fortunate. Especially in restructured electricity markets, where load obligations are uncertain and purchasers often have poor credit, retail suppliers have often been reluctant or unable to enter into longer term contractual arrangements for renewable electricity or unbundled RECs (for example Massachusetts, Connecticut). In these instances, the RPS is driven by short-term REC trade, often causing REC prices to rise considerably, and it remains to be seen whether such short-term trades can adequately support new project development. In many ways, this experience is similar to that with the UK's renewable obligation.

In recognition of this fundamental difficulty in achieving success with an RPS policy, several states have devised policies and programmes (from simple to elaborate) to overcome financing and contracting hurdles. In Connecticut, for example, a revised RPS was established that requires the electric utilities in that state to purchase at least 100MW of renewable electricity under long-term contract. Similarly, in several additional states including Nevada, New Mexico and California, the RPS requires that suppliers purchase all of their renewable electricity under long-term contract. Nevada also established a temporary electric rate surcharge dedicated to fulfilling renewable energy purchase requirements, to help mitigate concerns over utility credit. In Massachusetts, meanwhile, the state's renewable energy fund opted to offer long-term REC price insurance to help facilitate project financing. Finally, in New York, policy-makers decided to use a government-administered central procurement mechanism, at least initially, in part to overcome concerns with the ability and willingness of the private sector to enter into long-term contractual arrangements with renewable developers.

Whether and under what conditions these strategies will work effectively remains to be seen, but one thing is clear: a plain vanilla RPS cannot be assured to work at minimum cost and with maximum effectiveness without addressing the contracting and financing challenges faced by renewable developers.

Conclusions

Though the RPS has been hailed as a leading, 'market-based' approach to supporting renewable generation – and several countries have opted to replace traditional policy mechanisms with this new approach – experience is only now being gained on RPS implementation. What is becoming clear is that, like any renewable energy policy, an RPS can be designed well or it can be designed poorly. Experience in several US states and European countries shows that inadequate purchase obligations, overly broad renewable energy eligibility guidelines, unclear regulatory rules, insufficient enforcement and wavering political support can all doom an RPS to certain failure.

And yet the Texas policy shows that an RPS, if properly designed and carefully implemented, can deliver on its promise of offering a low-cost, flexible and effective support mechanism for renewable energy. The Texas wind rush drove half of all wind development in the US in 2001, and moderate growth has continued since that time, slowed primarily by the absence of more aggressive renewable energy targets. In this respect, Texas' renewable energy targets are arguably not ambitious enough to foster a lasting deployment of renewable energy technologies.

To be sure, this wind power boom is not solely an outgrowth of an effective RPS policy. A developing customer-driven market for green power and the wind power plans of electricity utilities not subject to RPS requirements have also driven some of the development. The federal PTC for wind, favourable transmission rules and an outstanding wind resource have additionally played important roles. Such complementary policy and market mechanisms are nearly always essential for effective renewable energy deployment. In fact, it should be re-emphasized that the Texas RPS is largely supporting the development of the lowest cost renewable energy technology – wind power. Other US states have developed additional policies to ensure a diversity of renewable energy supply options.

Nonetheless, it can be said with near certainty that, given previous development plans, the major driver in the resurgence of wind energy development in Texas has been the state's aggressive RPS. Other countries and US states would be well served to study carefully the successful efforts of RPS design in Texas.

Perhaps the most intriguing element of the Texas RPS is that it obliged electricity suppliers to deal with wind power and other renewable energy sources on a large scale and in a proactive fashion. Growing industry confidence in these technologies seems inevitable, and electricity suppliers are beginning to realize that sizeable wind projects in Texas, with the PTC, are often able to compete on an equal footing with other, more traditional generating sources. While the 2000MW purchase obligation established by the RPS will provide a good footing for initial development, a maturing wind industry able to compete at or near the cost of natural gas will surely offer more substantial market opportunities over the long term.

Notes

1 Denmark has abandoned its plans for the introduction of an RPS. The Netherlands is credited with being the first to develop an REC trading programme (in 1998), but that programme has not yet been used to meet mandatory renewable energy obligations under a RPS.
2 In addition, 15 states have developed renewable energy funds, often funded through a small surcharge on retail electricity rates.
3 §39.904 of the Public Utility Regulatory Act (PURA).
4 PUC Substantive Rules §25.173 Related to Goal for Renewable Energy.
5 The level of the standard was established in a political setting, and was viewed at the time as being an aggressive but achievable target.
6 Based on an assumed average capacity factor of 35 per cent. Assuming an average annual growth in demand of 3 per cent this translates to a renewable energy share of 2.2 per cent by 2009.
7 DOE (2000) estimates that wind power alone in Texas has the resource potential to deliver over 400 per cent of the state's present electricity consumption.
8 The factor was adjusted to 27 per cent for the years 2004 and 2005.
9 In mid-2005, Texas raised its RPS to 5800MW by 2015.
10 It should be noted that the PTC expired briefly after 2001, only to be re-extended in 2002 and then again in 2004. Were it not extended, RPS compliance costs in Texas would have increased and other (non-wind) renewable technologies may have more effectively competed for a share of the RPS market with wind power.
11 The security requirements imposed by retail suppliers favour renewable energy developers or development teams with strong financial backing.
12 In both the UK and California, a substantial number of the new renewable energy projects that won bids under the Non-Fossil Fuel Obligation (UK) and the production-incentive auction (California) have never been developed. This result is partly due to the design of each policy, where a certain degree of speculative bidding by renewable energy developers has been allowed.
13 The prices under these contracts are often fixed over the entire contract term, though a fixed annual escalation is sometimes applied. We note that the cost-competitive pricing offered relies on the availability of the PTC.
14 One reason for this strong commitment to the success of the policy is that earlier polling in Texas showed surprisingly strong support for developing renewable energy among the state's residents.
15 Though REC purchase obligations did not begin until 2002, to help ensure RPS compliance, RECs generated during the later half of 2001 could be used to meet 2002 compliance obligations.
16 In March 2002, the PTC was extended for another two years. In 2004, the PTC was extended again.

References

Berry, T. and Jaccard, M. (2001) 'The renewable portfolio standard: design considerations and an implementation survey', *Energy Policy*, vol 29, no 4, pp263–277

Bolinger, M., Wiser, R., Milford, L., Stoddard, M. and Porter, K. (2001) *Clean Energy Funds: An Overview of State Support for Renewable Energy*, Berkeley, California, Lawrence Berkeley National Laboratory

Bolinger, M., Wiser, R. and Golove, W. (2004) *Accounting for Fuel Price Risk When Comparing Renewable to Gas-Fired Generation: The Role of Forward Natural Gas Prices*, Berkeley, California, Lawrence Berkeley National Laboratory

Department of Energy (DOE) (2000) 'Database. Texas Wind Resources', www.eren.doe.gov/state_energy/tech_wind.cfm?state=tx, accessed in May 2001

ERCOT (2003a) 'Wind Generation Curtailment information', www.ercot.com/ercotPublicWeb/PublicMarketInformation/marketinformation/upliftedWind%20GenerationCurtailmentInfo.xls, accessed in December 2004

ERCOT (2003b) 'Report on existing and potential electric system constraints and needs within the ERCOT region', www.ercot.com/NewsRoom/MediaBank/2003TransmissionReport.pdf, accessed in December 2004

ERCOT (2004) 'Electricity Reliability Council of Texas: ERCOT's 2003 Annual Report on the Texas Renewable Energy Credit Program. Austin' www.texasrenewables.com/publicReports/2003_Report.doc, accessed in December 2004

Espey, S. (2001) 'Renewables portfolio standard: A means for trade with electricity from renewable energy sources?' *Energy Policy*, vol 29, no 7, pp557–566

Haddad, B. and Jefferiss, P. (1999) 'Forging consensus on national renewables policy: The renewables portfolio standard and the national public benefits trust fund', *The Electricity Journal*, vol 12, no 2, pp68–80

Helby, P. (1997) 'The devil is in the detail', *Windpower Monthly*, vol 13, no 11, p32

Langniss, O. (ed) (1999) *Financing Renewable Energy Systems*, Kiel, Oetker-Voges-Verlag

Langniss, O. (2003) *Governance Structures for Promoting Renewable Energy Sources*, Dissertation, Lund, Lund University

Lorenzoni, A. (2003) 'The Italian green certificates market between uncertainty and opportunities', *Energy Policy*, vol 31, pp33–42

Mitchell, C. (2000) 'The England and Wales non-fossil fuel obligation. History and lessons', *Annual Review of Energy and the Environment*, vol 25, no 1, pp285–312

Mitchell, C. and Anderson, T. (2000) *The Implications of Tradeable Green Certificates for the UK*, ETSU, TGC

Morthorst, P. E. (2000) 'The development of a green certificate market', *Energy Policy*, vol 29, no 15, pp1085–1094

Rader, N. (2000) 'The hazards of implementing renewables portfolio standards', *Energy and Environment*, vol 11, no 4, pp391–405

Rader, N. and Hempling, S. (2001) *The Renewables Portfolio Standard: A Practical Guide*, prepared for the National Association of Regulatory Utility Commissioners, www.naruc.org/associations/1773/files/rps.pdf, accessed 2 June 2005

Rader, N. and Norgaard, R. (1996) 'Efficiency and sustainability in restructured electricity markets: The renewables portfolio standard', *The Electricity Journal*, vol 9, no 6, pp37–49

Schaeffer, G. J., Boots, M. G., Mitchell, C., Anderson, T., Timpe, C. and Cames, M. (2000) *Options for Design of Tradeable Green Certificate Systems*, ECN-C-00-032.

Schaeffer, G. J. and Sonnemans, J. (2000) 'The influence of banking and borrowing under different penalty regimes in tradable green certificate markets: Results from an experimental laboratory experiment', *Energy and Environment*, vol 11, no 4, pp407–422

Timpe, C., Bergmann, H., Klann, U., Langniss, O., Nitsch, J., Cames, M. and Voß, J. (2001) *Realisation Aspects of a Quota Model for Electricity from Renewable Energy Sources* (in German), Study commissioned on behalf of the Ministry of Environment and Traffic Baden-Württemberg, Stuttgart

Wiser, R. and Hamrin, J. (2000) *Designing a Renewables Portfolio Standard: Principles, Design Options, and Implications for China*, Report prepared for the China Sustainable Energy Project, San Francisco, California, Center for Resource Solutions

Wiser, R. and Pickle. S. (1997) *Financing Investments in Renewable Energy: The Role of Policy Design and Restructuring*, Berkeley, California, Lawrence Berkeley National Laboratory

Wiser, R., Porter, K. and Grace, R. (2004) *Evaluating Experience with Renewables Portfolio Standards in the United States*, Berkeley, California, Lawrence Berkeley National Laboratory

9

European Union Policy towards Renewable Power

Volkmar Lauber

With regard to coal (with the European Coal and Steel Community) and nuclear power (Euratom), energy appeared quite early on the agenda of European integration. But coal was already on its way out in the 1950s and 1960s, and nuclear power never became as important as Monnet imagined it would (Dinan, 2004). In other areas, particularly those of oil, gas and electricity, the powers granted by the member states to European authorities remained very modest for a long time. Even the oil crises of the 1970s did not bring much change to this area, as many member states developed their own policy of how to deal with OPEC (Matláry, 1997). New impulses came from the emphasis on the internal market that developed in the 1980s, reflected in the Single European Act; this included the idea of an internal market for energy, to be characterized by greater competition and market integration. This idea was specified in the second half of the 1980s by Council and Commission proposals. In the area of electricity, the new passion for deregulation and privatization led the European Commission to the idea of taking on some of the last, largest and most powerful public monopoly structures that still existed, that is the electricity suppliers. The driving force in this area was Energy Commissioner Cardoso e Cunha, who on this issue formed an alliance with two commissioners who carried particular weight with their commitment to deregulation and competition: Martin Bangemann from Directorate General (DG) Industry[1] and Leon Brittan from DG Competition (Padgett, 2001). This led to the proposal of what became Directive 96/92/EC on the internal market in electricity, a proposal that was adopted after much resistance. Many other Directives were adopted in this area, with the emphasis clearly on increasing competition (Cini/McGowan, 1998; Palinkas and Maurer, 1997); this process is still continuing.

It is in this context that electricity from renewable energy sources appeared on the agenda. It was clear that it stood little chance to develop under intensified competition and therefore needed a special regime; Directive 96/92/EC only provided for priority dispatching of renewable power and contained no basis for market support programmes. Rules on this subject were to be found only in the Community framework on state aid, to be discussed later on. The initiative for a special regime for renewable power came from parliament. The Commission wanted above all to make sure that some form of competition should operate in the sector of renewable power as well. Also, climate change as well as considerations of security of supply (that is the prospect of growing import dependence on fossil energy resources in the first decades of the 21st century) made it seem desirable to the Commission to merge and harmonize national support policies, which already existed in quite a few member states, and to bring about international trade in renewable power which, in the understanding of the Commission at that time, would be the best way to accelerate its development.

In 1996, the European Commission submitted a Green Paper on renewable energy sources (European Commission, 1996); it was clearly deregulationist in inspiration and already proposed, for the renewable power sector, a quota system accompanied by tradeable certificates. The same applies to the subsequent Commission White Paper on the subject (European Commission, 1997). For several years the deregulationist approach was the main source of inspiration for the draft proposals issuing from DG Energy, which in this respect was under strong influence from DG Competition. The activities of this DG with regard to renewable energy focused at this time on creating a legal framework for renewable power in the form of a Directive, the preparation of which took about five years.

Actors in the politics of Directive 2001/77/EC

The central actors in the legislative process of this Directive were the European Commission and the Council of energy ministers, that is the member states. The European Parliament played a somewhat lesser role; so did the European Court. Outside actors providing important input to these official public actors were the renewables industry with its associations, the electricity supply industry of some countries (organized at the European level in EURELECTRIC), environmental non-governmental organizations (NGOs) and probably major electricity consumers such as IFIEC (International Federation of Industrial Energy Consumers).

The European Commission encompasses several sub-actors relevant for renewable energy politics (Lauber, 2001). While the college of Commissioners acts as a single body with majority voting (Hix, 1999), the DGs involved did not follow the same administrative traditions and paradigms. Most important was undoubtedly DG Energy which was in charge of the legislative dossier, first under commissioner Christos Papoutsis, then under commissioner Loyola de Palacio (DG Transport and Energy or DG TREN). Under Papoutsis this DG argued that only a system of tradeable certificates or a similar, 'competitive' system was compatible with electricity liberalization; such a system would necessarily be best at securing low prices, innovation and rapid expansion of renewable power (European Commission, 1999). De Palacio was more pragmatic from the beginning, at least for the preparation of the immediate proposal submitted in 2001, which left the choice of support systems to member states. However, the ultimate transition to 'competitive systems' was also prepared by a clause of the Directive, to be discussed below.

Another important actor within the Commission was DG Competition, in those years under the leadership of Commissioner Mario Monti. DG Competition has several instruments to participate in the political struggle around renewable energy policy. It formulates Community frameworks on environmental state aid (the last one was issued in early 2001, see European Commission, 2001), which contain clauses pertaining to renewable energy and which limit member states' ability to subsidize this activity. It may also review legislation of and initiate lawsuits against member states for violating the Community state aid regime. While such lawsuits are usually settled out of court, the instrument also serves to exert pressure on governments of member states to stay away from certain approaches, in this case from fixed feed-in tariffs. Such pressure was put on Denmark, Germany (at least until 2002), Ireland and probably some other governments as well.

DG Environment also played a role on occasion, especially in the legislative process regarding the Directive. Thus, Commissioner Wallstrøm came out forcefully in favour of permitting substantial financial support to renewable power and questioned the 'competitive' approach of Papoutsis, arguing that his proposal was likely to do more for competition than for the environment (Lauber, 2001).

Adoption of Directive 2001/77/EC was governed by the co-decision procedure. Despite the formally equal status between Parliament and Council under this procedure, the Council clearly played a bigger role. This is well illustrated by two central conflicts between Council and Parliament, that is the issue of mandatory versus indicative targets for member states and the recognition of household waste for incineration as (partly) biomass. On both issues the (Energy) Council prevailed.

In the European Parliament, the most important committee dealing with the Directive was the Committee on Industry, External Trade, Research and Energy, with Mechtild Rothe (a German social democrat strongly committed to renewable energy) as rapporteur. The Environment Committee provided an opinion, formulated by Hans Kronberger who took a similar stance. Claude Turmes, Green MEP from Luxemburg, also played a key role. The European Parliament wanted mandatory rather than indicative targets for the member states and a clause favouring fixed feed-in tariffs at a time when the Commission wanted to phase those out. Within parliament, support for renewable energy came most consistently from the Greens and the Party of European Socialists, with the European People's Party somewhat more likely to espouse the nuclear cause (but not all view renewables and nuclear as conflicting). Support for the Directive was however quite broad in the final vote.

The European Court of Justice played an important role due to its decision in a case involving the German feed-in scheme (*PreussenElektra* vs *Schleswag*), a decision handed down a short time before the submission of the final Directive proposal. In matters of state aid, the Court is the chief control on the Commission. It rejected the reasoning of DG Competition that the German system of fixed feed-in rates should be viewed as state aid even though it clearly did not fit the criteria of state aid applied so far (Advocate General, 2000). The final decision of the court meant that the Commission was not able to ban systems based on a fixed feed-in tariff by invoking the Treaty articles on state aid (articles 87 and 88).

An important role was played by the European associations of renewable energy producers, especially EREF (European Renewable Energies Federation), Eufores (European Forum for Renewable Energy Sources), EWEA (European Wind Energy Association), European Photovoltaics Association (EPIA) and European Biomass Association, supported by related organizations such as FEDARENE (European Federation of Regional Energy and Environmental Agencies). National associations also played an important role at times, as in the case of Britain and Germany. In the United Kingdom (UK), the British Wind Energy Association thought favourably of a quota system with certificate trading, initially influencing EWEA in this direction. By contrast, German wind energy associations were strongly opposed to such a system and successfully lobbied the federal government (based on a red–green coalition since 1998; see Jacobsson and Lauber, in this publication) to oppose such a system on the European level. While renewable energy associations acted more backstage, appeals to the general public and open political confrontations were often taken up by environmental NGOs, most notably by Greenpeace and WWF. They very resolutely espoused the cause of renewable power, conducting important campaigns at a time when the renewable power industry itself was still in an early phase of its development. In general they took a line similar to that of the European Parliament – in favour of mandatory targets and against phasing out fixed feed-in tariffs. The conventional electricity supply industry was mainly represented by EURELECTRIC, which sided with the Commission on the issue of tradeable credits, made concrete proposals and with Commission support organized such trade on a private basis (Renewable Energy Certificate System (RECS)).

Struggle over the contents of Directive 2001/77/EC

Overview

After defining renewable energy sources (see next paragraph), the Directive set national indicative targets for the consumption of renewable power by 2010, laid down some principles for national support systems and provided for a guarantee of origin of such power. It required member states to introduce simplified, transparent and non-discriminatory administrative and grid practices. Finally, it provides for a report on implementation as the basis for a potential harmonization Directive in the future, and for a minimum transition period should a new scheme be decided upon.

Definition of renewable energy for the purpose of Directive 2001/77/EC

The Directive defines renewable energy sources in its article 2 as non-fossil energy sources ('wind, solar, geothermal, wave, tidal, hydropower, biomass, landfill gas, sewage treatment plant gas and biogases'). The original Commission proposal (European Commission, 2000) limited hydro to 10MW; this provision was eliminated. Biomass is further specified to mean 'the biodegradable fraction of products, waste and residues from agriculture (including vegetal and animal substances), forestry and related industries, as well as the biodegradable fraction of industrial and municipal waste'. This was included upon the insistence of the Dutch, British, Italian and Spanish governments against the opposition of both Commission and Parliament (*European Voice*, 2001; *Solarthemen*, 2001). The proportion of electricity produced by renewable energy sources in such plants may now also be considered as renewable power if the waste treatment hierarchy is respected.

National targets

Historically, national commitments to renewable energy varied greatly with the vagaries of the oil market. Crash programmes in the 1970s and early 1980s were generally followed by a decline in activities following the oil price decrease in the mid-1980s. The diffusion of renewable energy technologies also differed significantly among member states. Table 9.1 shows figures for renewable power ranging from 1.1 per cent (Belgium) to 70 per cent (Austria); these figures also include large hydro. One of the ideas of the green and white papers – and of the Directive – on renewable energy sources was to reduce the cost of the new renewable power technologies by achieving mass production on a European level, reinforced by exports on the basis of Europe's leadership in this area; this in turn required a Community-wide effort. For such a collective effort, mandatory targets were seen as an appropriate instrument. The European Parliament argued this case with great perseverance. On the other hand, nearly all member states were unwilling to accept such targets; Denmark and later on Germany were the only exceptions.

Article 3 of the Directive sets only indicative targets, listed in an annex to the Directive (see Table 9.1). Targets for the new members acceding in 2004 were added in 2003. The annex also lists some qualifications by certain member states, which thus weakened their commitment. Member states are required to 'take the appropriate steps ... in proportion to the objective to be attained' and to document their efforts in regular reports, the first no later than 27 October 2003, the date by which the Directive was to be transposed into national law, and thereafter every two years. The Commission shall evaluate these reports and publish the results. If member states fail to live up to their targets without valid reasons, the Commission shall make appropriate proposals, which may include mandatory targets.

Table 9.1 *National indicative RES-E targets 2010 for member states*

	RES-E % in 1997	RES-E % in 2010
Austria	70	78
Belgium	1.1	6
Denmark	8.7	29
Finland	24.7	31.5
France	15	21
Germany	4.5	12.5
Greece	8.6	20.1
Ireland	3.6	13.2
Italy	16	25
Luxemburg	2.1	5.7
Netherlands	3.5	9
Portugal	38.5	39
Spain	19.9	29.4
Sweden	49.1	60
UK	1.7	10
Cyprus	0.05[a]	6.0
Czech Republic	3.8[a]	8.0
Estonia	0.2[a]	5.1
Hungary	0.7[a]	3.6
Latvia	42.4[a]	49.3
Lithuania	3.3[a]	7.0
Malta	0.0[a]	5.0
Poland	1.6[a]	7.5
Slovakia	17.9[a]	31.0
Slovenia	29.9[a]	33.6
EU 25	12.9[a]	21.0

Note: a Data for 1999.

Source: European Commission, 2004b, p4.

Support schemes

Apart from the issue of targets, this was one of the most hotly contested provisions of the Directive. At first, Energy Commissioner Papoutsis planned to progressively liberalize renewable power in order to arrive at a community-wide harmonized regime fairly soon, based on mistaken assumptions about the success of the British non-fossil fuel obligation in driving down prices, especially of wind power (European Commission, 1999; Lauber and Toke, 2005). This implied a conflict over systems of support for renewable power. The first draft proposals submitted by Papoutsis made clear that only support schemes which relied on 'competition' and international trade of renewable power were judged compatible with the principles of electricity liberalization laid down in Directive 96/92/EC. This referred to the tendering systems practised at that time in some member states and ignored that there is competition also under fixed feed-in tariffs (Hvelplund, 2001). These tendering schemes consisted of prospective renewable power generators submitting competitive bids for fixed-price contracts. Such a system was practiced in Britain (where the NFFO system could hardly be considered a success in expanding production of renewable power; Mitchell, 2000, and Connor, in this publication), France (where results were simply dismal) and Ireland (where results were fairly modest). Britain and France in particular had low (or simply slow) rates of completion of accepted projects, even though they dispose of the best wind resources in the European Union (EU). As Figure 9.1 shows, this contrasted with the successful market introduction of renewable power achieved by those states that relied on fixed feed-in tariffs, that is Denmark, Germany and Spain, responsible for about 80–90 per cent of wind power installations in the EU (the most important of the new renewable power sources).

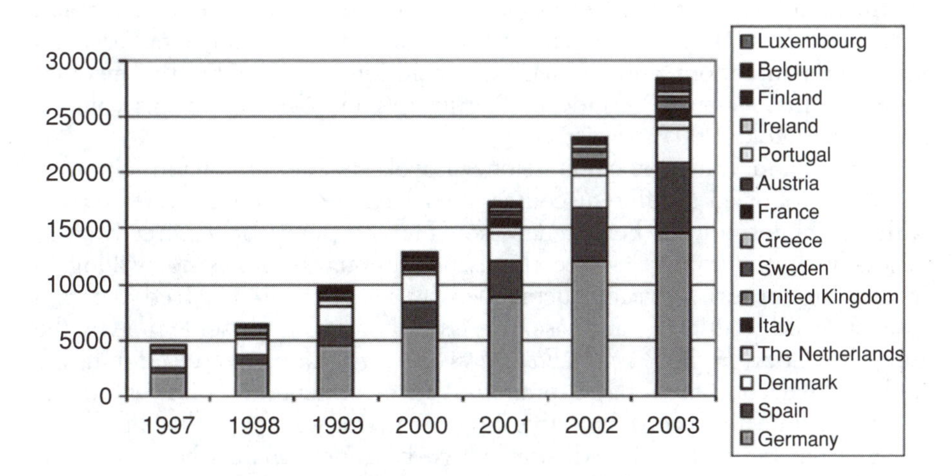

Source: EU Commission 2004a.

Figure 9.1 Growth of installed wind power capacity in EU-15, 1997–2003 – three leading markets

Later on, Papoutsis seems to have favoured a quota/tradeable certificate approach. Under this approach, generators of renewable power sell both the physical electricity they produce and certificates embodying the 'greenness' of that electricity; these certificates can then be traded on an exchange so that a market price results. The expectation was that such a system would promote trade of renewable power within the EU, encourage its development in those regions where conditions are most favourable and thus reduce costs. At the time the Directive was discussed, such a system was viewed favourably by the

UK, Italy and the Flemish part of Belgium; it also had supporters in Sweden and Austria. The Netherlands introduced tradeable certificates on a voluntary basis (without quotas, but with tax advantages as incentive). The Danish government was prepared to accept it after Commission pressure and promises that such a system would apply to the whole EU. EURELECTRIC was also supportive. The system was opposed most strongly by the German government, which was bent on preserving – and actually on improving – its own success story based on a feed-in tariff, and by environmental and renewable power associations.

Confronted with intense protests against his draft proposals, Papoutsis published the philosophy underlying his approach in a Commission Working Paper in April 1999 (European Commission, 1999). This paper argued rather ideologically the perspective of policy makers who had successfully deregulated transports, telecommunication, electricity and gas, achieving considerable efficiency gains in monopolistic or oligopolistic sectors that had grown lazy under 'friendly' regulation. But this was hardly the situation of the renewable power sector. The working paper claimed that 'according to economic theory' – without further supportive argument or evidence – competition-based systems were simply better with regard to innovation and reducing prices. With the benefit of hindsight, this claim appears based on mistaken assumptions and premature conclusions, at least for the medium term (see Lauber on RPS/tradeable certificates and REFITs, in this volume; Lauber and Toke, 2005).

Energy and Transport commissioner Loyola de Palacio – who took office in 1999 – followed a different course, after having suffered a severe setback with the philosophy of her predecessor. Her proposal for a Directive was generally neutral with regard to this issue, although she was not willing to include an express provision, demanded by some, to protect feed-in tariffs from state aid scrutiny. In any case, this issue was settled for some time by the European Court in the *PreussenElektra* vs *Schleswag* case. Fixed feed-in tariffs under which consumers pay a premium price for renewable power do not constitute state aid according to this judgement, nor was the exclusion of such power from international trade viewed as disproportionate when measured against the goal of sustainability as long as renewable power was still struggling and its share was still modest. This meant that REFIT systems could not, for the time being, be questioned from this angle (Nagel, 2001).

But DG Transport and Energy did not entirely give up on introducing a system of tradeable certificates in due course. Directive 2001/77/EC states in its preamble and in article 4 that while it was still too early to decide on a Community-wide framework regarding support schemes, it was necessary to adapt, after a sufficient transitional period, support schemes to the developing internal electricity market. For this reason the Commission should present a report on experience gained with the various support schemes – in particular on success in promoting consumption of renewable power in accordance with the national indicative targets – and, if necessary, make an appropriate proposal for a Community framework, which however, would have to provide for a transition time of no less than seven years from its adoption. The general expectation at the time of the Directive's passage was that this system would be the tradeable certificate system as described above.

Guarantees of origin

Article 5 of the Directive regulates the way in which member states shall set up systems of guaranteeing the authenticity of renewable power. The term 'certificates' was avoided as some member states viewed this as a first step towards introducing a system of tradeable certificates/quota-based support schemes. Such an interpretation was also excluded expressly by items 10 and 11 of the preamble to the Directive. Nonetheless EURELECTRIC, starting in 2004, favoured a certificate scheme based on such guarantees (see below).

Administrative procedures

Article 6 provides that member states shall evaluate their regulatory frameworks for renewable power (authorizations, permits, support decisions, and so forth) with a view to reducing regulatory barriers, streamlining procedures and ensuring that rules are objective, transparent and non-discriminatory, taking into account the particularities of the various renewable energy technologies. They had to submit reports on action taken to this effect on 27 October 2003. Even though such obstacles are considerable in at least a third of the EU 15 member states (European Commission, 2004a), no provision is made for future Commission proposals in this area, only for a report on best practices.

Grid access

Article 7 regulates the relation between renewable power generators and operators of the transmission and distribution system. Its purpose is to make sure that there is no discrimination against such generators taking into account all the costs and benefits of renewable power. This concerns grid access, connection costs, transmission and distribution fees and the like. Prospective connection costs must be communicated to RES-E generators and standard rules on costs published. The original Commission proposal – supported by Parliament – provided for priority access. The Council changed this to 'guaranteed access', although priority access may still be granted. In dispatching generating installations, transmission system operators shall give priority to producers of electricity from renewable sources insofar as the operation of the national electricity system permits. This clause is important wherever grid operators are hostile to renewable power and try to impede its deployment (not a rare instance by any means). Member states shall report on measures taken to facilitate access to the grid in the report referred to in the preceding subsection. Again there is no provision for future Commission action to be taken in this area, even though this is one of the major hurdles for renewable power.

External costs and subsidies/Summary report on implementation

Article 8 provides for the Commission to submit in late 2005 a summary report on implementation on the basis of the member state reports referred to above at the end of 2005, and thereafter every five years. Such a report 'shall consider the progress made in reflecting external costs of electricity from non-renewable sources and the impact of public support granted to electricity production' as well as progress on achieving national targets and discrimination between different energy sources.

The chief purpose of the Directive is to make renewable power competitive for the internal electricity market. Now the competitiveness of this form of electricity is greatly inhibited by the fact that the generation of electricity from conventional sources entails substantial social costs which remain 'external', that is which are not reflected in its price; on top of that, conventional generation often receives subsidies, and it benefits from poor competition in many member states. Research conducted for the EU in the ExternE project shows that the cost of electricity generated from coal and oil in the EU would on average double if external costs to environment and health were included; from gas, it would increase by 30 per cent (Massy, 2001; Milborrow, 2002; see also Lauber, section on external costs in Chapter 1 of this volume). In addition, fossil and nuclear generation are often subsidized by a variety of mechanisms. Without these market distortions, renewable power would not nearly need the same amount of support; some would already be fully competitive. The internalization of external costs is a highly controversial issue; in early 2002 the European Parliament only narrowly prevailed upon the Council with its demand for a provision, in the 6th Environmental Action Programme, aiming at the progressive elimination of environmentally negative subsidies (*Environment Watch Europe*, 2002a and 2002b). Also in the context of the adoption of Directive 2001/77/EC, DG Competition promised to look at subsidies to nuclear generation (low insurance coverage, insufficient preparation for waste disposal, tax advantages etc; *European Voice*, 2002a, 2002b, 2004c and 2004d). The value of article 8 is primarily programmatic; it should help to focus discussions of support on the comparative treatment of all forms of electricity generation and relate levels of support to external costs.

Controversy continued: Targets and support systems

Targets for 2010 and beyond

As early as 2002, evidence accumulated that the EU and its member states would not, without major policy changes, reach the targets of the Directive with regard to the share of renewable power contained in its annex. This was first stressed in a consultant report to the Commission prepared in mid-2002 (*Renewable Energy Report*, 2002). The same conclusion was voiced in 2003 by the European Environment Agency (*Environment Watch Europe*, 2003). But DG Energy was 'not alarmed' (*Windpower Monthly*, 2003).

The original target of 22 per cent for 2010 was lowered as a result of Eastern European enlargement. Individual targets for accession states were negotiated in mid-2003. Even though they were quite ambitious, they provide for an average of 11 per cent for this group (Table 9.1); this reduced the original 22 per cent target of the Directive by one point.

The question of targets stirred new controversy in the spring of 2004. Following the recommendation of a preparatory meeting for the Bonn Renewables Conference earlier that year, Parliament urged Commission and Council to set a strong target for 2020 – 20 per cent of all energy (not just power) from renewable sources by then – to support decision-making at the Bonn conference. The Commission declined to do so; Commissioner de Palacio declared that to do so without adequate information would not make sense, particularly as most member states were not even fulfilling their targets for 2010. In fact, powerful directorates general were opposing such targets: Economic and Financial Affairs (then under Almunia) plus Enterprise and Competition; only DG Budget (Schreyer) and DG Environment (Wallstrøm) demanded it strongly – and also Jacqueline McGlade, chief of the European Environment Agency (*European Voice*, 2004a and b; Environmental Watch, 2004a). As a result of these conflicting pressures, the Commission in May acknowledged the importance of providing a longer term perspective to an infant industry and announced how it would go about developing new targets: it would first make a thorough assessment of renewable energy sources, 'notably with regard to their global economic effects', review developments to ensure compatibility with sustainable development policy and consider issues such as the competitiveness of the European economy, security of supply, technical feasibility, contribution to climate change policy and other environmental priorities, and finally potential. In other words, it would reinvent the wheel, or so it seemed. After a debate on this review, targets for the period after 2010 would be set in 2007. This position was fully accepted by the Council (Environmental Watch, 2004b).

However, at the same time, in a communication to Council and Parliament, the Commission declared that achievement of the 2001 goals had become unlikely; the practical measures put in place by member states so far would deliver only a renewable power share of 18 to 19 per cent. It also proposed to take a series of initiatives to improve this situation as far as possible (for example use of structural and cohesion funds), but above all stressed the responsibility of member states to devote sufficient financial resources to this task (European Commission, 2004a).

Support systems

In 2001, it was almost universally expected that the Commission would propose, in 2005, harmonization of support policies on the basis of a quota/certificate system. In early 2005, the situation is far less clear. For one, only four member states have introduced such a system so far (the UK, Belgium, Italy and most recently Sweden), and up to now results are far from fulfilling the expectations that the Commission once placed in this policy. The British

Renewables Obligation, during its first three years – it started to operate in 2002 – paid compensations that were about 50 per cent higher than the German feed-in tariff, limited innovation and even so is likely to miss the self-set quotas for 2010 and 2015 by about one-third (Lauber and Toke, 2005). The other such systems have even less experience to show for. As to the experiment of RECS (renewable energy credit system to trade certificates on a voluntary basis between members of the electricity supply industry), financially supported by the Commission, it suffered a significant setback in 2003 when the Netherlands stopped credit trading because this had led to a very high level of imports subsidized by Dutch taxpayers (Lauber, 2004).

Even so, the renewables industry is clearly worried that the Commission, pushed by EURELECTRIC (EURELECTRIC, 2004), might propose such a system, which in turn might disrupt more substantial efforts to build up renewable power. EURELECTRIC seems to favour a voluntary scheme financed by consumers (Scheer, 2005). In anticipation of such an initiative, the European Wind Energy Association has proposed an approach that differs from harmonization while at the same time promoting a measure of international competition. The idea is to gradually increase the compatibility of support systems and thus to merge the markets of countries that have chosen the same type of support mechanism. In this way experience could be gathered in preparation of an internal market for renewable electricity. Possibly in this vein Commissioner Andris Piebalgs,[2] who became the successor to de Palacio in the fall of 2004, declared at his confirmation hearing before the European Parliament that he was more interested in finding viable national solutions than in a 'one size fits all' policy (Massy, 2004b). And as already stated earlier, the Commission in its 2004 communication on the share of renewables in the EU stressed the need to commit sufficient resources rather than the need to curtail subsidies, and cited Germany and Spain (two countries with a feed-in tariff or premium) as having particularly effective and efficient support systems, along with Denmark and Finland. Not one member state with a quota/certificate system received a similar distinction (European Commission, 2004a). However, other developments are less favourable to renewable energy. Most of the leaders of the new member states from Eastern Europe are likely to experience little public pressure to conduct a strong policy in favour of renewable energy. In addition, the European Parliament's Committee on Industry, Research and Energy, which is in charge of renewable power matters, is now – since the elections in June 2004 – headed by Giles Chichester, a British conservative who supports nuclear power and actively opposes wind power at home (Massy, 2004a). And when the Transport and Energy Council in late 2004 agreed on a general approach to a Directive to safeguard security of energy supply, it agreed to a wording of the Dutch presidency which scrapped all references to renewable energy and decentralization of supply as important contributions to security (Environmental Watch, 2004c). On the other hand, the entry into force of the Kyoto Protocol, recent evidence of stronger climatic change than expected so far and the doubt about oil and gas reserves lingering since the oil price increases of 2004 may well operate in favour of renewable power, although they may give an even bigger push to nuclear.

Notes

1 Directorates General are administrative subdivisions of the EU Commission.
2 Commission President Barroso first proposed Laszlo Kovacs for this office. At his hearing before the European Parliament, he suggested a harmonized quota/tradeable certificate system for renewables, without even pretending to await the outcome of the 2005 Commission report. He was strongly criticized by MEPs, so that Barroso decided to shift him to another portfolio.

References

Advocate general (2000) Opinion of advocate general delivered on 26 October 2000, Case C-379/98, *PreussenElektra* vs *Schleswag*

Cini, M. and McGowan, L. (1998) *Competition Policy in the European Union*, Basingstoke, Macmillan

Dinan, D. (2004) *Europe Recast*, Boulder, Lynne Rienner

Directive 2001/77/EC of the European Parliament and of the Council of 27 September 2001 on the promotion of electricity produced for renewable energy sources in the internal electricity market, *Official Journal of the European Communities*, 27.10.2001, L 283/33

Environment Watch Europe (2002a), 15 March

Environment Watch Europe (2002b), 29 March, p11

Environment Watch Europe (2003), 4 July, p23

Environment Watch Europe (2004a), 8 June, pp16–19

Environment Watch Europe (2004b), 25 October, pp20–21

Environment Watch Europe (2004c), 6 December, p13

Eurelectric (2004) *A Quantitative Assessment of Direct Support Schemes for Renewables*, first edn, January

European Commission (1996) *Green Paper for a Community Strategy – Energy for the Future: Renewable Sources of Energy*, COM(96)576, 19 November

European Commission (1997) *Energy for the Future: Renewable Sources of Energy. White Paper for a Community Strategy and Action Plan*, COM(97)599 final, 26 November

European Commission (1999) *Working Paper. Electricity from renewable energy sources and the internal electricity market*, SEC(1999)470, 13 April

European Commission (2000) *Proposal for a directive of the European Parliament and of the Council on the promotion of electricity from renewable energy sources in the internal electricity market*, COM(2000)279 final, 10 May

European Commission (2001) 'Community guidelines on State aid for environmental protection'. Document 301Y0203(02), *Official Journal* C 037, 03/02/2001, pp3–15

European Commission (2004a) *Communication to the Council and the European Parliament. The share of renewable energy in the EU*, COM (2004) 366 final, 26 May

European Commission (2004b) *Commission staff working document. The share of renewable energy in the EU. Country Profiles. Overview of Renewable Energy Sources in the Enlarged European Union*, SEC(2004)547, 26 May

European Court of Justice (2001) *Judgement in Case C-379/98, PreussenElektra* vs *Schleswag*, 13 March

European Voice (2001), 7–13 June

European Voice (2002a), 17–23 October, p29

European Voice (2002b), 28 November–4 December, p21

European Voice (2004a), 10–16 June, p23

European Voice (2004b), 17–23 June, p23

European Voice (2004c), 17–23 June, p24

European Voice (2004d), 2–8 December, p23

Hix, S. (1999) *The Political System of the European Union*, Basingstoke, Macmillan

Hvelplund, F. (2001) 'Political price or political quantities?', *New Energy*, Vol 5, pp18–23

Lauber, V. (2001) 'Regelung von Preisen und Beihilfen für Elektrizität aus erneuerbaren Energieträgern durch die Europäische Union', *Zeitschrift für Neues Energierecht*, vol 5, no 1, pp35–43

Lauber, V. (2004) 'REFIT and RPS: options for a harmonised Community framework', *Energy Policy*, vol 32, no 12, pp1405–1414

Lauber, V. and Toke, D. (2005) 'Einspeisetarife sind billiger und effizienter als Quoten-/Zertifikatssysteme. Der Vergleich Deutschland–Großbritannien stellt frühere Erwartungen auf den Kopf', Zeitschrift für Neues Energierecht, vol 9, no 2, pp1–8

Massy, J. (2001) 'External costs benchmark too low', *Windpower Monthly*, vol 17, no 10, p24

Massy, J. (2004a) 'Policy upheaval threatens in Europe', *Windpower Monthly*, vol 20, no 11, pp51–56

Massy, J. (2004b) 'Call for evolution not revolution', *Windpower Monthly*, vol 20, no 12, pp29–30

Matlary, H. (1997) *Energy Policy in the European Union,* Basingstoke, Macmillan

Milborrow, D. (2002) 'External Costs and the Real Truth', *Windpower Monthly*, vol 18, no 1, p32

Mitchell, C. (2000) 'The England and Wales Non-Fossil Fuel Obligation: History and lessons', *Annual Review of Energy and the Environment*, vol 25, pp285–312

Nagel, B. (2001) Rechtliche und politische Hindernisse bei der Einführung Erneuerbarer Energien am Beispiel Strom', *Solarzeitalter,* vol 13, no 4, pp14–20

Palinkas, P. and Maurer, A. (1997) 'Erneuerbare Energien als Teil der Energiestrategie der Europäischen Gemeinschaft: Entwicklung, Stand und Perspektiven', in Brauch, H. G. (ed) *Energiepolitik*, Heidelberg, Springer, pp197–220

Padgett, S. (2001) 'Market Effects and Institutions in Transnational Governance Formation: European Regulatory Regimes in Telecommunications and Electricity', Paper, ECPR Joint Sessions in Grenoble, 6–11 April

Renewable Energy Report (2002), vol 42, August, p20

Scheer, H. (2005) 'On the future of national support for renewable energy in Europe', *Photon International*, February, p80

Solarthemen (2001), vol 114, 21 June, p4

Windpower Monthly (2003), vol 19, no 2, p14

Part 3

Evaluation of Policies and Approaches

10

Comparing Support for Renewable Power

David Elliott

The renewable challenge

Renewable energy technologies are new entrants in the energy system. They face an uphill struggle against the well-established dominant electricity supply technologies – coal and gas, plus nuclear. Given increasing concerns about climate change, governments around the world have over the years tried to stimulate the expansion of renewable energy via a range of subsidies and other financial support systems. This chapter looks at the economic support mechanisms that have been used in the UK, comparing them with those adopted elsewhere.

Underlying the approaches to the development of renewable energy technology that have been adopted around the world is a basic distinction between supply side 'technology push' approaches and demand side 'market pull' approaches. It was perhaps inevitable that technology push dominated initially, in the mid 1970s, as the new technologies needed research and development (R&D) effort, with much of the funding coming from government in the form of grants to research teams. However, by the early 1980s, the emphasis shifted in most countries to a market pull approach.

This shift was only partly due to the successful emergence of viable technologies from the R&D phase. In fact many technologies still needed R&D support. It was arguably much more a result of a political shift away from government intervention and on to market led developments, with the emphasis being on achieving competitive success. This emphasis has continued to dominate in the UK and also in the US. As we shall see, however, in some other countries there has been less emphasis on the need to attain rapid commercial viability and a greater willingness to condone interim subsidies.

The UK 'market enablement' approach

The UK renewable energy programme was started in the mid 1970s, under a Labour administration, with a range of government funded research programmes, focusing in particular on novel wave energy technologies as well as tidal barrages and geothermal energy. However, a Conservative government was elected in 1979 on a tax cutting policy, and in 1982 drastic cuts were made in the wave and tidal research programmes. The Hot Dry Rock geothermal project also closed. The underlying approach was well reflected in a response, in January 1988, from Michael Spicer, then an energy minister, to a Parliamentary Question on the role of subsidies and grants. Spicer commented: 'the renewable technologies would not be best served in the long term by distorting the market by grant aid or other subsidies for their use.'

In parallel, the Conservative government initiated a programme of privatization of the energy sector, including, in 1990, electricity supply. Despite its relatively poor economics, nuclear power was initially to be included in this sell-off. To try to improve the economic position of nuclear power, a special subsidy was established to meet the gap between the price of nuclear and fossil fuel generated electricity, which involved putting a 10 per cent surcharge on consumers bills – the fossil fuel levy. Renewables were also allowed to benefit from this levy under the Non Fossil Fuel Obligation (NFFO). The NFFO/levy was subsequently presented as a way to help stimulate the market for renewables – as a 'market enablement' mechanism. However, the bulk of the money it raised went to nuclear – initially 98 per cent of the £1.2 billion the levy typically raised each year. The small and gradually expanding amount going to renewables did however support a growing number of renewable energy projects. By 2000, around 800MW had been installed, including around 300MW of mostly imported wind power plants.

The NFFO required electricity companies to buy in specified amounts of renewable energy, with a competition being held among rival candidate projects to meet this demand. A series of separate 'technology bands' was established, for wind, biofuels, and so on, to segment this competition – so wind would compete with wind, biogas with biogas and so on. The result of this competitive approach was that prices began to fall dramatically. Whereas wind projects got 11p/kWh in the 1991 NFFO competition, by 1998 wind projects were going ahead at 2p/kWh in some locations – cheaper than gas CCGTs.

Following the liberalization of the electricity supply market in 1998, and the introduction of new more competitive trading arrangements, the NFFO had to be abandoned. It was replaced, from 2002 onwards, under a Labour administration, by a new support mechanism, the Renewables Obligation (RO), which requires electricity supply companies to move in stages to obtain 10 per cent of their electricity from renewables by 2010/2011. If they do not, they are in effect fined 3p/kWh – the so-called 'buy-out' price. That means that renewable electricity is attractive to them even if it cost up to 3p/kWh more than conventional power, that is roughly 5p/kWh at the wholesale price (around 2p/kWh) current when the RO was conceived. This 3p/kWh buy-out price ceiling was imposed to stop prices to consumers escalating – the extra cost of using renewables is passed on to the consumers, as with the NFFO, but the aim was to keep the overall price increase down to 3–4 per cent by 2010 (DTI 1999, 2000, 2001). However, the RO system also includes a certificate trading system, which has allowed prices to rise above this level. The energy suppliers are given Renewable Obligation Certificates (ROCs) for each certified MWh of green power they purchase, and they can trade any extra ROCs they earn, over and above their set RO requirement, with companies who have not met their obligation. ROCs thus have a market value, and trading in them has, in effect, raised the ceiling price, although with no guarantee of the level.

For its part, the government seemed confident that the competitive mechanisms in the RO system would nevertheless lead to price reductions in the future. Table 10.1 shows estimates for prices for renewables and other sources by 2020 quoted in the 2002 Cabinet Office Energy Review. However, while these reductions might be achieved given the competitive nature of the RO, market pressures may not lead to enough capacity being installed to meet the 10 per cent by 2010 target. One study suggested that, unless policies were changed, at best the RO might only lead to a 7 per cent renewable contribution by 2010 (RPA 2002). Certainly, given that unlike NFFO, it does not have segmented 'technology bands', the RO focuses on the 'near market' options and does very little to stimulate investment in the *newer* renewables like wave and tidal current energy. This will be increasingly important given that, in 2004 the government announced a 15 per cent by 2015 target, on the way to achieving the 'aspiration', mentioned in the energy white paper published in 2003, of obtaining 20 per cent of UK electricity from renewables by 2020.

Table 10.1 *Estimated cost of electricity in the UK in 2020 (pence/kWh)*

On land wind	1.5–2.5
Offshore wind	2–3
Energy crops	2.5–4
Wave and tidal power	3–6
PV Solar	10–16
Gas CCGT	2–2.3
Large CHP/cogeneration	under 2p
Micro CHP	2.3–3.5
Coal (IGCC)	3–3.5
Nuclear	3–4

Source: Performance and Innovation Unit, 'The Energy Review' UK Cabinet Office, 2002 http://number10.gov.uk/files/pdf/piuf.pdf.

The RO system has provided prices that have allowed some offshore wind projects to go ahead, but in 2003 the government decided to provide some extra support in the form of a competitive series of grants to help new renewables like offshore wind, and also energy crop projects. In 2004 it also decided to allocate £50 million to wave and tidal projects. The decision to provide grant aid may be seen as an admission that the UK's competitive market enablement approach may not be effective at delivering sufficient capacity.

Table 10.2 illustrates the levels of support provided under the various mechanisms.

Table 10.2 UK support for renewables (£ million)

	Research grants[a]	RO	NFFO	Capital grants
1990–1991	21.3	–	6.1	–
1991–1992	24.8	–	11.7	–
1992–1993	26.6	–	28.0	–
1993–1994	26.8	–	68.1	–
1994–1995	20.5	–	96.4	–
1996–1997	18.5	–	112.8	–
1997–1998	15.9	–	126.5	–
1998–1999	14.4	–	127.0	–
1999–2000	14.9	–	56.4	–
2000–2001	15.9	–	64.9	–
2001–2002	24.0[b]	–	54.7	–
2002–2003	27.6[b]	282.0[b]	Unknown	60
2003–2004	29.0[b]	405.0[b]	Unknown	131

Notes: a Direct government funding for R&D on renewable energy through the DTI's Sustainable Energy Programme and through the Research Councils via the Science Budget.

b estimates.

Source: Hansard, 21 Nov 2001: Column: 300-01W.

Comparison with other approaches

The amount of renewable energy generating capacity that has been installed in the UK so far is small compared for example with Germany, where quite generous support was provided under the so-called Renewable Energy Feed in Tariff (REFIT) scheme and its subsequent variations. REFIT provided prices that initially were generally higher than offered by the RO (for example up to 9 Euro cents/kWh for wind projects), although, in subsequent phases, the level of subsidy has fallen. Indeed, when other elements in the UK support system are added in (including the revenue from the parallel Climate Change Levy) the UK subsidy has actually in some cases become larger than that offered in Germany, especially on a kW installed basis (Toke, 2005). However the key factor was that REFIT-type schemes provided *guaranteed* prices. The secure investment climate that REFIT-type schemes created contrasts strongly with the situation in the UK, where RO prices were subject to competitive market pressures. The result was that, by 2003, the UK had only around 530MW of wind capacity in place while Germany, which has a much worse wind resource, had over 20 times more – around 12,000MW – over ten times the total renewable capacity that had been installed in the UK (1GW by 2003).

The competitive system used in the UK not only created an insecure investment climate, it also meant that wind developers had to select very high wind speed sites (7m/s typically), most of which were in environmentally sensitive upland areas. This could deliver lower cost power on a pence/kWh basis. However, in some cases this invasive approach provoked a major local backlash against projects. Local objections have meant that around 70 per cent of wind farm proposals have been turned down by planning bodies (Elliott, 2003). The low take-up rate of renewables in the UK was also arguably one result of the adoption of a very competitive approach to electricity prices generally, following the adoption in 2001 of the New Electricity Trading Arrangements, which aimed to increase competitive pressures among energy suppliers so as to force prices down. It did that very successfully: wholesale electricity prices fell by 20–25 per cent in the first year of NETA's operation and more, subsequently. However, the impact on renewable energy schemes was disastrous – most were run by small companies that could not compete with the large conventional generation companies when prices were pushed down. The result has been that demand for their output fell and, despite the RO, the incentive for investing in more schemes was reduced. Part of the problem is that most renewables can only offer power intermittently and this is penalized heavily in the NETA structure – their environmental contribution is not recognized. So while the government says it is trying to push renewables forward via the RO, NETA pushed in the opposite direction (EAC, 2002).

The end result of the UK's competitive market approach as enshrined in the NFFO, the RO and also NETA, compared with the REFIT guaranteed price type approach adopted in Germany and elsewhere is fairly clear. Despite substantial subsidies and an excellent wind speed regime, not much capacity has been installed. By contrast REFIT-type schemes seem to have done much better. A report by the European Environment Agency, published in 2002, compared the successes and failures of EU renewable energy programmes between 1993 and 1999. It noted that three countries that guaranteed or fixed purchase prices of wind-generated electricity – Germany, Denmark and Spain – contributed 80 per cent of new EU wind energy output during the period (EEA, 2001).

A report by the World Wind Energy Association noted that more than 80 per cent (1144MW) of the 1388MW installed around the world in the first half of 2002 was installed in three countries with guaranteed minimum prices: Germany, Italy and Spain. In countries with quota/certificate systems, including the UK, the US and the Netherlands, only 75MW was installed. Perhaps unsurprisingly, subsequently, France and Brazil decided to introduce minimum price systems and the national environmental organization, the Sierra Club, has also been campaigning for the adoption of Feed-In tariff schemes in the US.

The REFIT system has not been without its opponents, for example in terms of the loading up utilities with extra costs. However, competitive pressures are not absent. Generators that can make use of cost-effective equipment will still be better placed than those that do not. Despite the REFIT-type schemes in Germany and, until recently, Denmark, there seems to have been no lack of technological innovation, quite the opposite. Both have been at the forefront of wind technology innovation. The same can hardly be said of the UK – there are no major UK wind manufacturers.

Nevertheless, the European Commission clearly wants to shift to a market-led certificate trading system. Certainly, REFIT type subsidies should not be needed to be retained across the board for ever. They are useful at the early stage of a technology's commercial history, but they can be progressively withdrawn as it matures. Some have argued that tradeable certificates, like the ROCs provided in the RO system, can be just as effective in providing support, but so far, as we have seen, this has not been the reality. The European certificate trading system, which is to be put in place in 2005, may begin to offer more success, and provide a common focus, but there is still some way to go before the various national support schemes in the EU can be harmonized, with REFIT-type schemes remaining popular. Given the situation for the moment, for most individual EU countries, stepped REFIT systems with subsidy levels falling in a phased way may be the better option (ElGreen 2001).

Markets or subsidies?

Clearly, new renewable energy technologies need subsidies to get established, but at some point they should be able to compete unaided. Wind has nearly reached that point in some parts of the UK, and some waste/biofuel combustion options have already passed it. So, arguably, for these technologies, market enablement has achieved its goal, even if it has not led to much capacity being installed.

However, there are new renewable energy options that need continued support, such as, in the UK in particular, wave and tidal power. With the large-scale wave and tidal programmes abandoned, and, in the new liberalized electricity market, the emphasis being on smaller scale plants, the focus among the surviving UK research teams had been on smaller scale in-shore and onshore wave system and on the more recent idea of extracting power from tidal flows. Projects like this, which were at best at the demonstration stage and more usually at the R&D stage, are not suited to support under the NFFO or the RO, which are meant for 'near market' technologies. By contrast the REFIT approach has provided support for technologies such as PV solar, which are still very expensive – on the assumption that costs will come down later as the market for the technology was expanded by subsidized lift-off. So far, as we have seen, the UK approach does not seem to have done enough to help much near-market technology take off. It is even less suited to less developed technologies. This may be one reason why, despite having a very large renewable potential, the UK lags well behind most other EU countries in terms both of developing capacity now, and in terms of targets for the future (see Table 10.3).

Table 10.3 *Renewable energy as a percentage of total primary energy in 2001*

Sweden	29.4
Finland	22.4
Austria	21.5
Portugal	13.7
Denmark	10.4
France	7.0
Spain	6.5
Italy	5.6
Greece	4.6
Germany	2.6
Ireland	1.7
Luxembourg	1.6
Netherlands	1.4
United Kingdom	1.1
Belgium	1.0

Source: International Energy Agency.

To be fair, the UK's lowly ranking is in part due to the fact that it started from a low level of renewable energy capacity, essentially a few hundred megawatts of hydro. By contrast Austria, Sweden and Finland enjoy very large hydro and biomass heat contributions. However, the competitive approach the UK has adopted does not seem likely to help it to catch up rapidly. The agreed EU targets for *electricity* from renewables by 2010 are shown in Table 10.4. As can be seen, despite its very large renewable resource, and the fact that these data leave out large hydro and biomass heat, the UK still occupies a lowly position. That seems to be, at least in part, the price paid for the emphasis on competition.

Table 10.4 *EU Directive: 2010 targets for electricity from renewables (%) excluding large hydro*

Denmark	29.0
Finland	21.7
Portugal	21.5
Austria	21.1
Spain	17.5
Sweden	15.7
Greece	14.5
Italy	14.9
Netherlands	12.0
Ireland	11.7
Germany	10.3
UK	9.3
France	8.9
Belgium	5.8
Luxembourg	5.7

Source: European Commission.

Conclusion

Competition can ensure that resources are used efficiently and, certainly, there is a need to increase energy productivity for environmental as well as economic reasons. But obsession with low prices and competition can be counter--productive when it comes to developing new energy technologies. There are well-established 'learning curves', showing how, if given support to continue to develop and build markets, technologies improve in performance and cost over time (OECD, 2000).

As has been illustrated, the competitive NFFO and RO approaches do provide some market enablement for near market technologies, but do not help new technologies to develop. In 2004, the Carbon Trust produced a report with the Department of Trade and Industry, which accepted that there was a 'funding gap' in the UK between the end of the R&D phase and the market up-take phase, which the RO could not bridge. It looked at various options for remedying this situation, including the expansion of the capital grants approach (Carbon Trust, 2004). The problem then is that the level of government grants that would be needed to support a significant development programme for new technologies could begin to dwarf the support being raised from consumers via the RO system. In effect the support system could default to a traditional technology push approach, thus undermining the aim of having a market oriented competitive system.

This would no doubt be opposed by OFGEM, the energy regulator, which is charged with stimulating competition in the energy sector in the belief that this would keep prices low. However, views on the need for low prices may change. In 2003, OFGEM argued that 'competitive energy markets give lower prices in the long run, and so help to mitigate the inevitable social impacts as necessary environmental measures raise energy costs' (OFGEM, 2003). Moreover, when the DTI announced an expansion of its offshore wind programme in July 2003, the minister, Patricia Hewitt, seemed to condone initially high prices and accept the 'learning curve' argument: 'the more we build of these offshore wind farms the more the price will drop as people learn to do it more efficiently' (Hewitt, 2002).

However, there may be some way to go before the UK abandons low prices as a key policy aim. The government still remains very nervous about potential consumer and voter reaction to price rises – and state subsidies. This may be short-sighted. Fossil fuel prices seem bound to rise, while, in the longer term, the new sustainable energy technologies should be able to expand to meet energy needs at reasonable costs. Indeed, the net cost to society could well be less, when the huge potential costs of dealing with the impacts of unmitigated climate change are included. That lesson seems to have been learnt elsewhere in the EU.

References

Carbon Trust (2004) *The Renewables Innovation Review*, DTI and the Carbon Trust, London

DTI (1999–2001) 'New and renewable energy: Prospects for the 21st century', consultation paper series, London, Department of Trade and Industry

EAC (2002) 'A Sustainable Energy Strategy? Renewables and the PIU review', House of Commons Environment Audit Committee, Session 2001–2002 Fifth Report, London, July

EEA (2001) 'Renewable energies: Success stories', Environmental issue report no 27: Ecotec Research and Consulting Ltd and A. Mourelatou, for the European Environment Agency, Copenhagen 2001, http://reports.eea.eu.int/environmental_issue_report_2001_27/en

ElGreen (2001) *Action Plan for a Green European Electricity Market*, ElGreen project report (WP7); Claus Huber, Reinhard Haas, Thomas Faber, Gustav Resch, John Green, John Twidell, Walter Ruijgrok, Thomas Erge; Energy Economics Group (EEG), Vienna University of Technology, Austria

Elliott, D. (2003) *Energy, Society and Environment*, London, Routledge

Hewitt, P. (2003) as reported by *BBC online*, 14 July

OECD (2000) *Experience Curves for Energy Technology Policy*, International Energy Agency. See also the review McDonald, A and Schrattenholzer, L. 'Learning rates for energy technologies', *Energy Policy*, vol 29, pp255–261

OFGEM (2003) 'Competitive markets and the low carbon economy', Boaz Mosell, OFGEM, paper to the annual conference of the Parliamentary Renewable and Sustainable Energy Group, London, July

RPA (2002) Submission to the Governments Energy Review, Renewable Power Association, London, Sept; see also the Lords Science and Technology Committee, 4th report Session 2003–04, London 2004

Toke, D. (2005) 'Are green electricity certificates the way forward for renewable energy?' Paper to 4th International Conference on Business and Sustainable Performance, Aalborg, Denmark, 14–15 April

11

Renewable Energy: Political Prices or Political Quantities

Frede Hvelplund

Two models: Political prices or political quantities

The two main renewable energy governance systems that are discussed in this chapter are:

1. the 'Political price/amount market' model, often called the 'feed-in model', which has politically set prices for RE (renewable energy) electricity, and where the produced quantity of RE electricity is determined on the market;
2. the 'Political quota/certificate price market' model, often called the 'certificate market model' in which the RE electricity quantity is politically fixed as a quota and the RE electricity prices are determined on the market.

International experiences

The 'political price/amount market' model has, among others, been successful in Germany and Spain. Those countries boasted around 65 per cent of the growth in world wind power production from 2001 to 2003. In the same period the 'political quota/certificate price' governance system in the US and the UK, with much better wind conditions, has resulted in an increase in wind power production equal to 17 per cent of the growth in world wind power production.

In 2000, the German parliament approved a new advanced political price/amount market governance model, and in 2001, the French Parliament accepted a similar model. We call the German model advanced, as it includes both a price differentiation between good and poor wind sites and an innovation pressure, decreasing the kWh price by 1.5 per cent annually in relation to year of plant construction.[1]

In the UK a political quota/certificate price market was introduced in 2002. This model has resulted in relatively high prices for wind power, 9–10

Euro cent per kWh (Mitchell et al, 2006), which are very high prices, when taking the exceptionally good UK wind sites into consideration. The high prices are caused by the fact that the investors in wind turbines in the UK cannot get any long term contracts, which forces them to require very short-term pay back periods, mainly due to high price risks. In Texas another version of this model has been implemented. However, it is very difficult to learn from the Texan experiences as the political quota/certificate price model is mixed up with a system of tax deductions, which plays a crucial role for the economy of a given project. In Sweden and Japan a political quota/certificate price model was introduced in 2003. In Japan it replaced a political price/amount market system, resulting in a decrease in the wind power market from 330MW in 2003 to 50MW in 2004. In Sweden the quota model so far has been very unsuccessful (see Lida and Kåberger, 2004).

A couple of years ago, the EU commission accepted the use of the political price/amount market model in the latest Directive proposal, which has been accepted by the Council of Ministers. This keeps the question of the future regulation framework open. Therefore, the political amount/certificate price market model is presently losing its foothold and is no longer 'the only possible future regulation model'. Furthermore, this development was supported by a European Court adjudication[2] which said that the German political price/amount market model is not to be regarded as illegal state aid, and is therefore acceptable as a way of regulating RE development (Meyer, 2003).

In 2004 the political quota/certificate price market model has been used in practice in for example Britain, Japan, Sweden and Texas, and its performance with regard to furthering a stable implementation of RE technologies at a reasonable price has been very disappointing.

Changing the Danish RE governance system

The 1999 electricity law included a change from a political price/amount market model to a political quota/certificate price market, which should be implemented at the end of 2003. It should consist of two price elements, namely an electricity market price plus a green certificate price. It was not implemented, and it is difficult to say whether it will ever be implemented.

This intended double price model should include the two price components below.

Price component 1: A spot market price for electricity

With reference to average 1999 and 2002 prices, the prices on Nordpool are oscillating between 1.3 Euro cents per kWh in 1999 and 3 Euro cents per kWh in 2002. The Nordpool spot market price is hampering any innovation, both because of its considerable fluctuations and because of its tendency to reflect the short-term marginal costs of fossil fuel technologies in periods of excess capacity. For instance, in 1999, 2000 and 2001, the spot market price was around 1.6 Euro cents per kWh, which is close to the short-run marginal costs of coal-fired plants.

Price component 2: A price for green certificates

A political quota/certificate price market should be established, where consumers are obliged to buy a certain quantity of renewable energy (RE), and where producers of RE obtain a green certificate, which is then sold to the consumers, so that they can fulfil their quota obligation. The quotas of RE for a given number of years should be set by the Danish parliament. The law established price limits, so that the certificates could not be cheaper than 1.30 Euro cents per kWh or exceed 3.6 Euro cents per kWh. The 'Green Certificate' price would therefore oscillate between 1.30 Euro cents and 3.6 Euro cents per kWh.

Although Denmark gets almost 20 per cent of its electricity supply from wind power, the green certificate market was not implemented, and in 2002 the new right-wing government introduced new rules, with the Nordpool electricity spot market price plus a fixed 1.30 Euro cents per kWh CO_2 premium and a price ceiling at 4.9 Euro cents per kWh. A probable average price will be around 3.5 Euro cents per kWh, which is close to the long term marginal cost at a large coal-based power plant. This was around 50 per cent of the price paid until 2002, and below the level where new windmills will be built. No new contracts have been signed under these new terms in 2002[3] and 2003, simply because the expected payment per kWh is 1–2 Euro cents per kWh lower than the wind power production costs. The outcome of this policy has been a halt for wind power development.

In 2004 a new agreement between a broad spectrum of parties in the parliament was reached. The main component of this agreement was the establishment of two 200MW offshore projects and the establishment of a further 350MW wind power onshore during the next five years. The price for the offshore plants is established in a tendering process, and will be a fixed price for 50,000 hours of full capacity productions. This equals a fixed price for around 12 years. After this the price will be electricity market price plus a CO_2 payment. The onshore plants are getting the electricity market price plus a CO_2 payment of 1.6 Euro cents per kWh for a period of five years.

In fact the pricing for offshore wind power has all the important features of the political price model, as it gives a fixed price for a relatively long period. The pricing for onshore plants also now has a fixed price element for at least a period of five years. So to a large extent the political price/amount market system has entered the Danish energy scene again.

The political quota/certificate price market model is now being practiced in an array of countries, and has generally shown results that are not satisfactory, when compared with the theoretically based promises of its proponents. The analysis on the following pages might explain why this model has worked out to be less successful and price efficient than the political price/amount market model. As a first step in this analysis we will discuss the RE characteristics and as a second step analyse the institutional context, which we see as adequate for the discussion of RE governance systems. But before doing this we need to do some ideological cleaning by finishing the discussion regarding which model is a real 'market model'.

Which governance model is a market model?

First of all we should be aware of the many markets shown in Figure 11.1 below, when discussing which governance model one should call a market model. So when we talk about the RE market, which market(s) are we then referring to? Are we talking about the electricity price market or the RE equipment market, or are we talking about the investor market and/or the market for flexibility and regulation? Or are we talking of a combination of these markets? In the Danish discussion up to the 1999 reform, the political quota/certificate price market system has persistently been 'sold' to the public as being more *market-oriented* than a political price/amount market system. First of all the inherent context in this discourse only included the price element. It did not seem to be included in the perceived reality that the amount of RE was determined as a politically set quota. This delusion has been so successful that it is now an almost undisputed 'fact', that Green Certificate trading on the basis of a price market plus quota regulation should be 'the genuine market' system. Table 11.1 illustrates why this is a delusion.

Table 11.1 *Political and market determination of price and quantity in two regulation models*

	'Political quota/certificate price market' model	'Political price/amount market' model
Price determination	Market and political	Political
Amount determination	Political	Market

Note: The price in the planned Danish political quota/certificate price market model is partly politically defined, since the law determines that the certificate price should not be below 1.30 Euro cents per kWh or above 3.57 Euro cents per kWh. Later on the new right wing government decided to put a ceiling on the total price paid to RE-based electricity.

As illustrated in Table 11.1, the political quota/certificate price market model shows the political interference on the market at the quantity as well as at the price levels in the Danish case with its price ceiling. The only political intervention in the political price/amount market model is at the price level.

The political quota/certificate price market model, therefore, is not more liberal or market-oriented than the political price/amount market model. On the contrary, the political quota/certificate price market model, due to its 100 per cent state-governed amounts, and partly state-governed prices, includes more state governance, than the political price/amount market model.

The political quota/price market model does not at all, even in the limited context of Box 1 in Figure 11.1 below, constitute any clean market model. This conclusion is reinforced, when extending the analytical context so that it includes the totality of the Figure 11.1 context, with its equipment and investor market, and its variation in natural resources from location to location, as we will see in the following.

Renewable energy characteristics and governance systems

When examining the various arguments in this debate, it is striking that there is no thorough discussion regarding the basic differences between 'the nature' of fossil fuel and RE technologies. These fundamental characteristics where RE technologies differ from fossil fuel and nuclear technologies are among others:

- Many RE techniques are *energy automations* having a cost structure with a relatively *high percentage as investment costs* and proportionally low annual running costs. Both these things implicate high sunk costs and therefore investor risks, and an increasing importance in keeping the competition at the market for RE technology alive, as this market contains by far the largest cost fraction
- Mostly RE technologies have *a site dependent* resource base, where the resource base varies from location to location. Consequently, there is a need for a governance system that furthers an EU-wide 'site efficiency'[4] generating process rather than a 'mono price' (one price on a European market) based price competition
- RE technologies are often *dispersed technologies*, being distributed around the country. Often close to residential areas, this makes it particularly important to involve neighbours and people from the region in the design, development and ownership of RE projects
- *Increasing market share results in growing visibility*, enhancing the need for local participation. In a mature phase of implementation,[5] especially wind-power becomes increasingly visible, making local resistance against wind-power projects increasingly frequent. In this phase RE has also become a serious competitor to the old energy companies, raising a tough resistance from these interests
- RE technologies have a value-added profile that is very different from the value-added profile of the fossil fuel based energy companies. This enhances the resistance from these companies, simply because it is relatively difficult to maintain the economical turnover, when replacing fossil fuel technologies with RE systems. When this is combined with the relatively weak *newcomer position of RE* it is important to develop and maintain a general political and public support for RE technologies. In the initial phases of RE development the costs are high, and a public support for the payment of these initial development costs is a must. In later phases the dispersed RE technologies will be seen in the landscape to an increased extent, and the public engagement and support become necessary due to this cause
- Wave, wind and photovoltaic energy are fluctuating forms of energy, requiring an infrastructure that is able to cope with these fluctuations in a cost-efficient way. This need for establishing a RE infrastructure has been increasingly clear in Denmark, where wind power in 2003 amounted to around 20 per cent of the total electricity consumption. The present electricity infrastructures are designed for large fossil fuel- and uranium-based electricity systems, and are therefore advantageous for these systems. An active political process is needed in order to establish an energy

infrastructure, which also develops and supports the often decentralized RE-based systems (Lund and Münster, 2004; Lund and Münster, 2003; Alberg Ostergaard, 2003).

Institutional context and the analysis of regovernance models

When analysing an RE governance system it is necessary to establish a context, which is adequate seen in relation to the energy policy goals, and the six special characteristics of RE technologies. The present Figure 11.1 analysis of the institutional context is focusing on the institutions in which the RE technology market/development is embedded, exemplified by a 'political quota-/certificate price market' governance system.

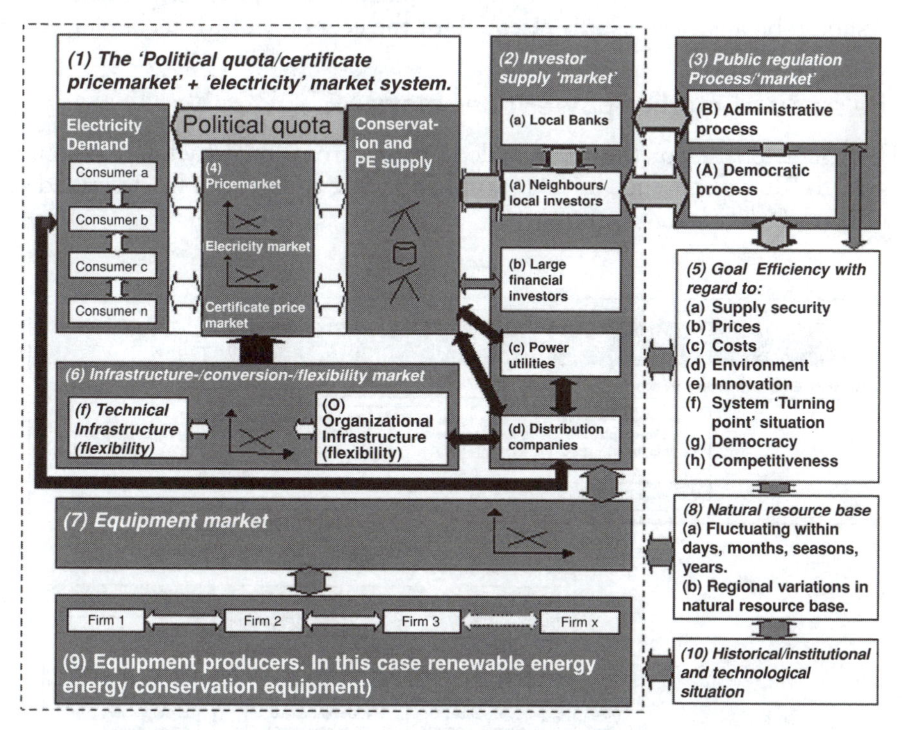

Note: The figure focuses on the interrelations between the policy goals (Box 5), the public regulation process (Box 3) the investor supply market (Box 2), the equipment market (Box 7), the conversion and flexibility market (Box 6), the natural resource base (Box 8) and the historical institutional situation (Box 10). The specific interest of this paper thus resides in an analysis of the links between the above areas and the specific characteristics of RE mentioned above seen in relation to different RE Governance systems. This analysis should be performed for both the initial phase of RE implementation and the next phase of RE implementation to which Denmark and Schleswig Holstein in Germany have reached.

Figure 11.1 A relevant institutional context around RE governance systems, and the case of the 'political quota/certificate market' model

It should be mentioned that the theoretical approach used by the Danish Ministry of Environment and Energy before proposing the 1999 change in the RE Governance system did not include any systematic description of its perception of the institutional context of the reform.

The context, which was implicit in the analysis of the Ministry of Environment and Energy only included some considerations concerning the market described in box 4 in Figure 11.1, and not any systematic description or knowledge regarding any other parts of the Figure 11.1 institutional context.

The most important messages in the Figure 11.1 context are the following:

- the study of market power does matter;
- the RE equipment market should be included in the context (Box 7);
- 'RE-investor supply' market and its composition does matter (Box 3);
- the regional variation of the natural resource base makes a difference, when designing an RE Governance system (Box 8);
- the infrastructure for electricity conversion and production flexibility should be analysed, as it is essential for the economy of RE technologies.

The study of market power does matter

The characteristics of the value-added change from fossil fuel and uranium based to RE systems such as solar, wind and wave energy[7] can be described as in Figure 11.2 below.

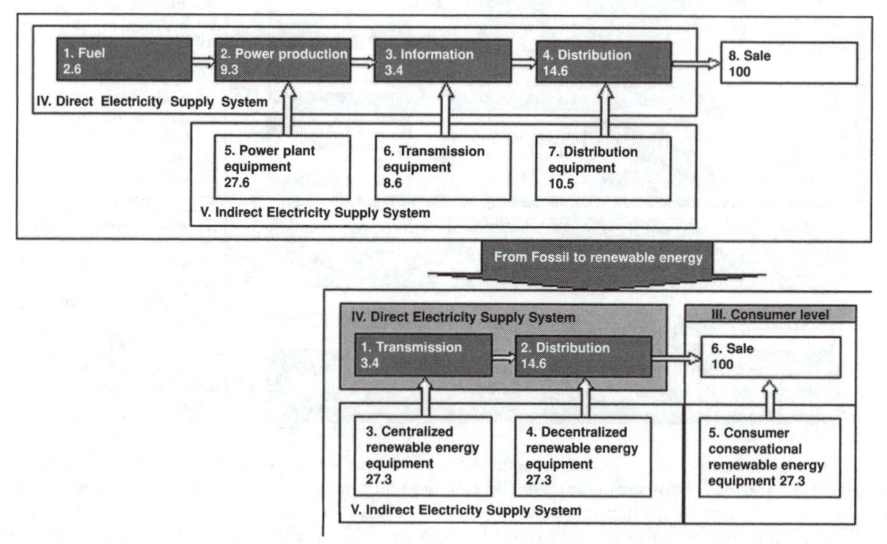

Note: In the old fossil fuel-based system a 100 Euro sale at consumer level will have the value added divided between the different levels of vertical integration as shown in the upper figure. The figure at the bottom shows a possible value-added distribution in an energy conservation and an RE system.

Figure 11.2 The change in value-added profile connected to the change from uranium and fossil fuel, to RE and energy conservation systems

From Figure 11.2 we can observe that the value-added chain of RE and conservation (REC) technologies differs clearly seen in relation to the value-added chain in a fossil fuel-based system within two areas:

1. In the REC value-added chain, the fossil fuel value-added part has disappeared, and is replaced by investment in RE capital equipment
2. In the REC value-added chain, the power production value added in a specific direct electricity supply system organization has been replaced by 'RE system automation', where it is probable that the maintenance, at least at the decentralized and consumer level, will be performed by the producers and suppliers of the wind turbines, the photovoltaic cells, the hydrogen production system, the electricity battery charging system, and so on. The need for a specific power production organization might decrease or disappear, as the day-to-day work on a power plant has been replaced by automatons requiring maintenance from, for instance, the RE equipment producers.

The result is that a transformation to RE systems *decreases the value added on the market for electricity from the present power companies, and increases the value added on the market for RE equipment*, seen as a proportion of the sales price at the consumer level. This basic condition constitutes an important problem for the existing fossil fuel and uranium companies at the market, as RE innovation makes them lose turnover and profit. As a result they tend to be financially 'forced' to oppose any transformation to a system based upon RE technologies.

Due to the different value-added profiles of RE and fossil fuel/uranium technologies, one therefore can expect that fossil fuel/uranium companies in general will oppose RE technologies.

It is also crucial to analyse monopolistic and/or oligopolistic tendencies at the different markets. Figure 11.1 illustrates by means of the black arrows the ownership relationship between the demand and supply sides. The electricity distribution companies in Denmark are the owners of the power utilities, which again can also build and own RE plants. There are similar ownership links in all the northern European electricity systems. The above ownership links naturally cause huge problems with market control carried out by large fossil fuel- and uranium-based actors, enabling these to squeeze out investors[8] who for instance do not have a strong capital background and ownership relations to the distribution companies. This 'factor' should be seriously taken into consideration when defining the selection criteria presiding over the choice of an appropriate RE institutional framework.

The RE equipment market should be included in the context (Box 7)

It is necessary to establish a systematic analysis of the contents and inter-relationships between the market for electricity and its Green Certificates (Box 4), and the market for energy equipment (Box 7).

As a consequence of the Figure 11.2 change from fossil fuel and nuclear energy to RE systems, the direct electricity supply system organization might decrease its size until it only consists of the transmission organization and the distribution network organization.

Consequently, a main characteristic of technological change can be that the part of the indirect electricity supply system that directly relates to equipment for power production, transmission and distribution might increase from 46.7 per cent of the total value added in the fossil fuel system to 81 per cent of the added value in an RE system. This is mainly due to the fact that fuel import is replaced by RE equipment/capital.

Concretely, the change to some types of RE systems, such as wind power, *represents an automation of electricity production*, with 80–90 per cent as investment costs and the rest as maintenance costs. Once the wind turbine, solar cell, wave power plant is built, hardly anybody works on it. It just produces electricity for 20–40 years, and is usually maintained by service units linked to the wind turbine factories in the case of windpower. Consequently, once built, the wind turbine – photovoltaic – or wave power plant will not work more efficiently just because of the introduction of competition at the electricity market. In a traditional electricity service supply system, the situation is totally different. At least in theory, one might expect that market competition on the electricity market might put pressure on the power utilities, which will then rationalize and dismiss some of the people employed at the power plant. Once it is built, a wind turbine can dismiss nobody. Any potential personnel compression can then only happen at the level of the wind turbine factory.

At present, fossil fuel back-up systems are still being used. But in the future, a system with different types of storage techniques, such as for instance hydrogen storage, will be developed. These systems also appear to be 'automatic storage systems', which hardly require any maintenance to be performed by employees in an energy organization.

Thus, when introducing RE systems, the market for electricity is decreasing in importance, whereas the market for energy equipment is becoming increasingly important.

In Table 11.2, the relative importance of the market for equipment is compared within a fossil fuel system and a RE/electricity conservation system.

Table 11.2 *Moving from fossil fuel to RE means a change in value added from the electricity market to the equipment market*

	Equipment market	Electricity market
Fossil fuel systems	47%	53%
Renewable energy systems (electricity)	81%	19%

Placing this discussion in the context of Figure 11.1, it is especially important to include the interplay between an RE governance system and the competition at the market for RE equipment (Box 7 in Figure 11.1). One should be aware that there is not only an electricity market to discuss, but also an equipment market – especially in a situation of technological change, where there is a transformation to RE automations.

In a political price/amount market system, the wind turbine factories are, and historically have been able to increase their total turnover and profit by decreasing wind turbine costs and prices. It is due to this system that the wind power costs per kWh have decreased by 80 per cent since 1980. In a political quota/certificate price market system, the quantity of wind power is politically decided several years ahead. Consequently, the wind turbine producers, as a group, can only increase their turnover by increasing prices. This motivates the wind power firms to establish 'strategically collaboration' or mergers, in order to achieve increased market control.

In the present situation of technological change, the 'political quota/certificate price market' system ends up introducing price competition on a dwindling market and abolishing market competition on the expanding equipment market. The advanced 'political price/amount market' system supports market competition on the growing market for equipment, and therefore, is especially well suited to the present type of technological change.

'RE-investor supply' and its composition does matter (Box 3)

(Boxes 3A and 2a are connected to each other displaying the importance of the link).

To include neighbours and local banks in the context is a must, and is also linked to the discussion of governance systems and the need for *price stability* over a longer period of time. Long term price stability is necessary for the strategically and politically important investor group – neighbours and local communities. Without long term price stability the neighbours and local communities are not able to participate as viable investors at the investor market.

One of the main historical secrets behind the Danish wind power success was that a system of cooperatives and neighbour and local ownership was furthered by the public regulation, resulting in more than 120,000 wind turbine owners in Denmark. People seem to like wind turbines, when they own them, and are not annoyed by the noise and visual inconveniences; especially when receiving a fair compensation. However, with a system of distant utility or shareholder ownership, the local inhabitants are only getting the disadvantages without any compensation. This is seen as unjust and results in increasing local political resistance against wind power. It is as simple as that.

Neighbours and local communities are not only important investors due to their political importance, but also as a result of their cultural and economic independence of fossil fuel companies. Therefore, historically these groups have also been the initiators and innovators within the investor group in Denmark during these past 25 years.[9] Thus, there is also a link back to the descriptions of both the character of political processes furthering the innovation potential within this field, and the policy influence on the composition of the 'investor supply'.

The new, not yet implemented Danish political quota/certificate price market system results in very fluctuating prices due to a range of different factors. The cost structure of wind turbines results in a very vertical supply curve. Once wind turbines are built, they will not close down production, as the majority of costs are fixed. Annual wind resources vary by up to 30 per cent, making it impossible to govern by quotas, as the annual change in wind resources will surmount the size of any quota increase. Furthermore, the market will be characterized by large actors who will have the ability to manipulate market prices. Altogether, this causes the certificate prices to fluctuate heavily and often in a manipulated way, making it impossible to draft trustworthy wind power project budgets. Consequently, the old procedure of financing a wind turbine project together with the local bank is no longer possible. Only large financial investors and power utilities are left over in the market. This means that the number of investors and consequently the competition between investors is decreasing, resulting in higher project prices. Furthermore, it results in increased local and regional political resistance against wind power.

It is important to underline that the need for public participation in RE projects should not be limited to an introduction phase. Due to the fact, that RE technologies are dispersed around an area/country, an increased market share means that more and more people will see and hear the RE units. Consequently, it is, concurrently with increased RE market share, increasingly important to achieve acceptance and participation of local people living in the neighbourhood of RE plants. Methodologically, it is important to be aware that this type of conclusion can only be developed when placing the RE development, and its specific characteristics of being geographically dispersed, in the context of Figure 11.1, Boxes 2a and 3a and b.

The regional variation of the natural resource base makes a difference when designing an RE governance system (Box 8)

Any context describing the RE governance systems should consider that the RE resource base varies from location to location. And the use of RE is characterized by 'harvesting' and transforming to electricity by means of the same plant, where the fossil fuel and uranium technologies are characterized by 'harvesting' the oil, coal and uranium in one phase and in another phase processing it to electricity at power plants.

This means that the market for RE technologies should take care of both the 'harvest' procedure and the transformation to electricity in one market process. Therefore, it is not possible just to use the market mechanisms from the fossil fuel technologies upon the development of RE technologies.

The RE resource base variations, site efficiency and RE governance systems

As mentioned initially, a very important context characteristic (Box 8, Figure 11.1) is that RE technologies have different natural resource capacities from location to location. A wind turbine on an inland site in Germany only produces around 50 per cent of the quantity produced on a very good coastal site in Ireland or Scotland. When dealing with nuclear, natural gas or coal-fired

power plants, variations from location to location will mainly depend on differences in cooling facilities, with a coastal site being slightly cheaper than an inland site that needs cooling towers.

Due to the declared EU goal[10] of increasing the percentage of RE-based electricity production (not including large hydro) from 3.2 per cent to 12.5 per cent during the 1997–2010 period, it is necessary not only to exploit the best coastal sites for wind power but also to use good inland wind sites all over Europe. With a political quota/certificate price market system for the EU, there would be only one certificate price for wind power in the EU.

Regarding wind power, Figure 11.3 shows the different production prices in a 'model union' consisting of three case countries.

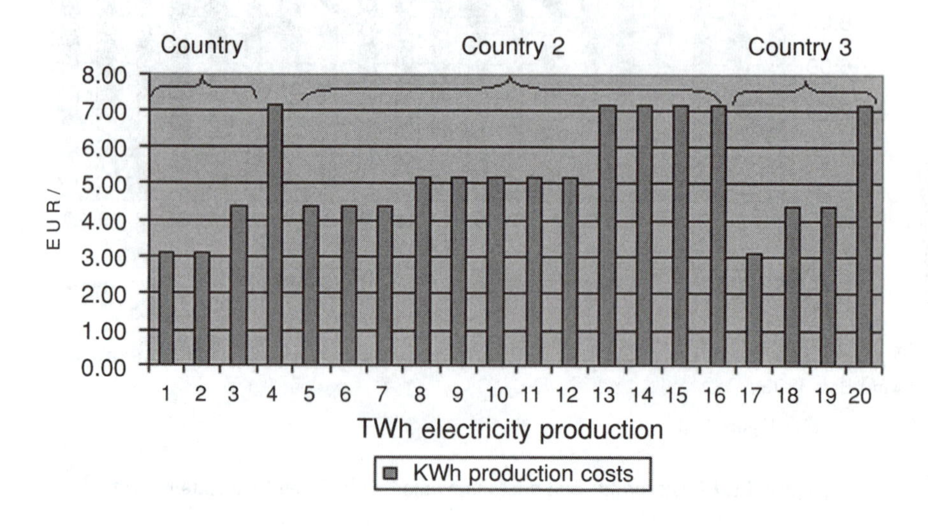

Source: Hvelplund, 2001.

Figure 11.3 Wind power production costs in three countries

The costs of producing wind power vary from around 3 Euro cents per kWh on a very good coastal site, in Ireland for instance, to around 7 Euro cents per kWh on good inland sites in central Europe. As wind power production on inland sites is also required if climatic and security of supply goals are to be met, and there will only be one price for Green Certificates in Europe, the price level needed in order to produce wind power on inland sites, especially in Central Europe, will be around 9 Euro cents per kWh. This price is required because some profit is necessary to stimulate investment. This price would result in very high profits on the good coastal wind sites, with between 90 and 160 per cent profits on these (wind class 0 and 1) sites.

RE resources and the 'political quota/certificate price market' model

In the political quota/certificate price market model, a quota politically regulates the amount of RE-electricity. The price is determined on a market for electricity and a market for certificates.

In Figure 11.4, the three countries have introduced a common political quota/certificate price market system. Linked to their different wind resources, this governance system entails the following wind power cost functions and profits for wind site and wind turbine owners:

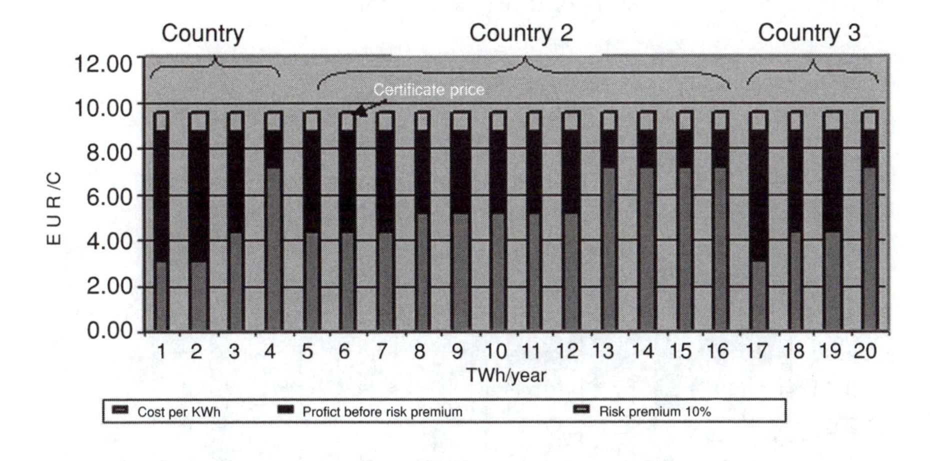

Note: Assumptions: 10 per cent risk premium due to fluctuating prices. 20 per cent profit demand on a wind class 3 site.).

Source: Same as Figure 11.3.

Figure 11.4 Costs, profits and prices in a union-wide Green Certificate market

Figure 11.4 shows that on this market, there is one price for wind power all over the Union, namely the one developed in the EU certificate market.[11] Politicians have established a quota system that ensures that an annual production of 200TWh RE-electricity is implemented. In order to reach this TWh goal, it is necessary that the kWh price on the market is at least so high that it becomes profitable to use wind class 3 sites, which concretely translates into a price slightly above 8 Euro cents per kWh. Additionally, the fluctuating prices on the certificate market imply that the investors demand a 10 per cent risk premium, increasing the price to an average of 9.8 Euro cents per kWh.

RE-resources and the advanced 'political price/amount market' model

We call the model 'advanced' because of its ability to foster a competition process, which increases 'site efficiency' in a non-bureaucratic way.[12] In Figure 11.5, the effects of this type of regulation are illustrated.

Cost and price efficiency of the advanced political price/amount market' model

The price performance of the advanced political price/amount market model is shown in Figure 11.5. The figure displays exactly the same cost structure as for the three countries in Figures 11.3 and 11.4. The only difference is that the advanced political price/amount market model has a politically defined, site-dependent price framework, which makes it possible, as shown in the figure, to decrease the profit on good wind sites without destroying the economy of inland wind sites.

The infrastructure for production flexibility should be analysed, as it is essential for the economy of RE technologies

The economy of RE sources and especially wind power is a function of the technical infrastructure in which the particular RE is embedded. Wind power, photovoltaic and wave energy production varies according to the amount of wind, sun and wave energy. When the RE proportion of total electricity production is lower than around 10–15 per cent, these variations will be absorbed by the present electricity system. When the proportion of RE rises above this percentage, and there is, as in the Danish energy system, a very high proportion of cogeneration, it results in a surplus electricity production due to wind and heat bound electricity production. With almost 50 per cent of the

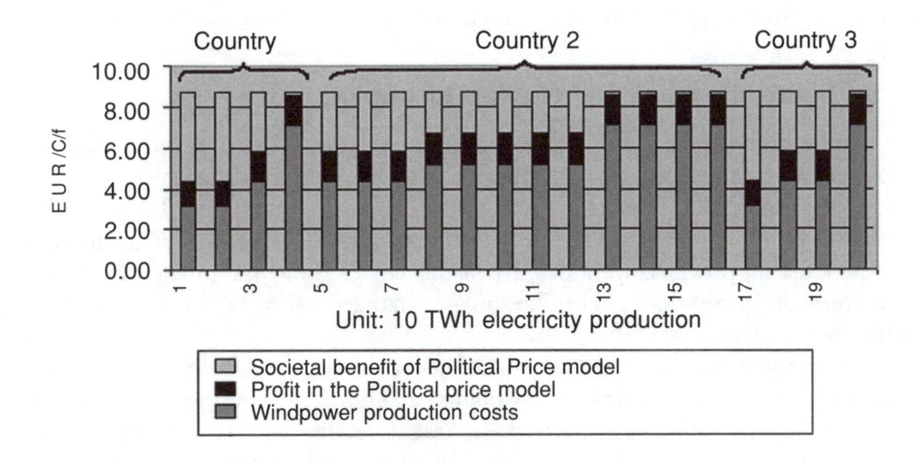

Note: Profits are a percentage of costs: 40 per cent on wind site 0; 35 per cent on wind site 1; 30 per cent on wind site 2 and 20 per cent on wind site 3. These profit percentages are approximations of the profits that we have calculated on the basis of German wind power prices in 2001. Since prices are politically guaranteed, there is no need for any risk premium.

Source: Same as Figures 11.3 and 11.4.

Figure 11.5 Price, profit and costs in the political price/amount market model

heat consumption based on cogeneration of heat and electricity and around 20 per cent of total electricity production coming from wind power, the Danish electricity system has reached a boundary with several periods of 'excess' wind- and heat bound electricity production. This 'problem' will increase in parallel with the fulfilment of the official Danish goal of producing at least 50 per cent of electricity by means of wind turbines around 2030.

The present official strategy to solve this problem, that is huge electricity exports during windy periods, entails the risk of very low sales prices for wind power. A scenario built on an increasing proportion of forced export at low prices will probably result in political resistance against wind power and make a CO_2 reduction policy increasingly difficult to champion. The present successful expansion of wind power capacity in northern Germany, producing electricity during the same periods as Denmark, reinforces this problem.

Another strategy is to use wind power in regional systems, consisting of heat pumps linked to cogeneration, flexible regulation of cogeneration units and the establishment of cogeneration and wind turbine systems, which, on their own, can already stabilize the grid. This strategy will make it possible for Danish system operators to decide when to import and when to export electricity. At present, this type of system is being discussed in Denmark and seems to be much more promising economically than the current official export strategy (Lund and Münster, 2004).

As a consequence of the above discussion, it is important to analyse to what extent a well-functioning competition exists between various technical solutions. This is crucial with regard to the establishment of a flexible infrastructure, which results in a politically acceptable economy for RE technologies in general and for wind power in particular. The 'Infrastructure, conversion and flexibility market' should therefore imperatively be included in the analysis (Box 6 in Figure 11.1).

RE is a politically weak 'newcomer' technology

The competition between RE technologies and existing fossil fuel and uranium-based power companies is very often a win/lose situation. If wind power production increases, then the profit of the power companies, ELSAM, in Denmark, E.ON in Germany, and so on decreases. Due to excess capacity linked to these existing power companies, RE technologies are, when owned by these companies, often competing with the short-term marginal costs of existing plants within these companies.

Hence, these old fossil fuel and uranium-based companies do not have real economic interest in investing in RE plants. Therefore, it is important that the politicians are establishing development tracks facilitating the conditions for investors without 'stranded costs' in fossil fuel or nuclear plants. As described above, the political quota/certificate price market system tends to hamper the possibilities of such independent neighbour and local investors. Therefore, this governance system is leaving the economically unmotivated uranium and fossil fuel utilities alone with regard to investments in the RE market. This is not the case with the political price/amount market system, which, with its foreseeable prices, makes it possible for independent 'neighbour and local' investors to establish wind turbine projects.

Conclusion

The political quota/certificate price market system introduces an inefficient competition between the RE automatons, and weakens the increasingly important competition between RE equipment producers.

It hampers competition between investors by making it difficult for neighbours and local investors to invest in wind turbines.

Due to its mono price character, it gives too high profits to wind turbine owners at very good wind sites.

Table 11.3 *A comparison of the political price/amount market model with the political amount/certificate price market model*

	'Political price/amount market' model	*'Political amount/certificate price market' model*
(a) Is it a market model?	The price is political; the amount is decided upon a market	The amount is political; the price is partly decided upon a market, partly politically set
(b) Is it furthering competition between equipment producers?	The equipment producers as a group can expand sales and profit by lowering production costs	The equipment producers are facing a 6–8 year politically set annual production quota. They can expand profit by lowering costs and and especially by increasing sales prices
(c) Can it price differentiate in space?	Yes, as is done in the German model	No. In this 'mono-price' model, the same price is paid to the very good coastal sites, as to the good inland sites
(d) Can it price differentiate in time?	Yes, as is done in the German model	No. The same price has to be paid during the whole lifetime of an RE plant
(e) Can it lower the price in parallel with productivity improvements?	Yes, as is done in the German model. 2002 wind turbines are getting 1.5% lower KWh prices, than 2001 wind turbines	No. The quota has to be set for a 6–8 year period, and new improved wind turbines are getting the same certificate price as less efficient wind turbines built earlier
(f) Does it support neighbour and local investors?	Yes. The foreseeable prices make it possible for local groups to get loans from local banks	No. The very fluctuating and possibly manipulated prices make it too risky to invest, and difficult to obtain loans from local banks
(g) Does it put a cost pressure upon equipment producers?	Yes. Almost the same cost pressure is put upon investors at good wind sites as on investors at bad sites	In general, no. The mono-price system gives very high profits to owners of good coastal sites. This weakens the cost pressure upon equipment producers
(h) Does it support independent investor groups?	Due to the above (f), yes	Due to the above (f), no

The political quota/certificate price market system is very far from being a market model, as the RE amount is politically decided and the certificate market price is also politically influenced.

Table 11.3 summarizes our conclusions.

Therefore, the conclusion is that it is time to find an RE governance model that considers the specific needs and characteristics of RE technologies and to analyse this model in an adequate context. The present analysis strongly indicates that a political price/amount market model in the Figure 11.1 context is far better than the political quota/certificate price market model with regard to cost, price as well as implementation efficiency.

Furthermore, a common EU model, based on the principle of site efficiency, would be much more flexible, cheaper and easier to pursue than the political quota/certificate price market, or mono price model, which is designed for uranium and fossil fuel technologies, and represents a governance model designed for the technologies of yesterday. The present disappointing practical experience with the political quota/certificate market model underlines this conclusion.

Notes

1 By this is meant that a wind turbine built in 2003 – during its lifetime – will obtain a kWh price, which is 1.5 per cent lower in running prices, than the kWh price of a wind turbine built in 2002. A wind turbine built in 2004 will – during its lifetime – obtain a price which is 1.5 per cent lower than the price of a wind turbine built in 2003, and so on.

2 13 March 2001: Judgement of the Court, Case C-379/98.

3 Windmills built in Denmark in 2002 are all linked to the rules where it is possible to obtain a relatively good price, if you can document that a certain capacity of wind turbines smaller than 150kW has been scrapped.

4 By 'site efficiency' is meant efficiency with regard to the exploitation of a specific regional renewable energy resource.

5 Like wind power in Denmark, supplying around 20 per cent of total electricity consumption in 2003.

6 In relation to energy policy goals, and the RE characteristics described above.

7 When dealing with biomass-based RE technologies, the value-added chain will also include biomass fuel procurement.

8 Due to the cost structure of RE production (for example wind power), characterized by both high capital costs (85–90 per cent of costs) and low running costs (10 per cent), low price periods will drive investors with a weak capital background to bankruptcy. This state of affairs may then potentially result in small investors being bought up by big investors with a strong capital background, such as the large utilities.

9 On 1 February 1999 in Copenhagen, S. Auken, the Minister of Energy, said at a meeting with Tage Dræby, Hans Bjerregård, Frede Hvelplund, and the chief of his department, Leo Bjørnskov, as well as others from the department, that we had now reached a level where development could no longer be 'driven by 12 idealistic schoolteachers'. With this colourful metaphor, he hinted that it was now time for big capital and large companies especially, to invest in 'offshore wind power' development. The assumption behind this statement, whereby technological development is now out of the reach of local people and their

creativity, does not seem to be well grounded, especially when one considers the big problems linked to the present utility-driven (ELTRA and ELKRAFT) model of exporting surplus wind power to neighbouring countries. The technological alternative to this export model is to integrate the fluctuating RE sources locally. This strategy might well prove much more efficient than the utility-driven export model and the creativity and support of the 'twelve schoolteachers' is certainly required to develop and implement this local integration.

10 See Annex 1. In 'Draft directive of the European Parliament and of the Council on the promotion of electricity from renewable energy sources in the internal electricity market', December 2000.

11 In order not to complicate the argument, we assume that the basic price for electricity and the 'Certificate price' is the same in all three countries.

12 It is worthwhile mentioning that the UK tendering system could, in theory, be called an ideal system, in the sense that it potentially fosters the pursuit of site efficiency in a detailed way, as the auction is linked to a specific wind site. Meanwhile, unfortunately, the UK system seems too bureaucratic and unable to ensure sufficient wind power capacity. The advanced 'Feed-in' system might secure site efficiency without the bureaucratic disadvantages of the UK system.

References

Alberg Ostergaard, P. (2003) 'Transmission-grid requirements with scattered and fluctuating renewable electricity- sources', *Applied* Energy, vol 76, nos 1–3, pp247–255

Hvelplund, F. (2001) *'RE Governance Systems' – A Comparison of the 'Political Price/amount Market' Model with the 'Political Quota/certificate Price Market' System (the German and Danish cases)*, Institute for Development and Planning, Aalborg University

Lida, T. and Kåberger, T. (2004) 'One year after RPS Introduction', presentation at the conference: Climate Change Policy, National Allocation Plans and Emissions Trading in Comparison, Schloss Leopoldskron/Salzburg September 2004. Presentation at www.fu-berlin.de/ffu

Lund, H. and Münster, E. (2004) 'Integrated energy systems and local energy markets', *Energy Policy*, in press

Lund, H. and Münster, E. (2003) 'Management of surplus electricity-production from a fluctuating renewable-energy source', *Applied Energy*, vol 76, nos 1–3, pp65–74

Meyer, N. I. (2003) 'European schemes for promoting renewables in liberalised markets', *Energy Policy*, vol 31, no 7, pp665–676

Mitchell, C., Bauknecht, D. and Connor, P. M. (2006) 'Quota versus subsidies–risk reduction, efficiency and effectiveness – A comparison of the renewable obligation and the German feed-in law', *Energy Policy*, in press

12

Tradeable Certificate Systems and Feed-in Tariffs: Expectation Versus Performance

Volkmar Lauber[1]

Introduction

Currently two main instruments are dominating the discussion on how to promote renewable power most efficiently. The first historically is the feed-in tariff (here abbreviated as REFIT, for renewable energy feed-in tariff), set in advance for a considerable length of time. The second is the system of tradeable green certificates combined with minimum quotas for renewable power, here abbreviated as RPS (renewable portfolio standard, the term used in the United States where this system was first introduced) or TGC (tradeable green certificates, an essential feature of such a quota system).

This chapter will describe concrete examples of these two main schemes. It will look at their philosophical origin, the historical and political context of their introduction, the expectations regarding their performance derived from theoretical arguments (particularly with regard to promoting growth, fostering a renewable power equipment industry and technological innovation, and reducing costs and prices) and finally at the actual experiences with these instruments thus far. These will be studied during two time periods: the period before 2002 (when assumptions about RPS/TGC systems were more important than actual practice in Europe, since such regimes were only being introduced then), and the period since then, for which actual practical experience allows for a better comparison. Although each instrument was implemented in several different versions across the European Union (EU) member states, the main focus will be on comparing the German feed-in system with the British tradeable certificate system, in order to introduce a stronger empirical element.

REFIT Schemes

The first successful REFIT was introduced in Germany in 1990, at the apex of a phase of high environmental concern, which had led to the appointment of a special parliamentary committee to study the impact of industrial society – and in particular of energy consumption – on the earth's atmosphere, and to come up with recommendations for action. Renewable energy had been an important subject in the 1970s, with considerable federal research funding of wind and photovoltaics (PV). However, funding stagnated in the early 1980s and was scheduled to decline in the second half of that decade. With Chernobyl and the emergence of climate change as a political issue, it was back at the centre of attention, and government and parliament were looking for ways to create a market for the innovations resulting from earlier research, which they thought was among the forefront worldwide. The work of the Enquete Commission, supported by all political parties, helped to create a political consensus on this issue, at least until German reunification led to new priorities (Kords, 1996). The German parliament unanimously adopted strong reduction goals for carbon dioxide – minus 30 per cent by 2005 – and a series of recommendations to the government on legislative initiatives in the area of energy. One of these was the electricity feed-in law sponsored by all-party co-operation (Ganseforth, 1996). The first proposal gave the utilities the option to avoid legislation by negotiating a voluntary agreement with renewable power generators. When the economically powerful but unpopular utilities failed to do that (it seems that they thought the whole business was too insignificant to warrant much effort), the feed-in tariff was passed into law.

The majority parties (conservative CDU/CSU and liberal FDP) pointed out that the law was designed to compensate generators for the lack of internalization of external costs in fossil fuel plants and therefore did not introduce a subsidy (Bundestag, 1990a). These external costs were estimated at DM 0.06–0.10 per kWh (3–6 Euro cents) not including climate costs. The feed-in tariff was fixed at a certain percentage of the current tariff to final electricity consumers (90 per cent of those rates for wind and PV generation). The parliamentary representative of the ministry of economic affairs (in charge of the matter) pointed out that in a market economy such a fixed tariff had to remain an exception and should not be expanded. At the same time he promised additional measures such as special tax deductions for RES-investments. The chief criticism of the opposition social democrats and greens was that the new piece of legislation was dramatically insufficient to rise to the task of averting climate change (Bundestag, 1990b). The final vote of the Electricity Feed-in Law or *Stromeinspeisegesetz*, in October 1990, was nearly unanimous; there was just a number of abstentions (for a more detailed history of German renewable power legislation see Jacobssson and Lauber, in this publication).

In terms of public philosophy, the Feed-in Law fits well with German *Ordo-Liberalismus*, that is the neo-liberal philosophy of public management dominant in Germany in the second half of the 20th century. This philosophy was a reaction to the laissez-faire liberalism of the inter-war period, but also to state interventionism and the growth of regulation since the end of the 19th century and during the Third Reich. It stresses the need for the state to create a market framework for competition and to prevent monopolistic or oligopolistic power, and accordingly to prevent restrictions of market access, promote competition and take measures against negative external effects. Appropriate framework conditions should create market incentives for achieving public policy goals (Brockhaus, 1991). In keeping with these priorities, the Feed-in Law excluded from its terms for compensation renewable power generated by facilities under public or utility ownership. Utilities were supposed to increase their own renewable power generation by voluntary commitments (Salje, 1998). Generators qualifying under the law were expected to be newcomers to the business. In fact, by the mid-1990s farmers made up the majority of the new electricity producers (Bundestag, 1997); in 2000 they still amounted to 30 per cent (Gipe, 2002). Producers of renewable power equipment would receive a boost, but over the medium term, competition would bring prices down. A stable feed-in tariff would give them an extra incentive to accelerate cost reductions in order to maximize profits.

Even though renewable power costs declined substantially in the years after the law was adopted, the tariffs of the 1990 law were not lowered. The utilities intensified their opposition to the law when renewable generation kept increasing in volume, partly because the extra cost was not spread evenly across the different firms. In 1996 they complained to the EU Commission about violation of state aid rules, especially in the area of wind where expansion was particularly strong. The Commission proposed a reduction of the wind rates from 90 per cent to 75 per cent of the final consumer tariff (Advocate General 2000, #19–31). Confronted with uncertainty, wind power investments stagnated in 1996 and 1997 (Connor, in this publication). A government proposal taking up the Commission suggestion failed narrowly in a Bundestag committee after strong political mobilization by the wind power community (Bundestag, 1997). Thus the 90 per cent rate was left intact in 1997; however, due to the liberalization of electricity markets introduced that year, this rate meant effectively declining compensation.

In 2000, the Feed-in Law was replaced by the Renewable Energy Sources Act (EEG) of the red–green coalition that had come to power in 1998. Once again the commitment to compensate for non-internalized external costs of coal and nuclear generation was specifically mentioned (Nagel, 2001). The law set rates in monetary amounts rather than percentages, differentiated the tariff for wind-generated electricity according to location quality (wind speeds) and provided for a steady decline of the tariff over time for each year's new installations. These features were also taken up by the new French REFIT based on the decree of June 2001. It too differentiates according to location and features declining compensation, directly inspired by the German example (Chabot/Kellet/Saulnier, 2002). Wind speeds are also taken into account in the Portuguese REFIT of December 2001 (McGovern, 2002). The Renewable Energy Sources Act also set a special tariff for PV. It finally addressed the conflict with the utilities and provided for a mechanism distributing the extra cost equally (that is proportionately) among all German electricity consumers.

Finally, utilities too were allowed, under the new act, to obtain this compensation for renewable power they generated. The opposition parties (conservatives and liberals) protested against the law but did not formulate a clear alternative. In the course of the 2002 election campaign they became increasingly critical of the REFIT. In the opposition, some in the CDU demanded a RPS/TGC model; the FDP came out in favour of coal and nuclear altogether (*Solarthemen*, 2002a and 2002b). By early 2003, CDU/CSU seemed to renew its support to the REFIT (Knight, 2003). But when the Renewable Sources Act was amended in 2004, the leadership of both opposition parties in the Bundestag (the lower chamber of parliament) proclaimed that they would phase out this act should they come to power at the next election (see Jacobsson and Lauber, in this publication).

RPS/TGC schemes: US and UK

Origins of the concept

The renewable portfolio standard – or obligatory minimum quota – was conceived in a context that differed very radically from the conditions surrounding the Feed-in Law in Germany. The US and particularly California had been crucial in starting wind power as an industry. The instruments for reaching this goal were the Public Utilities Regulatory Policies Act (PURPA) of the Carter Administration in 1978 (providing for feed-in rates at the level of avoided costs), investment tax credits introduced in 1979, and a rather fruitless and expensive federal programme for developing large-scale wind turbines that cost $0.5 billion (Asmus, 2001). The tax credits 'jump-started' the industry, with effects reaching as far as Denmark. They also bankrupted much of it – including Vestas and Nordtank – when they were abolished in the mid-1980s (van Est, 1999).

Political fortunes changed with the Reagan administration and a similar change at the state level in California (governor Brown being replaced by Deukmejian, 1983–1991). Regulation to promote independent power producers and renewable power gave way to a new approach – deregulation. This wave reached a high point with the 1994 elections, which resulted in Republican control of both houses of Congress. The prevailing outlook behind deregulation was summed up in 1995 by Carl Weinberg as 'market machismo' in these terms:

> What we really want is cheap electricity. We care less about the environment. All this global climate change is just a bunch of scientists' pipe dream that want research money. What we really want, and what really is good for everybody is to have cheap electricity. Therefore we ought to move to a very market-driven, price-driven, short-term price-outlook market, which is the spot market (van Est, 1999, p292).

If anyone wanted to buy renewable energy, it should of course be available ('consumer choice'), but that was all. Confronted with this new situation, the public interest coalition that supported renewable energy in California (which included the American Wind Energy Association) had to devise a new and more modest strategy in favour of renewables that might be supported by public authorities. At the very least it wanted to protect existing renewable power installations, many of whom were going out of business as old PURPA contracts expired (Rader, 2001). For this it needed to develop a rationale that made concessions to the dominant value system of deregulation. This gave rise to the formulation of the renewable portfolio standard by Nancy Rader, then working for the AWEA. Under this concept, a certain percentage of energy consumption would be required to come from renewables. Market competition via a system of tradeable certificates would make sure that the cheapest options would prevail – a clear concession to the spirit of the times. However, in the political climate of 1995 California even this proposal was too ambitious, and the first RPS proposal was defeated (van Est, 1999).

Obviously the RPS was also inspired by another market instrument – the bubble concept in environmental economics, first developed in the US for sulphur dioxide policy as an alternative to command-and-control regulation. It sets an upper quantitative limit for emissions during a given time period. Polluters are issued certificates (pollution rights) whose volume declines over time. They may reduce their own pollution and/or trade in certificates. This arrangement should make sure that the cheapest ways of reducing pollution are used first.

The authors of the RPS concept extolled the advantages of this concept by comparing it – naturally enough – to other US policy measures that had been resorted to previously. They stressed in particular that it was bureaucratically simple (unlike some instruments developed by public utility commissions with their long and complex procedures often followed by litigation) and economically efficient (in the sense of securing renewable power at marginal cost, whereas under some PURPA contracts prices were pegged to natural gas costs resulting at one time in an astronomical level of compensation; later on, avoided costs fell so low that no independent power producers could meet them (Klass, 2003). At the same time it overcame the imperfections of the market which otherwise handicap renewable power: the public goods/free rider problem (citizens like clean energy but as consumers will go for the best deal since they cannot reserve for themselves any environmental benefits flowing from renewable power), the need to finance capital-intensive installations with ten-year payback periods at a time when 'deregulated' investors want short paybacks and can achieve six-year paybacks with natural gas plants, and the problem of marketing costs for small generators (Rader and Norgaard, 1996). Renewable power generators – expected to be mostly utilities and similarly large investors – are compensated by two income streams, one from the sale and physical delivery of electricity, the other from the sale of certificates. Competition among generators should automatically put pressure on prices, without any need for further public intervention.

From the US the RPS concept came to Europe, where it found particular support in the British Wind Energy Association and in the EU Commission. There it was supported by Directorate General (DG) Competition and former Energy Commissioner Christos Papoutsis, and its virtues were extolled in a working paper as superior in reducing costs, stimulating innovation and promoting deployment (European Commission, 1999). Its most outspoken and eloquent support was probably on the pages of *Windpower Monthly*, the leading trade publication in the area of wind power. It produced visions of wind farms developed with due diligence procedures and much more cost efficient than small plant financed and operated 'like a tractor'. The emphasis on profitability, combined with considerable independence from governmental policy, would make those investments interesting to the financial community (banks, pension funds, and so on, that is large and very large private investors) – an essential step since soft state money would not be sufficient to meet the financing needs of the sector (estimated to be in the order of Euro 100 billion in the first decade of this century). Minimal political risk is portrayed to appeal more strongly to the financial community than a combination of a tariff with tax privileges and soft state loans exposed to changing political fortunes (Harrison, 2000; Massy, 2000; Probyn, 2000; Harrison and Milborrow, 2000). It is also assumed that this scheme will facilitate international trading of renewable power due to the separation of physical electricity from certificates. It should be noted however that the editorial line of *Windpower Monthly* has been highly critical of the implementation of the TGC/quota model in the UK via the Renewables Obligation (Harrison, 2004).

Implementation of the concept

The political fortunes of the RPS concept in the US have been quite mixed so far. In 1998, four major electric industry restructuring bills in Congress included an RPS; the strongest was already then by Senator Jeffords of Vermont (Rader and Short, 1998). In 2001, Jeffords persuaded the Senate to include this measure in its Energy Bill. It would have required an increase from about 15,000MW of renewable power in 2002 to about 74,000MW in 2010 (Union of Concerned Scientists, 2002). However, the bill failed in House–Senate negotiations after the Republicans acquired control of both houses of Congress in the 2002 elections (*Renewable Energy Report* 46, December 2002, pp2–3. In the meantime, nearly two dozen states have adopted such a measure. Together they should increase renewable power generation in those states several times over (Union of Concerned Scientists, 2003). On the level of the states, the Texas RPS is generally held up as a model. It was introduced in 1999 under Governor George W. Bush during restructuring of the electricity sector. Even though it is limited to a tiny fraction of Texas's potential, it was strongly resisted by large industrial customers. However, renewable energy proved overwhelmingly popular with the general public, and its cause was well represented by public interest groups; this explains why it was included in the end after all (Langniss and Wiser, 2003, and this volume).

In Europe, a Danish coalition government led by social-democrats was first to plan the introduction of an RPS/TGC scheme in the course of the electricity reform of 1999; the chief reasons were the growing burden of the earlier REFIT subsidized by tax revenues, pressure by the EU Commission (Morthorst, 2000; *Windpower Monthly* 1999, pp34–36) and the expectation that this system would be introduced at EU level fairly soon. However, the reform was postponed at first in the face of strong opposition, fuelled by uncertainty about its real consequences; in 2004 it was given up for the time being (Meyer and Koefoed, 2003; Nielsen, in this publication). The first government to implement such a system in Europe was the New Labour government of Tony Blair, which was looking for an alternative to the existing system of bidding (NFFO; see Connor in this publication), followed by Belgium (or rather, its territorial sub-units) where this reform was pushed by the greens (Lovinfosse and Varone, 2002 and 2004). In Italy, an expensive REFIT system was also abandoned in favour of an RPS by a centre-left government in 2001; for some years it was not fully developed, and in particular lacked a clear penalty, a long-term commitment and a provision to increase renewable power demand (Lorenzoni, 2003). The British system is interesting because of the country's early commitment to electricity privatization and liberalization inspired by Thatcherism. When the sector was privatized, an exception had to be made for nuclear generation due to its high risks and insufficient profitability. A levy was then introduced to subsidize it, called the non-fossil fuel obligation or NFFO. This mechanism was later used to also create a market for renewable power by several rounds of competitive bidding, all of them for fairly small amounts of power. Even though the generators received good prices for their contracts, this – combined with difficulties in the planning system, that is in obtaining site approvals for wind farms – resulted in one of the lowest levels of renewable power in the EU and failed to create a domestic renewable power equipment industry (Mitchell, 2000; Dinica, 2003) – all this in one of the best-endowed countries with regard to wind resource availability.

Under Tony Blair a new policy was defined, and finally put into practice in 2002: the Renewables Obligation or RO, an RPS scheme with tradeable certificates. It is supposed to combine continued emphasis on market competition with ambitious growth targets for renewable power capacity (from 1.7 per cent of total consumption in 1997 to 10 per cent by 2010). Against the dismal background of NFFO, which had allowed so little use of Britain's excellent renewable resources (especially in wind), this was an enormous step for electricity generation from renewable sources.

Early expectations regarding RPS/TGC performance

Prior to 2002, there was no practical experience in Europe with RPS (del Río and Gual, 2004). The UK had just adopted the Renewables Obligation (RO), but it was still unclear how (and in fact whether) it would work out in practice. The comparison in this subsection rests on real-life systems cited as models, mainly the German REFIT (the laws of 1990 and 2000), the 1999 Texas RPS and the emerging UK RO. It strives to gather the evidence that was available at that time regarding the performance of the two support systems with regard to the following policy goals: promoting growth of renewable power; fostering a renewable energy equipment industry and technological innovation; and reducing costs and prices.

Promoting rapid growth and achieving ambitious targets

REFITs with a generous tariff have regularly surpassed the targets set by their authors. This was the case in both Denmark and Germany, where national goals for wind power were achieved years ahead of schedule (in Germany, installed capacity rose from 50MW in 1990 to about 12,000MW at the end of 2002). In Spain, another country with a REFIT system, wind was the only renewable source of energy to meet the growth targets of government policy (McGovern, 2001). By the end of 2001, the three classic REFIT-countries together made up 84 per cent of installed wind capacity – the most important of the new renewables – in the EU. Favourable REFITs introduced in France and Portugal in 2001 were expected to produce a veritable rush towards wind power projects. France, which in 2001 had an installed capacity of about 90 MW (Inman, 2002) and rather modest growth expectations (Forster, 2002), now expected 7000–14,000MW by 2010 (*Renewable Energy Report*, 2002b). In Portugal, with 125MW on line at the end of 2001, the announcement of a favourable tariff produced wind power applications for over 7000MW, more than double the target set by the government for 2010 (*Renewable Energy Report*, 2002a, p1; McGovern, 2002). It is true that in both countries, administrative delays in the permitting system – in France also a change in government, and to some extent in policy – greatly dampened this impulse (Inman, 2003; Knight, 2003). By contrast, the announced shift to a RECs/quota system in Denmark brought construction of wind plant almost to a halt; this was the main reason why the scheme was first postponed and then withdrawn in late 2001. The British RO, in the first two years of its implementation, showed only a modest increase of renewable power projects. In Italy, a similar shift also led to a significant slowdown at first (Jackson, 2002; Knight and Harrison, 2003). Such slowdowns could of course prove to be temporary.

So much for very dynamic early growth. There is another side to this anarchic growth, a side that Germany was soon to face. After the rapid increases that lasted until 2002, reaching over 3000MW in that year, new construction of turbines was expected to decline to below 1000MW around the middle of the decade, with interesting onshore sites running short and offshore not yet making a significant contribution (Deutsches Windenergie-Institut, 2002). Such results would not be quite as radical as the on/off policies of the production tax credit in the US, but they could prove dramatic enough for German turbine makers.

RPS systems have no similar record of inducing such dramatic growth so far. In some cases – UK and Belgium – the targets were very ambitious, and this was also true of the now defunct US Senate Energy Bill of 2001. There was an impressive wind rush in Texas (about 900MW) in 2001, but it was undoubtedly due in part to the availability of the US production tax credit (PTC, with a value of 1.7 cents/kWh) whose extension beyond 2001 could not be taken for granted. But the RPS was also crucial: without its existence the utilities would not have had any great interest in signing up this capacity for the long term (Langniss and Wiser, 2003). It should be added that the 2000MW targeted in Texas for 2009 (in addition to the 880MW already on line at the time the RPS was introduced) amount to a mere 3 per cent of present electricity consumption and represent less than 1 per cent of a Texas wind potential estimated at over 400 per cent of present consumption. There is one very desirable thing, however, that a well-functioning RPS was expected to deliver for a sector shaken by ups and downs, provided it is designed accordingly and no other factors interfere (such as an off–on PTC in the US, or the lack of an obligation to conclude long-term contracts in the UK): 'a steadily growing market for renewable energy' according to a 'predictable fixed schedule' (Rader and Hempling, 2001, pxii). It is curious to find this aspect of planning in a neo-liberal policy instrument.

Fostering a renewable power equipment industry and technological innovation

The very purpose of REFIT introduction in Germany and Denmark was to promote a domestic wind power industry. In the late 1980s, wind power technology was under development in a variety of European countries. Significant efforts were underway not only in Denmark, Germany and Spain, but also in the UK, the Netherlands and Sweden. Today the market is clearly dominated by turbine producers from the rapid growth/REFIT countries Denmark, Germany and Spain. The decisive step towards success was probably the successful market introduction programme in those countries (Johnson and Jacobsson, 2002). France was trying to follow with its new REFIT; Jeumont could benefit as domestic turbine producer, and other firms were building subsidiaries. REFITs help to create markets by combining relative security to entrepreneurs with encouraging a long-term perspective. As Menanteau et al (2003) argue, in case of a guaranteed price level 'producers keep the surplus created by technical change', that is by increasing their profits through investing in R&D. This is probably more favourable to 'radical innovations' that require a longer payback period (Dinica, 2003). To be sure, the existence of a generous REFIT is no guarantee that a domestic renewable power industry will develop; this was the Italian experience, where a tariff more generous than the German, introduced in 1992 (Piacentino, 2002), had comparatively little impact on the Italian renewable power equipment industry.

Arguments in favour of RPS schemes stressed that they could promote renewable power while minimizing costs (Nielsen and Jeppesen, 2003); for this very reason they tend to limit technological variety, although they could theoretically provide for technology bands. The pervasive pressure to reduce costs was expected to discourage high R&D investments on the part of equipment producers. As generators are compelled to remain competitive, they must however try to benefit from technical progress as well. 'In an open economy, this situation may encourage them to turn to foreign technology' (Menanteau et al, 2003). Utilities sometimes argued that if a government is interested in promoting such R&D, it should pay for it directly and not burden electricity customers. This missed the fact that with deregulation and liberalization, public R&D expenditures on energy went down along with private ones (Dooley, 1998).

RPS countries showed little concern for developing a renewable power equipment industry. In the US, the chief support for wind power is the production tax credit. Created with the Energy Policy Act of 1992 (Kahn, 1996), it was in recent years always reinstated for short time periods only. This led to boom–bust cycles in the industry, which deterred European investors – in particular Vestas, which had quite advanced plans – from setting up production facilities there (*Renewable Energy Report*, 2002c).

Table 12.1 *Wind power investment cycles in the United States and availability of PTC*

	1999	2000	2001	2002	2003	2004	2005
PTC status	Expired end of year	Reinstated late	Expired end of year	Reinstated late	Expired end of year	Reinstated in October until end of 2005	
Capacity installed (MW)	800	53	1695	410	1687	400	Over 2000 expected

Source: Bailey and O'Bryant, 2004; *Windpower Monthly*, 2005

Reducing costs and prices

REFITs were expected to contribute to cost reduction by encouraging higher growth rates early on, and by better encouraging long-term technological innovation (see above). It is a common assumption that a doubling of market volume will lead to a 15–20 per cent reduction of costs. Of course, even if there is intense competition among turbine producers, there is little incentive for developers to pass on these cost reductions to the consumer unless the tariff is adjusted accordingly. And equipment producers and developers can be expected to make tariff adjustments politically difficult, as Germany experienced in 1997. One of the key goals of RPS/TGC is to achieve renewable power at at least cost (Rader and Hempling, 2001; Connor, in this publication; Elliott, in this publication; Langniss and Wiser, 2003; Nielsen and Jeppesen, 2003). With their emphasis on immediate competition between projects, RPS schemes were generally argued to be more efficient in driving down prices since there is no safe floor in the form of a feed-in tariff, so that generators are likely to put greater pressure on equipment producers for lower prices and on developers for the best available locations.

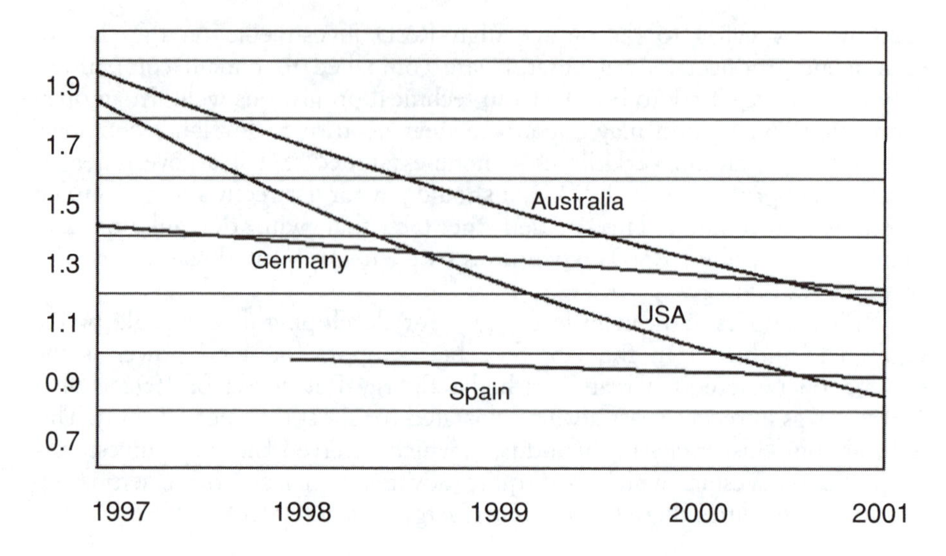

Source: Meier, Milborrow and Wiesenthal (2002).

Figure 12.1 Wind: installed cost reduction over time (in million US$/MW installed)

Figure 12.1 shows impressive differences between cost reductions for new wind power installations in four countries. The fact that the decline is considerably faster in Australia and the US than in Germany or Spain led to the research hypothesis that competitive regimes such as RPS/TGC lower costs faster than REFITs (Meier et al, 2002). There could also be a different explanation. In the US, particularly in Texas, where more than half of all US wind power installations were built in 2001 (O'Bryant, 2002) as well as in Australia, there is much open and – by European standards – uninhabited land, a condition that cannot be replicated in the EU. Also, with more extensive use of wind power, less attractive resource locations (wind speed, connection to transmission lines) and more expensive land would in turn be used, raising the cost of plant again after an initial downturn (Energy Information Administration, 2002). Finally, there would be another problematic consequence in the case of a pan-European RPS. Wind power developers would necessarily go for the best available locations first, most of them in Britain and Ireland. One can imagine the logjam of projects and the amount of local resistance resulting from such a concentration, which would result from the search for an advantage that after all would only be temporary, as less favourable locations will also be needed if wind power is to make a significant contribution (Hvelplund, 2001 and in this publication; Menanteau et al, 2003).

Experience with comparative performance since 2002

The recent performance concerns above all, RPS regimes, due to the experience accumulated in recent years, primarily in the UK, but also in other states. However, there are also new experiences in the area of REFITs. It is clear that the expectations derived from theory, ideology and some early experiences were not always borne out by actual developments. In fact, some results are the very opposite of what was expected.

Promoting rapid growth and achieving ambitious targets

Clearly REFITs have far outperformed RPS so far. However, even generous REFITs experienced some difficulties in recent years. In Germany, new wind power development, while still the highest of any country at present, is declining due to a lack of appropriate onshore sites; offshore development is slow due to special permitting procedures. In 2004, new construction of wind capacity was just above 2000MW and expected to decline further in 2005. REFITs were however very successful in this regard in Spain (in a modified version, primarily as a premium added to market prices), in Austria (where this led to legislative efforts in 2004 to curtail renewable power) and in Portugal (after administrative difficulties in a first phase). In France, permission problems and administrative ambiguity about wind power led to a lowering of the growth expectations that had resulted from the introduction of this instrument a few years ago. Also, some new REFITs introduced recently have such low rates that they hardly led to any expansion of renewable power at all (Denmark, Netherlands). With regard to RPS, there are still few success stories with regard to rapid growth. In the UK, the RO has run into a series of obstacles so far and has not yet delivered, though this may be about to change. One reason for the problem in Britain is the special design of the RO. Unlike the Texas RPS is rarely implemented by long-term contracts with utilities (in Texas: 10–25 years). Without such contracts developers can always be undercut by later installations. In order to obtain long-term contracts, they have to hand over to the utility part of the extra income they expect to derive from the sale of certificates (Connor, in this publication). However, the RO has certainly led to a considerable acceleration of renewable power construction in the UK since 2002 and holds promise for further development in this direction. This is partly the result of a different outcome of permission procedures (positive conclusions increased every year since 2002), but also of the new incentive structure and its acceptance by the financial community. It is quite conceivable that the RO will lead to a very rapid growth of installed capacity within a few years, but even so it may well be that RO targets – 10 per cent by 2010, 15 per cent by 2015 – will be missed by a margin of one-third to one-quarter (House of Lords Technology Committee, 2004). Substantial growth is also likely to take place in Texas if the target of 2000MW of installed capacity by 2009 is increased to 10,000MW by 2020, a development that seemed well on its way in 2005 (*Windpower Monthly*, 2004). In Italy, substantial growth resumed in 2004. In Sweden on the other hand, the introduction of this type of system in May 2003, but combined with a renewable mandate that extended only until 2010, led to a standstill of new construction (Mortimore, 2004).

Fostering a renewable power equipment industry and technological innovation

REFIT systems still seem more successful today in this respect, both with respect to large and small equipment producers. Neither Texas nor the UK developed wind power industries of their own; instead, they rely on imported technology (with at best local assembly). The absence of technology bands in both cases – that is reserving specified market shares for different technologies – means that technological diversity is not encouraged. In the area of wind turbines, large firms – Danish Vestas (which now also includes NEG-Micon), German Enercon and Spanish Gamesa – still dominate the scene, together with General Electric Wind, which took over the former Enron Wind, which in turn merged American and German manufacturers; also, Siemens took over the Danish producer Bonus. Firms are becoming ever more international though as markets are being more open to competition. The smaller firms are also mostly from REFIT countries, with few exceptions. The UK, despite its substantial new growth in the area of wind power, still has trouble in developing an industry of its own (bankruptcy of Cambrian engineering, a tower producer, in 2004; withdrawal of FKI from the turbine business after poor results with its takeover of a German producer). There is only a Vestas subsidiary in Britain. As expected, there is little diversity of renewable power technologies on the market as there is little incentive to invest in technologies other than the cheapest, that is wind power. In the case of solar power, the leading markets are those of Japan and Germany, which both had stable support schemes during this period (of the REFIT type in Germany, and fixed subsidies in Japan).

Reducing costs and prices

This is the area with the most surprising results. Here, REFITs were expected to perform poorly, due first to the lack of direct competition between developers to drive down prices, and second due to political opposition to such reductions on the other. But the new REFIT introduced in Germany in 2000 built in an annual decline of rates (and no compensation for inflation), a policy imitated by several other countries and continued in the 2004 amendment. Rates go down more or less in line with the learning curve of particular technologies. By contrast, the TGC/quota systems were expected to drive prices down more efficiently. However, the British RO, in combination with other regulations such as the climate levy, has led to prices for wind power, which are substantially above German rates by about 50 per cent, and this despite a wind resource that is far superior in the British case (Toke, 2005; Lauber and Toke, 2005). At the same time, competition in the turbine market was less intense than in Germany, possibly due to the smaller size of the market (Butler and Neuhoff, 2004). The reason why high renewables obligation payments do not translate into more deployment is that part of the money goes not to developers but to distributors, who make developers pay for granting them long-term contracts; in the UK such contracts are not required by legislation. Another reason is the oligopolistic structure of the distribution sector and the fact that fulfilling RO targets would mean lower income from certificates; the distributors have every interest in limiting fulfilment. Italy has even higher certificate prices, which since 2004 have attracted new renewable power capacity. In the US, the PTC is paradoxically indexed to inflation, which means that this premium is growing instead of declining. All this is the very opposite of what the European Commission expected in the document that laid down its philosophy on renewable power support schemes (European Commission, 1999).

Conclusions

On the basis of current experience, it is possible to draw several conclusions. First, it seems clear that so far, at least in Europe, best results on all three dimensions – capacity deployment, developing a renewable power industry and bringing down costs and prices – were clearly achieved by countries with REFIT or related systems (such as Spain, where the dominant system is that of a premium added to the market price). Second, it is also clear that there are some goals that cannot be integrated into an RPS – hence the development of an industrial policy in favour of renewable power. External costs also cannot be integrated into this instrument, but there are other ways to do that (for example a tax or trading regime for carbon and for pollutants).

On the other hand, some aspects of success do not depend on the choice of the instruments as such, but on the way they are designed and implemented – and on how they are received by the many actors involved in renewable power development. It is easy enough to imagine poorly designed instruments of both types; examples of these abound. Success is less easy to come by, and to upset a success story should be viewed as a very serious matter. The stability of an instrument has a substantial value of its own; stakeholders need to become familiar with it before accepting it fully, and this takes time. At a time when popular support for renewable power is an important asset in order to bridge the gap to full competitiveness, any instrument that is likely to create resistance (such as excessive regional concentration of wind turbines in a few areas of Europe) or to produce windfall profits should be viewed with great scepticism (see also Hvelplund, in this publication).

The peculiarities of renewable power (high upfront costs, low running costs, small external costs) – the opposite of fossil power with its high fuel and high external costs must be taken into account by any instrument if it is to become successful. It is also important that an instrument does not discourage the variety of technologies in renewable power development. REFITs can solve this problem quite easily, while RPS systems with their emphasis on encouraging the cheapest technologies already available on the market have greater difficulty in financing new developments; technically though it can be done just as easily by technology banding.

There is one criterion stressed particularly by the European Commission in this context: which instrument is more compatible with international trade in renewable energy, a feature judged by the Commission as central in advancing competition and thus in reducing prices and driving innovation. It is this criterion that led the Commission to prefer RPS systems, even if this preference was suspended temporarily in the formulation of the 2001 directive on renewable power. But this is hardly a reason to insist on a single instrument, epecially one that proved so problematic in the course of implementation. Even in the electricity directive of 1996, the Commission accepted three instruments for competition: regulated third-party access, negotiated third-party access, and the single buyer model. Why should it not develop different models for the trade with renewable power?[2]

The stark fact is that that up to now, there has been no experience in Europe with a quota/TGC system that can claim a performance with regard to cost-efficiency, innovation or deployment, which is superior to a good REFIT system. Nor are the existing systems easily compatible and thus do not encourage international trade in renewable power. It may not be impossible to design a quota/TGC system that lives up to the criteria listed above, but it would be quite unwise to impose a system for harmonization that has not shown its worth beforehand.

Notes

1 This chapter is based in part on an article by the author published in *Energy Policy* (Lauber, 2004).

2 On this and other aspects of EU policy, see Lauber in this volume, Chapter 9.

References

Advocate General (2000) Opinion of advocate general Jacobs delivered on 26 October 2000 in Case C-379/98 PreussenElektra AG vs Schleswag AG (proceedings before European Court of Justice)

Asmus, P. (2001) *Reaping the Wind*, Washington, DC, Island Press

Bailey, D. and O'Bryant, M. (2004) 'American roller coaster market sets off again', *Windpower Monthly*, vol 20, no 10, p19

Brockhaus Enzyklopädie (1991) 19th edn, vol 15, pp429–430, Mannheim, Brockhaus

Bundestag (1990a) Drucksache 11/7169

Bundestag (1990b) Parliamentary record, 17750–17756

Bundestag (1997) Parliamentary record, 18966–18994

Butler, L. and Neuhoff, K. (2004) 'Comparison of feed in tariff, quota and auction mechanisms to support wind power development', Cambridge Working Papers in Economics

Chabot, B., Kellet, P. and Saulnier, B. (2002) 'Defining advanced wind energy tariffs systems to specific locations and applications: Lessons from the French tariff system and examples', paper presented to Global Windpower Conference, Paris, 2–5 April

Del Río, P. and Gual, M. (2004) 'The promotion of green electricity in Europe: Present and future', *European Environment*, vol 14, pp219–234

Deutsches Windenergie-Institut (DEWI) (2002) 'WindEnergy-Study 2002. Assessment of the wind energy market until 2010', www.dewi.de

Dinica, V. (2003) *Sustained Diffusion of Renewable Energy*, Enschede, Twente University Press

Dooley, J. J. (1998) 'Unintended consequences: Energy R&D in a deregulated energy market', *Energy Policy*, vol 26, no 7, pp547–555

Energy Information Administration EIA (2002) *Impacts of a 10-Percent Renewable Portfolio Standard*, Washington DC

European Commission (1999) 'Working Paper: Electricity from renewable energy sources and the internal electricity market', SEC (99) 470, 13.4.1999

Forster, S. (2002) '"Oui!" to wind power', *New Energy*, vol 1, pp26–29

Ganseforth, M. (1996) 'Politische Umsetzung der Empfehlungen der beiden Klima-Enquete-Kommissionen (1987–1994) – eine Bewertung', in Brauch, H. G. (ed) *Klimapolitik*, Berlin, Springer

Gipe, P. (2002) 'Soaring to new heights – The world wind energy market', *Renewable Energy World*, vol 5, no 4, pp33–47

Harrison, L. (2000) 'A new class of finance', *Windpower Monthly*, vol 16, no 9, p6

Harrison, L. (2004) 'High time to strike back', *Windpower Monthly*, vol 20, no 11, p6

Harrison, L. and Milborrow, D. (2000) 'Commercial realities kept at bay in most of Europe', *Windpower Monthly*, vol 16, no 9, pp39–42

House of Lords Science and Technology committee (2004) *Renewable Energy: Practicalities*, 4th Report of Session 2003–04, London, Stationery Office

Hvelplund, F. (2001) 'Political price or political quantities?', *New Energy*, vol 8, pp18–23

Hvelplund, F. (2005) 'Renewable energy – Political prices or political quantities', in this publication

Inman, N. (2002) 'France gathers strength for a boom year', *Windpower Monthly*, vol 18, no 3, p51

Inman, N. (2003) 'France still in waiting. Permitting system still a major barrier', *Windpower Monthly*, vol 19, no3, pp47–48

Jackson, J. (2002) 'Green credit trading anticipation', *Windpower Monthly*, vol 18, no 3, p50

Johnson, A. and Jacobsson, S. (2002) 'The emergence of a growth industry – The case of the German, Dutch and Swedish wind turbine industries', in Metcalfe, S. and Cantner, U. (eds) *Transformation and Development: Schumpeterian Perspectives*, Heidelberg, Physica/Springer

Kahn, E. (1996) 'The production tax credit for wind turbine power plants is an ineffective incentive', *Energy Policy*, vol 24, no 5, pp427–435

Klass, D. (2003) 'A critical assessment of renewable energy usage in the USA', *Energy Policy*, vol 31, no 4, pp353–367

Kords, U. (1996) 'Tätigkeit und Handlungsempfehlungen der beiden Klima-Enquete-Kommissionen des Deutschen Bundestages (1987–1994)' in Brauch, H. G. (ed) *Klimapolitik*, Berlin, Springer

Knight, S. (2003) 'Portugal promise – Wind industry steadily fighting clear of red tape', *Windpower Monthly*, vol 19, no 3, pp38–40

Knight, S. (2003) 'German wind slowing – Record year as market peaks', *Windpower Monthly*, vol 19, no 3, pp35–36

Knight, S. and Harrison, L. (2003) 'The attraction of Italy – Renewables mandate creates buzz of activity', *Windpower Monthly*, vol 19, no 3, pp44–45

Langniss, O. and Wiser, R. (2003) 'The renewables portfolio standard in Texas: An early assessment', *Energy Policy*, vol 31, no 3, pp527–535

Lauber, V. (2004) 'REFITs and RPS in a harmonised Community framework', *Energy Policy*, vol 32, no 12, pp1405–1414

Lauber, V. and Toke, D. (2005) 'Einspeisetarife sind billiger und effizienter als

Quoten/Zertifikatssysteme. Der Vergleich Deutschland–Großbritannien stellt frühere Erwartungen auf den Kopf', *Zeitschrift für Neues Energierecht*, vol 9, no 2, forthcoming

Lorenzoni, A. (2003) 'The Italian Green Certificates market between uncertainty and opportunities', *Energy Policy*, vol 31, no 1, pp33–42

Lovinfosse, I. and Varone, F. (2002) 'Renewable energy policy in Belgium', in Reiche, D. (ed) *Handbook of Renewable Energies in the European Union*, Frankfurt, Peter Lang, pp49–62

Lovinfosse, I. and Varone, F. (2004) 'From private self-regulation to public renewable electricity policy: a paradigmatic change in Belgium', in Lovinfosse, I. and Varone, F. (eds) *Renewable Electricity Policies in Europe*, Louvain, Presses Universitaires de Louvain, pp73–119

McGovern, M. (2001) 'Attacks on wind subsidies in Spain', *Windpower Monthly*, vol 17, no 7, p24

McGovern, M. (2002) 'Portugal staggers under wind rush', *Windpower Monthly*, vol 18, no 3, pp49–50

Massy, J. (2000) 'From a commercial perspective', *Windpower Monthly*, vol 16, no 11, pp38–40

Massy, J. (2003) 'New transparency first step to reform', *Windpower Monthly*, vol 18, no 2, p41

Meier, P., Milborrow, D. and Wiesenthal, T. (2002) 'Statistical analysis of wind farm costs and policy regimes', paper for Asia Alternative Energy Programme (ASTAE) of World Bank, www.worldbank.org/astae/windfarmcosts.pdf

Menanteau, P., Finon, D. and Lamy, M.-L. (2003) 'Prices versus quantities: Choosing policies for promoting the development of renewable energy', *Energy Policy*, vol 31, no 8, pp799–812

Meyer, N. and Koefoed, A. L. (2003) 'Danish energy reform: Policy implications for renewables', *Energy Policy*, vol 31, no 7, pp597–607

Mitchell, C. (2000) 'The England and Wales non-fossil fuel obligation: History and lessons', *Annual Review of Energy and the Environment*, vol 25, pp285–312

Morthorst, P. E. (2000) 'The development of a green certificate market', *Energy Policy*, vol 28, no 15, pp1085–1094

Mortimore, G. (2004) 'Green credit market without a cap', *Windpower Monthly*, vol 20, no 12, p19

Nagel, B. (2001) 'Rechtliche und politische Hindernisse bei der Einführung Erneuerbarer Energien am Beispiel Strom', *Solarzeitalter*, vol 13, no 4, pp14–20

Nielsen, L. and Jeppesen, T. (2003) 'Tradable Green Certificates in selected European countries – Overview and assessment', *Energy Policy*, vol 31, no 1, pp3–14

O'Bryant, M. (2002) 'Promising market stopped in its tracks', *Windpower Monthly*, vol 18, no 3, pp33–34

Piacentino, D. (2002) 'Italy', in Reiche, D. (ed) *Handbook of Renewable Energies in the European Union*, Frankfurt, Peter Lang, pp157–169

Probyn, S. (2000) 'New discipline in the grown up world', *Windpower Monthly*, vol 16, no 9, pp36–39

Rader, N. (2001) 'Renewable energy in California: A brief history and current prospects', paper accessed on internet in 2003, but removed since

Rader, N. and Hempling, S. (2001) 'The Renewables Portfolio Standard – A practical guide' prepared for the National Association of Regulatory Utility Commissioners

Rader, N. and Norgaard, R. (1996) 'Efficiency and sustainability in restructured electricity markets: the renewables portfolio standard', *The Electricity Journal*, vol 9, no 6, pp37–49

Rader, N. and Short, W. (1998) 'Competitive retail markets: Tenuous ground for renewable energy', *The Electricity Journal*, vol 11, no 3, pp72–80

Renewable Energy Report (2002a) vol 36, February, p1

Renewable Energy Report (2002b) vol 37, March, p23
Renewable Energy Report (2002c) vol 43, September, p3
Renewable Energy Report (2002d) vol 46, December, pp2–3
Schindler, J. and Zittel, W. (2001) 'Die künftige Verfügbarkeit von Erdöl und Erdgas', *Solarzeitalter*, vol 13, no 4, pp5–7
Salje, P. (1998) *Stromeinspeisungsgesetz*, Cologne, Heymann
Solarthemen (2002a) vol 140, 25 July, p1
Solarthemen (2002b) vol 141, 8 August, p1
Toke, D. (2005) 'Are green electricity certificates the way forward for renewable energy? An evaluation of the UK's Renewables Obligation in the context of international comparisons', *Environment and Planning C*, forthcoming
Union of Concerned Scientists (2002) 'Clean Energy fact sheet. The Senate Renewable Electricity (Portfolio) Standard', www.ucsusa.org/clean_energy/renewable_energy
Union of Concerned Scientists (2003) 'Clean Energy fact sheet. Renewable Energy Standards at Work in the States', www.ucsusa.org/clean_energy/renewable_energy
Van Est, R. (1999) *Winds of Change*, Utrecht, International Books
Windpower Monthly (1999) vol 15, no 12, pp34–36
Windpower Monthly (2004) vol 20, no 12, p14
Windpower Monthly (2005) vol 21, no 4, p66

Index